THE COMPETITIVE
INTERNET SERVICE
PROVIDER

WILEY SERIES IN COMMUNICATIONS NETWORKING & DISTRIBUTED SYSTEMS.

Series Editor: David Hutchison, *Lancaster University*
Series Advisers: Harmen van As, *TU Vienna*
                     Serge Fdida, *University of Paris*
                     Joe Sventek, *Agilent Laboratories, Edinburgh*

The 'Wiley Series in Communications Networking & Distributed Systems' is a series of expert-level, technically detailed books covering cutting-edge research and brand new developments in networking, middleware and software technologies for communications and distributed systems. The books will provide timely, accurate and reliable information about the state-of-the-art to researchers and development engineers in the Telecommunications and Computing sectors.

Other titles in the series:

Wright: *Voice over Packet Networks* 0-471-49516-6 (February 2001)
Jepsen: *Java for Telecommunications* 0-471-49826-2 (July 2001)
Sutton: *Secure Communications* 0-471-49904-8 (December 2001)
Stajano: *Security for Ubiquitous Computing* 0-470-84493-0 (February 2002)
Martin-Flatin: *Web-Based Management of IP Networks and Systems*, 0-471-48702-3 (September 2002)
Berman, Fox, Hey: *Grid Computing. Making the Global Infrastructure a Reality,* 0-470-85319-0 (March 2003)
Turner, Magill, Marples: *Service Provision. Technologies for Next Generation Communications* 0-470-85066-3 (April 2004)
Welzl: *Network Congestion Control: Managing Internet Traffic* 0-470-02528-X (July 2005)

THE COMPETITIVE INTERNET SERVICE PROVIDER

NETWORK ARCHITECTURE, INTERCONNECTION, TRAFFIC ENGINEERING AND NETWORK DESIGN

Oliver Heckmann
Technical University Darmstadt, Germany

John Wiley & Sons, Ltd

Other Wiley Editorial Offices

John Wiley & Sons Inc., 111 River Street, Hoboken, NJ 07030, USA

Jossey-Bass, 989 Market Street, San Francisco, CA 94103-1741, USA

Wiley-VCH Verlag GmbH, Boschstr. 12, D-69469 Weinheim, Germany

John Wiley & Sons Australia Ltd, 42 McDougall Street, Milton, Queensland 4064, Australia

John Wiley & Sons (Asia) Pte Ltd, 2 Clementi Loop #02-01, Jin Xing Distripark, Singapore 129809

John Wiley & Sons Canada Ltd, 22 Worcester Road, Etobicoke, Ontario, Canada M9W 1L1

Wiley also publishes its books in a variety of electronic formats. Some content that appears
in print may not be available in electronic books.

Library of Congress Cataloging-in-Publication Data

Heckmann, Oliver, 1974-
  The competitive Internet service provider : network architecture,
interconnection, traffic engineering, and network design / Oliver Heckmann.
     p. cm.—(Wiley series in communications networking & distributed
systems)
  Includes bibliographical references and index.
  ISBN-13: 978-0-470-01293-2 (cloth : alk. paper)
  ISBN-10: 0-470-01293-5 (cloth : alk. paper)
1. Computer networks—Design and construction. 2. Internet. 3. Internet
service providers. I. Title. II. Series.
 TK5105.5.H4245 2006
 004.67′8—dc22
                                                         2006000988

British Library Cataloguing in Publication Data

A catalogue record for this book is available from the British Library

ISBN-13: 978-0-470-01293-2
ISBN-10: 0-470-01293-5

Typeset in 10/12 Times by Laserwords Private Limited, Chennai, India
Printed and bound in Great Britain by Antony Rowe Ltd, Chippenham, Wiltshire
This book is printed on acid-free paper responsibly manufactured from sustainable forestry
in which at least two trees are planted for each one used for paper production.

Acknowledgements

Writing a book is never an easy task and always involves more people than just the author. Therefore, I would like to thank a number of people who helped me in finishing this project.

Most importantly, I would like to thank Bibiana for her invaluable help and patience. I also want to thank Angelika, Klaus, Berit and the rest of my family for their support.

Large parts of this book are based on my PhD thesis at the Technische Universität Darmstadt, Germany. Therefore, special thanks go to Prof. Jens Schmitt, TU Kaiserslautern, from whom I learned so much. I would also like to thank my PhD supervisors Prof. Ralf Steinmetz from TU Darmstadt and Prof. Jon Crowcroft from the University of Cambridge. I also want to acknowledge the help and inspiration of my (ex-)colleagues, especially Nicolas Liebau, Vasilios Darlagiannis, Andreas Mauthe, Martin Karsten, Rainer Berbner, Ivica Rimac, Matthias Hollick, Michael Zink and Utz Roedig. I would like to also thank my students, especially Ian, Martin, Enis, Nikola and Axel.

I am grateful to John Souter, CEO of LINX (London Internet Exchange); Eike Jessen, director of DFN e.V. (German research network); Arnold Nipper, chief technical officer of DE-CIX; Gerhard Hasslinger from T-Systems (Deutsche Telekom); Raza Rizvi, technical manager of the UK ISP REDNET (www.red.net); Frank Kelly, Richard Gibbens, Damon Wischik and Gaurav Raina – all from the Cambridge University – for their valuable input and fruitful discussions.

Last but not the least, I would also like to thank open source software and the open source community.

Contents

Part III   Interconnections   193

Part IV   Traffic and Network Engineering   235

Foreword

The Internet is a fact of everyday life for vast numbers of people who use it explicitly, or even implicitly, as they go about their business or leisure – doing web searches, downloading files for their next meeting, making Voice-over-IP telephone calls, accessing e-mail, booking holidays, ordering goods, and so on. Users and usage alike are increasing dramatically, all over the globe, and the Internet is firmly established as a critical infrastructure. The user-view of the network comes through the medium of the web browser, and although many factors are jointly responsible for the end-to-end performance experienced by the user, there is no doubt that the Quality of Service (QoS) of the underlying network is one of the most important elements. The role of Internet Service Providers (ISPs), via which most people access the network, is therefore crucial.

In this book, Oliver Heckmann focuses on the world of the ISP and in particular on the efficiency of their network operation and on the QoS that they (aim to) provide. He structures the book into four parts, namely the ISP market, network architecture, interconnection issues, and traffic and network engineering. In that way he situates descriptions of the technology issues and emerging solutions in their proper context. New strategies and insights are presented that can help ISPs realise their QoS targets.

It is an undeniable fact that QoS-engineering has been difficult to 'sell' ever since the early work on QoS architectures began to appear in the early to mid 1990s. Partly this was because the proposed solutions seemed too complicated, and partly because it was felt that the Internet could cope without them – especially where over-engineered and where excess bandwidth is available (nearly always the case in the network core). However, perhaps the main reason was the lack of a compelling commercial reason for network operators to deploy QoS solutions. Now, or soon, the time may be right to do so, because of new and increasing competition amongst Internet Service Providers and the commercial pressures that surely dictate the urgent need to assure customers of an excellent level of service at all times.

This book gives a comprehensive coverage of the technical components needed by Internet Service Providers in the world in which they are all competing to succeed. I welcome its addition to the series, and highly recommend it to anyone with an interest in Quality of Service and the efficient operation of networks.

David Hutchison
Lancaster University
January 2006

List of Figures

List of Tables

List of Abbreviations

General Abbreviations

1GE	1-Gigabit Ethernet
10GE	10-Gigabit Ethernet
ABE	Alternative Best-effort
AC	Admission Control
ACK	Acknowledgement (Packet)
ADSL	Asymmetric Digital Subscriber Lines
AdSpec	Advertisement Specification
AF	Assured Forwarding
AH	Authentication Header
AIMD	Additive Increase - Multiplicative Decrease
AISP	Access Internet Service Provider
ANSI	American National Standards Institute
AOL	America Online
API	Application Programming Interface
APS	Automatic Protection Switching
AQM	Active Queue Management
AS	Autonomous System
ASP	Application Service Provider
ATM	Asynchronous Transfer Mode
B&B	Branch & Bound
B2B	Business to Business
B2C	Business to Consumer
BA	Behaviour Aggregate
BB	Bandwidth Broker
BE	Best-effort
BGP	Border Gateway Protocol
BRITE	Boston University Representative Internet Topology Generator
BSD	Berkeley Software Distribution
BSP	Backbone Service Provider
CAIDA	Cooperative Association for Internet Data Analysis
CBQ	Class Based Queueing
CBR	Constant Bit-rate

CD	Compact Disc
CDN	Content Delivery Network
CIDR	Classless Inter-domain Routing
CL	Controlled Load Service
COPS	Common Open Policy Service
CoS	Class of Service
CP	Content Provider
CPU	Central Processing Unit
CR-LDP	Constraint-based Routing Support for LDP
CSFQ	Core-stateless Fair Queueing
CSMA/CD	Carrier Sense Multiple Access with Collision Detection
CS	Class Selector
DCCP	Datagram Congestion Control Protocol
DE-CIX	Deutscher (German) Commercial Internet Exchange
DFN	Deutsches Forschungsnetz (German Research Network)
Diffserv	Differentiated Services
DNS	Domain Name System
DP	Dynamic Programming
DPS	Dynamic Packet State
DPT	Dynamic Packet Transport
DRR	Deficit Round Robin
DS	Diffserv, Differentiated Services
DSCP	Differentiated Services Codepoint
DSD	Duplicate Scheduling with Deadlines
DSL	Digital Subscriber Lines
DVD	Digital Versatile Disc
ECN	Explicit Congestion Notification
EDF	Earliest-deadline-first
EF	Expedited Forwarding
EIGRP	Enhanced Interior Gateway Routing Protocol
ENO	End-user Network Operator
ESP	Encapsulating Security Payload
FCFS	First Come First Serve
FEC	Forwarding Equivalence Class
FF	Fixed Filter
FFQ	Frame-based Fair Queueing
FIFO	First In First Out
FilterSpec	Filter Specification
FlowSpec	Flow Specification
FPS	First Person Shooter
FTP	File Transfer Protocol
GCRA	Generic Cell Rate Algorithm
GE	Gigabit Ethernet
GMPLS	Generalised MPLS
GS	Guaranteed Service
GT-ITM	Georgia Tech Internetwork Topology Models

HDLC	High-level Data Link Control
HFSC	Hierarchical Fair Service Curve
HPFQ	Hierarchical Packet Fair Queueing
HTML	Hypertext Markup Language
HTTP	Hypertext Transfer Protocol
ICMP	Internet Control Message Protocol
IEEE	Institute of Electrical and Electronics Engineers, Inc.
IETF	Internet Engineering Task Force
IGMP	Internet Group Management Protocol
IGRP	Interior Gateway Routing Protocol
IKE	Internet Key Exchange Protocol
ILEC	Incumbent Local Exchange Carrier
IMAP	Internet Message Access Protocol
InfoSP	Information Service Provider
INSP	Internet Network Service Provider
Intserv	Integrated Services
IOS	(Cisco) Internet Operating System
IOTP	Internet Open Trading Protocol
IP	Internet Protocol
IPng	Internet Protocol, Next Generation (=IPv6)
IPsec	IP Security Protocol
IPv4	Internet Protocol Version 4
IPv6	Internet Protocol Version 6
IS	Intserv, Integrated Services
ISDN	Integrated Services Digital Network
IS-IS	Intermediate System to Intermediate System Routing Protocol
ISP	Internet Service Provider
IT	Information Technology
ITU-T	International Telecommunications Union – Telecommunications Standardization Sector
IXP	Internet Exchange Point
JUNOS	Juniper Network Operating System
LAN	Local Area Network
LDP	Label Distribution Protocol
LFVC	Leap Forward Virtual Clock
LINX	London Internet Exchange
LP	Linear Programming
LSP	Label Switched Path
LSR	Label Switching Router
MAC	Media Access Control
MAN	Metropolitan Area Network
MBone	Multicast Backbone
MIP	Mixed Integer Programming
MLD	Multicast Listener Discovery
MMORPG	Massive Multiplayer Online Roleplaying Game
MPEG	Motion Pictures Experts Group

MPLS	Multi-protocol Label Switching
MPλS	Multi-protocol Lambda Switching
MPRASE	Multi-period Resource Allocation at System Edges
MSS	Maximum Segment Size
MTU	Maximum Transmission Unit
NAP	Network Access Point
NAT	Network Address Translation
NNTP	Network News Transfer Protocol
NS2	Network Simulator 2
NSF	National Science Foundation
NSFNet	National Science Foundation Network
OC	Optical Carrier
OLO	Other Local Operator
OSI	Open Systems Interconnection
OSPF	Open Shortest Path First
OXC	Optical Cross-connect
P2P	Peer-to-peer
PCBE	Price-controlled Best-effort
PDB	Per Domain Behaviour
PDH	Plesiochronous Digital Hierarchy
PDU	Protocol Data Unit
PGPS	Packetised General Processor Sharing
PPP	Point-to-point Protocol
PHB	Per Hop Behaviour
PI	Proportional Integrator
PLC	Packet Loss Concealment
PLR	Packet Loss Recovery
PMP	Paris Metro Pricing
POP	Point of Presence
POTS	Plain Old Telephone Service
PPP	Point to Point Protocol
QBSS	QBone Scavenger Service
QoS	Quality of Service
RAM	Random Access Memory
RED	Random Early Detection
REM	Random Exponential Marking
RFC	Request for Comments
RIP	Routing Information Protocol
RP	Retail Provider
RSpec	Reservation Specification
RSVP	Resource Reservation Protocol
RSVP-TE	RSVP Traffic Engineering Extensions
RTP	Real-time Transport Protocol
RTS	Real-time Strategy (Game)
RTT	Round-trip Time
SACK	Selective Acknowledgement

SCFQ	Self-clocked Fair Queueing
SCORE	Stateless Core
SCP	Strategic Consultant Provider
SDH	Synchronous Digital Hierarchy
SDL	Simplified Data Link
SFQ	Start Time Fair Queueing
SIP	Session Initiation Protocol
SLA	Service Level Agreement
SLO	Service Level Objective
SLS	Service Level Specification
SMS	Short Message Service
SMTP	Simple Mail Transfer Protocol
SNMP	Simple Network Management Protocol
SONET	Synchronous Optical Networking
SP	Service Provider
SSH	Secure Shell
SSP	Storage Service Provider
STM	Synchronous Transfer Mode
TCAM	Ternary Content Addressable Memories
TCP	Transmission Control Protocol
TCS	Traffic Conditioning Specification
ToS	Type of Service
TSpec	Traffic Specification
TTL	Time to Live
UDP	User Datagram Protocol
UMS	Unified Messaging Service
URL	Uniform Resource Locator
US	United States
VBR	Variable Bit-rate
VC	Virtual Clock
VoIP	Voice over IP
VPN	Virtual Private Network
VPOP	Virtual POP (Point Of Presence)
VQ	Virtual Queue
WAN	Wide Area Network
WDM	Wavelength-division Multiplexing
WFQ	Weighted Fair Queueing
WF2Q	Worst-case Fair Weighted Fair Queueing
WRR	Weighted Round Robin
WWW	World Wide Web
YESSIR	Yet another Sender Session Internet Reservations

Abbreviations of Models and Algorithms

Chapter 8	see also Table 8.8
IS-α_{GS}	Intserv QoS System with parameter α_{GS}
sDS-c-p	Standard Diffserv with Central Bandwidth Broker and Parameter p
sDS-d-p	Standard Diffserv with Decentral Bandwidth Broker and Parameter p
sDS-n	Standard Diffserv without Bandwidth Broker
oDS-c-p	Olympic Diffserv with Central Bandwidth Broker and Parameter p
oDS-d-p	Olympic Diffserv with Decentral Bandwidth Broker and Parameter p
oDS-n	Olympic Diffserv without Bandwidth Broker
BE-OF	Best-effort System with Overprovisioning Factor OF
Chapter 10	
OPT	Minimal Cost Interconnection Model (Model 10.1)
H TR	Transit Heuristics
H PA	Peer-with-all Heuristics
H PS	Peer-at-selected-IXPs Heuristics
H EV	Evolution Heuristics
MT	Minimum Number of Transit Providers Policy (Model 10.2)
MC	Minimum Free Capacity Policy (Model 10.3)
AF	Anticipating Failure Policy (Model 10.4)
MCAF	Combined MC and AF Policy
PB	Peering Bonus (Model 10.5)
HC	Hop Constraint (Model 10.6)
HP	Hop Count Penalty Costs Policy (Model 10.7)
PC	Penalty Costs Policy (Model 10.8)
LC	Limiting Change Policy (Model 10.9)
Chapter 12	see also Table 12.3
SP	Shortest Path Routing
CC	Path Selection Strategy Minimising (Weighted) Congestion Costs
CC$_{uw}$	Path Selection Strategy Minimising Unweighted Congestion Costs
U$_{max}$	Path Selection Strategy Minimising Max. Utilisation
U$_{max}$**L**$_{av}$	Path Selection Strategy Minimising Max. Utilisation and Av. Load
U$_{max}$**P**$_{av}$	Path Selection Strategy Minimising Max. Util. and Av. Path Length
U$_{max}$**U**$_{av}$	Path Selection Strategy Minimising Max. Utilisation and Av. Utilisation

$\mathbf{U}_{av}$	Path Selection Strategy Minimising Av. Utilisation
$\mathbf{U}_{av}\mathbf{P}_{av}$	Path Selection Strategy Minimising Av. Util. and Av. Path Length
$\mathbf{p}_{av}\mathbf{L}_{av}$	Path Selection Strategy Minimising Av. Path Length and Av. Load
$\mathbf{L}_{av}$	Path Selection Strategy Minimising Av. Load

Chapter 13

CE	Capacity Expansion
TMCE	Combined Traffic Engineering and Capacity Expansion
T	Threshold-based Capacity Expansion Strategy

Part I

Introduction and Basics

1

Introduction

1.1 Motivation

The Internet is a large network formed out of more than 30000 autonomous systems (AS), in which each AS is a collection of IP networks sharing a common routing strategy. These networks are operated by thousands of **Internet service providers** (ISPs). On the one hand, the ISPs compete with each other for customers and traffic, on the other, they have to cooperate and exchange traffic, otherwise the worldwide connectivity would be lost. In contrast with the traditional telecommunication markets, there are almost no central instances in the Internet enforcing cooperation and regulation of the market.

The ISP market is characterised by serious competition and is currently in a phase of consolidation: according to Access ECommerce (University of Minnesota Extension Service) in the three years after the dot.com crash in 2000, at least 962 Internet companies have either shut down or have declared bankruptcy. In 2002 WorldCom, one of the largest tier-1 ISPs, filed for bankruptcy protection, the largest such filing in US history. A year ago, in the US Internet access market, the top 10 providers accounted for less than half of the Internet users; today the market is consolidated more so that the top 10 providers account for almost three out of four Internet users, see Boardwatch (2004) (i). The situation was aggravated by peer-to-peer applications like Napster, Gnutella and Kazaa that led to extreme traffic growths, causing additional costs and challenges for access providers.

Never before had ISPs to be as competitive as they have to be today. The goal of this book is to help ISPs to be more competitive. The focus of this book is on network operations. It is highly important for ISPs to operate their network **efficiently**. In addition, they have to strive for successful business practices. Traditional successful business practices in a competitive environment are cost leadership, market segmentation and differentiation, see Porter (1980). Market segmentation and differentiation depend on measures to offer different products to different markets. A central service of ISPs consists of forwarding IP packets; this service can be differentiated by price and quality. Therefore, besides efficiency it is important to investigate the **quality of service** (QoS) for ISPs too. The latter is important for a second reason: many emerging multimedia applications such as voice-over Internet Protocol (VoIP) and video communication can greatly benefit from QoS support in a network, see Bhatti and Crowcroft (2000); Crowcroft *et al.* (1999); Steinmetz and Nahrstedt (2004). QoS support in a network therefore opens further possibilities

The Competitive Internet Service Provider: Network Architecture, Interconnection, Traffic Engineering and Network Design
Oliver Heckmann © 2006 John Wiley & Sons, Ltd

for value-added services with which providers can differentiate themselves and target new markets.

To summarise the motivation of this book and to be competitive, an ISP has to operate its network efficiently and it must also control the level of QoS offered to its customers. Efficiency and QoS are discussed in more detail before we make an overview and a short summary of the different chapters of this book are given.

1.2 Efficiency and Quality of Service

1.2.1 Network Efficiency

Merriam-Webster defines *efficiency* as an *effective operation as measured by a comparison of production with cost (as in energy, time, and money)*. In the context of network services, the "production" of an ISP's network can be described by the amount of traffic transported by the ISP. Therefore, we define the efficiency of a network as

$$\text{Network Efficiency} = \frac{\text{Transported Traffic}}{\text{Costs}}$$

Depending on the level of abstraction the traffic can be measured by

- the volume of traffic carried through the network, or
- the number of flows or sessions transported through the network, or
- the number of customers served.

The network costs can be monetary or non-monetary. Typical monetary cost factors for an ISP are

- costs for leasing communication lines,
- interconnection fees,
- costs of the network hardware (e.g. routers, switches and line-cards), and
- costs for the technical and administrative staff.

The examples for non-monetary costs are

- the complexity in computation time or the memory of managing and scheduling a packet,
- the amount of state necessary in a network to provide a certain QoS, or
- the technical effort of changing resource allocations.

In this book, many optimisation problems are presented and solved to maximise efficiency. In many circumstances, one can assume that the amount of traffic is given and constant on the timescale of the investigated problem. In that case, the efficiency is maximised if the costs are minimised.

1.2.2 Network Quality of Service

Quality of service (QoS) is defined in Schmitt (2001) as

> *the well-defined and controllable behaviour of a system with respect to quantitative parameters.*

Typical QoS parameters on the network layer are packet loss, packet delay, jitter (delay variation), throughput etc. Different applications have different QoS requirements: real-time applications are more sensitive to delay and jitter, while elastic bulk-transfer applications are relatively insensitive to the delay and jitter of individual packets but are sensitive to the overall achievable throughput.

For this reason, measuring the QoS of a specific flow or session directly with technical parameters like loss or delay can be misleading. Preferably, *utility functions* should be used; they transform the technical QoS parameters and traffic flow experiences, into a utility value depending on the requirements of that flow's application. Examples for this can be found throughout this book.

1.2.3 Trade-off between Efficiency and Quality of Service

If one looks at certain aspects of an ISP (e.g. the interconnection mix or the QoS system), a **trade-off** between the QoS and the network efficiency can be observed. This trade-off is depicted in Figure 1.1 in which the grey area depicts the **solution space** marking the points where feasible solutions exist for an ISP. Consider for example, interconnections – the connections of the network of one ISP with other networks. Typically, many different possibilities for an individual interconnection exist, resulting in a larger number of possible interconnection combinations as a provider typically has interconnections with many other providers. These interconnection combinations differ in their costs, efficiency and in the QoS they support (see Part III of this book for details); the solution space is formed by all feasible interconnection combinations. Offering a very high QoS usually leads to a lower network efficiency because either less traffic can be supported to provide the QoS or the costs for handling the traffic increase. The same holds true the other way around, leading to the shape of the solution space depicted in Figure 1.1.

It is important to stress that the solution space only contains *feasible* solutions. In the example above, if a specific interconnection mix violates the requirements of the ISP with respect to other criteria – e.g. security – it is not considered feasible and therefore not a part of the solution space. This book takes the position of an ISP and optimises the efficiency and QoS of its network, taking the trade-off between the two goals into account.

The **optimal performance boundary** is marked by the upper right border of the solution space (see Figure 1.1). As long as an ISP does not operate at the optimal performance boundary, it can improve either QoS or efficiency without having to reduce the other goal. It is clear that the goal of a competitive ISP is to operate at the optimal performance boundary. A major contribution of this book is that it investigates how the optimal performance boundary can be found for different aspects from building, operating and managing a network.

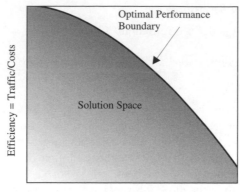

Figure 1.1   Trade-off between Network Efficiency and Quality of Service

A different question is where along that boundary should an ISP operate. Internet is a heterogeneous network of networks; the ISPs that control these networks differ from very small regional niche providers to huge multinational backbone providers. Where along the optimal performance boundary an individual ISP should operate, is basically its own decision. And that decision, of course, depends on the requirements of the market and the ISP's customers. The customer requirements and market structures are constantly developing. This book does not make any limiting assumptions in this respect as, for example, only looking at one point of the performance boundary (e.g. providers with high QoS but also high costs) this would make its results only applicable to a small subset of all ISPs. Even for those ISPs, the results would only be valid until the next change occurs in a fast evolving business like the Internet service provisioning business.

Instead, the book strives for generic solutions, if possible. It tries to investigate the entire boundary. For example, a large variety of different QoS systems are evaluated in this book, each with distinct advantages and disadvantages in which none of the systems is clearly better than another. They rather represent different points on the performance boundary. Which of these systems a provider should employ, depends on the specific situation it is in: its customers, its financial situation and many more factors .

1.3 Action Space and Approach

The next important point to discuss is the action space investigated in this book. The book focuses on *technical* operations, not on marketing or other measures. Of all technical measures, the focus is on building, running and interconnecting the *network* of an ISP. In Chapter 2, the term *INSP* for *Internet Network Service Provider* is therefore introduced to describe the entity responsible for these actions. The different actions can be grouped into the following three areas (see also Figure 1.2):

- **Network Architecture**
  The network architecture defines the properties of the INSP's network itself. Four sub-architectures can be distinguished:
  ○ the quality of service architecture,

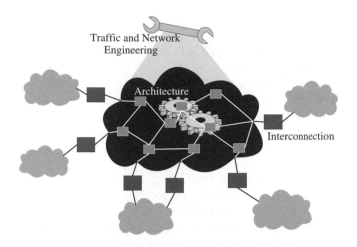

Figure 1.2 The Different Parts of this Book

○ the forwarding architecture,
○ the signalling architecture, and
○ the security architecture.
For the purpose of this book, the most important part of the network architecture is the QoS architecture as it directly determines the QoS and considerably affects the other sub-architectures.

- **Interconnection**
  The network of an INSP has to be connected to other INSPs' networks; the connection between two INSPs' networks is called an *interconnection*. There are different types of interconnections and typically many possible interconnection partners. The interconnection mix strongly influences the amount of traffic an INSP can carry, its costs and the achievable QoS, as this book will show.

- **Traffic and Network Engineering**
  Apart from operating a network on the basis of a selected network architecture and interconnecting it with other networks, the INSP constantly has to manage its own network: the INSP can use *traffic engineering* methods, e.g. to avoid bottlenecks by rerouting traffic and on a larger timescale it can use *network engineering* to update the topology and upgrade the capacity of the network.

Solutions obtained for one goal or in one category while ignoring the other goals or categories respectively can easily result in an overall suboptimal system: a QoS solution that offers good QoS but would result in unacceptable (monetary or non-monetary) costs or could not cope with the required amount of traffic would be inefficient and probably useless. Also, if the network architecture is highly optimised for efficiency and QoS, that advantage would be easily lost if the interconnections are not optimised for the same goal. Further, it would be lost after a few months if the network engineering process fails to upgrade the network to increasing traffic. To avoid this, a system-oriented view on building, operating and managing a network is necessary.

1.4 Overview

The book is structured along three areas: network architecture, interconnection and traffic and network engineering. Every single problem area has a large influence on the overall QoS and efficiency of the provider as the book will show. From this it follows that considering only one of these aspects – e.g. the network architecture – is not enough because the gain of e.g. QoS by carefully optimising the network architecture is immediately lost if the interconnection mix is not adapted to support this level of QoS. Also, it is lost soon if the capacity expansion process (part of the traffic and network engineering problem area) fails.

Part I presents the introduction and an overview of the ISP market. In addition, different basic methods for network performance analysis are presented. They are employed in the rest of the book to analyse the different facets of networks. The Internet protocol stack and the most important protocols are discussed. Finally, the most important applications for the present and the future of Internet are discussed with focus on the QoS requirements and traffic behaviour.

Part II investigates the network architecture, starting with a discussion of the state-of-the-art in Chapter 6. In this part the focus lies on the QoS architecture as it is the foundation for the QoS achievable in a network and also because it largely influences the other aspects of the network architecture and thus indirectly the architectural costs and efficiency. Using analytical methods, two aspects of QoS architecture (admission control and service differentiation) are investigated in Chapter 7. For both the aspects, the over-provisioning factors are derived with different analytical methods. An over-provisioning factor is the relation of capacity (mainly bandwidth) between a plain best-effort system and a QoS system at the point in which both systems offer the same QoS. It captures the benefit of the QoS system.

The benefit of admission control largely depends on the adaptivity of the application and the load distribution. In a well-dimensioned network, the over-provisioning factor is usually less than 300%; for adaptive applications it is significantly even smaller than 150%.

We derive a novel network model that – contrary to the existing approaches – allows us to analyse service differentiation. The over-provisioning factors resulting from service differentiation are significantly higher, typically between 200% and 500%, depending on the traffic assumptions.

Different QoS systems based on the QoS architectures that are in the standardisation process of the Internet Engineering Task Force (IETF) are analysed in a simulative study (Chapter 8). The study sheds light on the quantitative trade-offs of the different approaches to QoS, e.g. per-flow versus per-class scheduling and central versus decentral admission control. One of the conclusions of this chapter is that Diffserv networks can be over booked by at least a factor of three to increase efficiency. The over-provisioning factors determined in this chapter are similar to those determined with our novel analytical models for the service differentiation in the previous chapter. The book also shows that contrary to common belief, the utilisation of a network with Expedited Forwarding (EF) traffic can be higher than a few per cent. For the basic experiment in Chapter 8, the Charny bound predicts a maximal utilisation of 7.98%. A bandwidth broker described in this book can raise the utilisation to over 27%.

In **Part III,** interconnections are investigated starting with an overview and the discussion of the state-of-the-art in Chapter 9. Chapter 10 shows that the interconnection mix has significant impact on efficiency and QoS. Reliability is important in this context too and therefore is also discussed. Different strategies for optimising the interconnection mix with respect to efficiency are presented and evaluated by a series of simulations.

Interconnection related costs are one of the highest cost factors of an INSP. The results show that from 5% to more than 30% interconnection related costs could be saved. The analysed strategies can be easily extended to control reliability and QoS too. Chapter 10 presents strategies that allow to explicitly adjust the desired trade-off between QoS and efficiency. The result is the optimal interconnection mix for an INSP.

Traffic and network engineering is discussed in **Part IV** of this book, starting with the discussion of the state-of-the-art in Chapter 11. Several traffic-engineering strategies are presented in Chapter 12. The chapter shows that the maximum utilisation criterion, which is typically used in related work, is not a good objective function for traffic engineering. The chapter derives a congestion function that should be used instead as an objective function. The impact of traffic engineering on the network efficiency and QoS is evaluated in a series of experiments. The results show that traffic engineering can decrease the congestion in a network during times of high load. A traffic-engineered network can therefore offer higher QoS and/or higher efficiency (because more traffic could be carried than a network without traffic engineering). The absolute benefit of traffic engineering, however, strongly depends on the traffic-engineering strategy. On the basis of the experiments, the book gives recommendations of which strategies to use and also how to use them. However, for certain topologies and traffic distributions, the benefit of traffic engineering can be rather small. In this case, it is very doubtful whether a traffic-engineering solution can amortise its costs.

Network engineering with focus on capacity expansion is finally discussed in Chapter 13. Capacity expansion is an important and a frequent task in today's IP networks because the traffic volume is increasing steadily. First, the influence of capacity expansion on the performance of the different QoS architectures of Chapter 8 is discussed. If capacity is abundant, the differences between the QoS architectures diminish. If capacity is scarce, the systems with a strict admission control manage to maintain QoS while the other systems suffer to different extents. The best effort systems are the ones most sensible to in-time capacity expansions.

Besides this, different capacity expansion algorithms are presented and evaluated in Chapter 13. The best strategy takes the mutual influences of traffic engineering and capacity expansion into account. The rule of thumb often used by today's INSPs shows acceptable performance only if their parameters are set correctly; our strategy performs significantly better and is robust against uncertain traffic predictions. Finally in that chapter, the effects of elastic TCP traffic-on-traffic matrices and on capacity expansion are discussed with some analytical models. The elasticity of the traffic influences the capacity expansion measures if the network is highly utilised before the expansion. The presented models can be used to predict this effect and react accordingly.

To understand the terminology used in the book, we recommend every reader to have a look at Chapter 2. Readers who are technical experts in IP network technology and performance analysis can probably skip the introductory Chapters 3 and 4, possibly also 5. Readers who are interested in the basics should focus on Chapters 2 to 5 and the first chapters of each part (6, 9 and 11). Moreover, in Chapters 6, 9 and 11 most of the related scientific works are summarised. Finally, readers who are already very familiar with the topics and who look for results might want to focus on the chapters describing the experiments (7, 8, 10, 12 and 13).

2

Internet Service Providers

Today, the Internet service provider (ISP) market is characterised by a huge diversity of offered services and business connections, differing significantly from the traditional telecommunications market. The further progression of Internet technologies, business innovations and regulatory and policy factors are adding to the complexity. The diversity of ISP market services and interactions is also reflected by the companies involved, ranging from niche market ISPs to global players. Their business portfolios vary from one to multiple services.

In this chapter, we investigate the Internet service market and the different types of ISPs that exist. We identify a subset of ISPs that we call Internet *network* service providers (INSPs). INSPs provide packet-forwarding services and operate Internet Protocol (IP) networks; they form the main focus of the rest of this book.

The term *Internet Service Provider*, or *ISP* for short, is commonly found in literature, with a lot of different definitions. However, most of the classification models discussed in the literature are not detailed enough for technical analysis, or they cover only a limited sub-market of the whole Internet service market as we understand it.

A common perception of the Internet service provider is that of *"an organisation that sells access to the Internet"* – an access provider, see Norton (2002). Huston (1998) defines the ISP also as an access provider that may additionally *"provide various value-added services, such as email, bulletin board services and others"*.

This notion can be found in the work of Greenstein (1999) as well. According to his study, ISPs are selling basic (smallband) Internet access and some optional services. The services provided by ISPs fall into five broad categories: *basic access, frontier access, networking, hosting and Web page design;* see Table 2.1 for the results of the survey in Greenstein (1999).

A general classification of service providers from an industry point of view shows, for example, the service provider's initiative from Sun Microsystems (2000); it is summarised in Table 2.2. This is a more complete approach but is quite unstructured for technical analysis, as it is mixing Internet access and hosting, which are technically very different. Also, it basically ignores that many companies act in many different roles.

Lakelin *et al.* (1999) give a different classification of ISPs and their services. It is based on the size of the company and its business model, see Table 2.3.

The Competitive Internet Service Provider: Network Architecture, Interconnection, Traffic Engineering and Network Design
Oliver Heckmann © 2006 John Wiley & Sons, Ltd

Table 2.1 ISP Services as in Greenstein (1999)

Service	No. of Companies
Basic (Smallband) Access	3816 (100%)
Frontier (High Speed / Broadband) Access	1059 (27,8%)
Networking	789 (20,6%)
Web Hosting	792 (20,7%)
Web Design	1385 (36,3%)

Table 2.2 Service Provider as in Sun Microsystems (2000)

Role	Services	Companies
Internet Service Provider	Access, Hosting, Email	AOL, Mindspring, @Home
Network Service Provider	High Bandwidth, Backbone Services, VoIP, VPN	Level (3), Concentric, Qwest, UUNet
Application Service Provider	Storefront, Help Desks, Enterprise Resource Planning, ...	Digex, GTE, Savvis, Vantive, Siebel, Oracle, Corio
Full Service Provider	Turnkey Enterprise Services, Supply Chain, IT Services	EDS, AT&T Worldnet, Exodus
Portals	Aggregate Content, Destination	Yahoo, Excite@Home, AOL

Table 2.3 Service Providers as in Lakelin et al. (1999)

By Business Model		By Size
Online Service Providers	Cable Operators	Local ISPs
Incumbent Telecoms	IT companies	National ISPs
New entrant Telecoms	Brand driven ISPs	International ISPs

The approaches shown above are not comprehensive and structured enough to express the variety of the diverse ISP business – from access providers over Content Delivery Networks (CDN) to Communication Service Providers (CSP) – and at the same time the technical functions of the ISPs. Most of them focus too strongly on the Internet access market, neglecting other important Internet services, for example, caching and hosting services.

We propose a role model that describes the different technical roles that can be found in the business portfolio of actual ISP companies. Each role is used to provide a well-defined set of services; therefore, the individual roles and their relationship form a solid basis for scientific, technical or economical studies. Real-world ISPs typically act in more than one role of the role model. Therefore, the role model can also be used to classify existing ISPs, to describe the differences between two ISPs or to describe and analyse market trends.

Our role model and definition of ISPs is introduced in the next section, it is then used in Section 2.2 to classify a selection of well-known real-world ISPs.

2.1 A Classification Model for ISPs

We will now describe the relationship between a (real-world) company with its services and/or support functions and a role. It is essential to separate a role and a (real-world) company. A **company** is defined as a real-world entity that is taking on one or more roles. A **role** represents a group of functions to provide a set of related services to customers (see Figure 2.1). **Functions** can be divided into internal functions, which do not directly affect the customers and **services** (external functions) that are offered to the customers.

Example: Real-world ISPs like AOL (America Online) are often engaged in more then one role. AOL, for example, provides Internet access to its subscribers. This is its core role. However, AOL also offers web space, email, online games and a marketplace for other companies to sell their products.

One role, for example, that of offering Internet access, consists of several services, like the Internet dial-up access by modem, ISDN (Integrated Service Digital Network), cable or DSL (Digital Subscriber Lines). Internal functions in that context involve authentication and accounting actions by the provider or the operation of a routing protocol; they are transparent to the customer.

Using Constantiou and Altmann (2003) as basis, we derive the following detailed role model.

2.1.1 Definition of Internet Service Providers

The term **Internet service provider (ISP)** is used as an umbrella term for information providers, server service providers, Internet network service providers (INSP):

- An **ISP** is a company whose core business consists of at least one of the ISP roles of Figure 2.2.
  Similarly, an **Information Service Provider** is defined as a company whose core business consists mainly of one or more of the information provider roles of Figure 2.2. **Server service providers** and **Internet Network Service Providers (INSPs)** are defined accordingly.
- **Information Service Provider roles** provide services that offer different kinds of information via the Internet. **Storage Service Provider Roles** provide basic and sophisticated services for storing and distributing this information. **INSP roles** provide services to forward this information via IP packets towards their target.

Figure 2.1   General Role Model

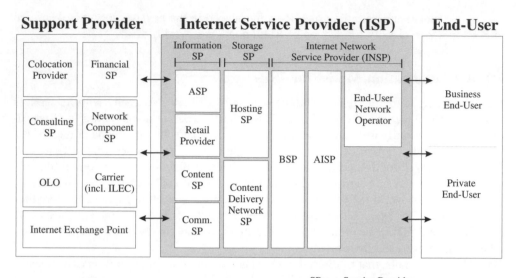

SP	Service Provider
AISP	Access Internet Service Provider
BSP	Backbone Service Provider
OLO	Other Local Operator
ASP	Application Service Provider
ILEC	Incumbent Local Exchange Carrier

Figure 2.2   ISP Role Model

Examples:

• AOL is an ISP because by offering content and communication services, it acts in the information provider role. By offering access service, it acts at the same time in the INSP role.

• Deutsche Bank offers online banking service and is therefore acting as content provider, which is a role of information providers. But as the core business of Deutsche Bank includes different roles (banking and brokering roles), Deutsche Bank itself is not an ISP. For more examples, see Section 2.2.

Next, a detailed description and a short discussion of the different roles found in Figure 2.2 is presented.

2.1.2 Internet Service Provider Roles

The ISP roles can be classified as follows (see Figure 2.2).

• The **Internet Network Service Providers (INSPs)** are responsible for the Internet connectivity; they operate a network and offer packet forwarding services. There are three types of INSPs:
  ○ The **ENOs (End-user Network Operators)** operate end-user network edges,

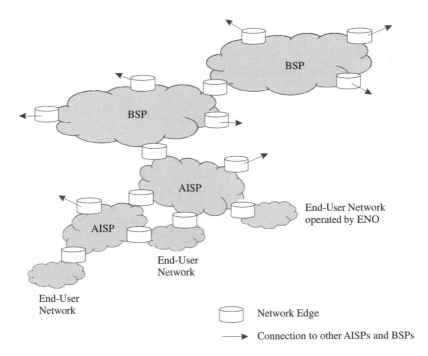

Figure 2.3   INSP Roles

○ the **AISPs (Access ISP)** aggregate and forward the traffic of network edges and
○ the **BSPs (Backbone Service Providers)** forward traffic without direct contact to
  end-user network edges (see Figure 2.3).
● The **Storage Service Providers** offer server and storage space in the Internet. Other
  ISPs might depend on this to be able to offer their own service or employ the storage
  services (e.g. caching) to improve the performance of their own services.
● **Information Service Providers** offer information. They cover the higher Internet layers. Their information is carried by INSPs to the end-user.

How the information flow is affected by different ISPs is described exemplarily in
Figure 2.4:

● End-users, for example, access the access provider's network via a modem or DSL
  connection to a point-of-presence (POP) or a virtual point-of-presence (VPOP) of the
  access provider.
  A POP can be described as a node in the INSP's network topology. The routers,
  switches, servers and other equipment of an INSP are located at its POPs. Typically,
  these POPs are geographically distributed to keep the distances to customers and interconnection partners short. The size of an INSP is often measured by the number of
  POPs it is operating.
  The difference between a POP and VPOP is that the latter does not actually belong to
  the ISP; it is only a leased access to another company's POP.

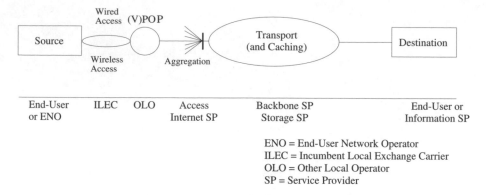

ENO = End-User Network Operator
ILEC = Incumbent Local Exchange Carrier
OLO = Other Local Operator
SP = Service Provider

Figure 2.4   Information Flow from Source to Destination (Example)

- The access medium (e.g. voice line, cable, radio transmission) from end-user to the (V)POP is usually owned by an Incumbent Local Exchange Carrier (ILEC).
- The AISP aggregates the data from many end-users and transports it via its own network and that of other connected AISPs' and BSPs' networks to the destination. The destination can be other end-users or information service providers.
- Depending on the type of the application used, storage service providers (SSPs) may provide caching space.

2.1.2.1 Internet Network Service Provider (INSP)

End-user Network Operator (ENO)

An *End-user Network Operator (ENO)* manages the network of an end-user and the network edge[1].

The services include forwarding the packets within the business end-user's network to the access point of the access ISP (AISP) as well as administration and support services. In the layer reference model (Figure 4.1), most of these services would be considered as layer 3 services. Supplementary services include the administration of an IP address pool, the operation of DNS servers and web caches as well as managing security issues.

The role of the ENO is usually filled out by a department of the business user's company or by a facility within a university.

Access Internet Service Provider (AISP)

The *Access Internet Service Provider (AISP)*[2] connects end-user networks with the Internet and forwards their IP packets toward their destination.

---

[1] The network edge is a connection between two networks. For a detailed definition, see Section 9.

[2] As shown above, in related works, the term *ISP* is often used as a synonym to *AISP*, to the combined portfolios of AISPs, OLOs, ILECs or sometimes to application service providers. In this book, we distinguish between the AISP role and the generic term ISP.

AISPs aggregate traffic from network edges and forward it directly to the destination host, if that host is reachable from within their network. Otherwise, the traffic is forwarded to other AISPs and BSPs (see Figure 2.4).

The ownership of local infrastructure is not a prerequisite to accomplish this as they can lease the needed infrastructure from OLOs and ILECs. This has been the case, particularly in the past, and for small AISPs. As the tendency of today's AISPs is to build up their own local networks to minimise costs and increase profit (see Lakelin and Wood (2000)), the OLO support role becomes more and more irrelevant.

The AISP market is going through a consolidation phase, in which business failure, mergers and acquisitions have all been important factors. Today, the top 10 AISPs in the United States account for almost three out of four Internet users. Only a year before, the top 10 accounted for less then half of them, see Boardwatch (2004), (i). The stiff competition in the market forces the AISPs to search for possibilities to stand out from the crowd, by acting in other roles as a content provider, for example.

Backbone Service Provider (BSP)

A *Backbone Service Provider (BSP)* provides packet forwarding services without direct contact to the end-user networks; typically across long distance.

BSPs operate large Internet backbone networks, aggregate the traffic from AISPs and transport them over their networks. A backbone network is supposed to have large capacities while concurrently spanning large geographical areas. There are only a few BSPs that operate worldwide, but almost every country with former state monopolies has at least one big national backbone provider (e.g. Deutsche Telekom AG in Germany).

The difference between AISPs and BSPs is that BSPs do not offer services directly to end-users. The most clearly defined BSPs, in the sense of our definition, are the *tier 1* BSPs. They operate large backbones that interconnect solely by peering (see Section 9.2.2) and do not need to purchase transit (see Section 9.2.3) from any other backbones. We will further discuss these thoughts in Chapter 9. Major worldwide BSPs include MCI[3], AOL, Qwest and Sprint.

Another difference between AISPs and BSPs is their revenue drivers. AISPs have to make revenue with a relatively small number of users, whereas BSPs' revenue driver is the volume transferred over their network. The distinction between both categories is becoming more and more blurred, however, as BSP companies are also starting to act in the end-user market by offering Internet access and added services to business users, see Huston (1999a).

2.1.2.2 Storage Service Providers

Content Delivery Networks

A *Content Delivery Network (CDN)* provides a platform based on overlay networks operating on top of the actual IP infrastructure. This allows information service providers to distribute their content without having to manage infrastructure.

---

[3] Previously called WorldCom respectively MCI WorldCom.

CDNs use specialised web-caches or video-caches to push replicated content close to the end-users. This service can be important for information providers if their services depend on very short retrieval times. For example, content providers who have to transfer large data volumes like videos to their customer in a short time could use this type of service.

CDNs get their revenue in most cases directly by charging the information provider that produces the data, not the end-user. A famous example for company offering CDN services is Akamai Technologies Inc. (2006).

As new services are offered in the CDN role, the market is further developing; one new service is enterprise CDN (eCDN). Different from content delivery or basic caching in that it goes beyond the traditional methods, eCDN allows enterprises to preposition specific content in certain caches for specific users and user groups. Akamai, Sprint, Qwest, Equant and IBM offer eCDN services, see, for example, Boardwatch (2003), (iii) or Akamai Technologies Inc. (2002).

Hosting Service Provider

The *Hosting Service Provider* offers housing, serving and maintaining storage space and files for customers. The services range from maintaining pure storage space over offering shared file systems to hosting Web sites and maintaining FTP servers. The services typically use layer 5 protocols.

The services also include periodic backup and archiving as well as consolidation of data from multiple customer company locations. This enables efficient data sharing. To realize the offered services, most hosting service providers use server farms[4] and the services of colocation providers (see below).

The hosting services can be divided into virtual and dedicated hosting. Often, "hosting" and "virtual hosting" are used as synonyms. *Virtual hosting* is the provision of hosting services so that a company does not have to buy servers with permanent connection to the Internet. Some virtual hosting service providers make it possible for customers to have more control of their files and Internet connection by providing a virtual server. *Dedicated Hosting* on the other hand provides customers with a dedicated server. The dedicated server can be rented at the provider's location or a customer can place his own equipment at the provider's location.

The spectrum of the offered services ranges from subscriber free space (as value-added service of AISPs like AOL for private Web pages) to extensive business solutions for other ISPs and business customers. For a complete hosting service, fast connections to the web pages are needed. Therefore, more and more hosting service providers engage in the CDN business field, see Boardwatch (2003), (iv).

---

[4] A server farm is a group of computers acting as servers and housed together in a single location, often under the control of a colocation provider. Server farms need a huge amount of power, typically 10 to 20 megawatt of power, to keep their servers running and cooled (see Abreu (2001)).

2.1.2.3 Information Service Provider

Application Service Provider

> The *Application Service Provider* offers access over the Internet to applications and related services that would otherwise have to be executed and managed locally.

The applications are normally accessed via a Web browser interface. The service of an application service provider[5] enables companies to move applications from desktops to dedicated application servers, having now only a centralised server to maintain instead of a larger number of workstations. The applications offered range from high-end enterprise resource planning and supply-chain management systems (such as those offered by Oracle and PeopleSoft) to simpler groupware and officeware applications. Service level agreements, covering bandwidth availability, software mechanisms and technical support, are also typically offered, see Lakelin and Wood (2000).

Essentially, the application service provider business model works by reducing infrastructure and management costs (using economies of scale) as it aggregates the infrastructure of multiple IT companies.

Retail Provider

> The *Retail Provider* can be seen as a merchant that offers its products or provides a marketplace for other companies' products over the Internet.

Often, retail providers simply use their Internet presence to increase the sale of their regular business. This method is often referred to as *multi-channel retailing*. However, there are a lot of companies that are only engaged in the retail provider role (e.g. Amazon or Ebay).

Internet marketplaces can be classified by defining whether the two involved parties are of the business type (B) or of the customer type (C) (see Table 2.4).

Content Provider

> The *Content Provider* creates or augments content. That content can be news, audio and video content, etc.

Table 2.4   Classification of Internet Markets

		Demand	
		B (=business)	C (=consumer)
Supply	B (=business)	**B2B**	**B2C**
	C (=consumer)	**C2B**	**C2C**

[5] A list of the top 25 application service providers from January 2004 can be found at ASPnews (2004).

Usually, content providers operate a central server to store their content or use the service of storage service providers. The services of this role range from offering company and product information on web pages to offering video-on-demand. Another type of content providers is comprised of information services like search engines (e.g. Google) and encyclopaedias (e.g. WhatIs?com) that are published on the Internet. Typically, content providers either charge the end-user directly or more commonly try to finance themselves by advertising.

Communication Service Provider

The *Communication Service Provider* offers Internet-based communication service like email, chat, e-cards and voice over IP (VoIP).

For example, companies such as GMX are offering unified messaging services to enable their customers to combine the vast communication options in one service; a customer can thus combine non-Internet-based services like fax or SMS with the Internet service platform.

Communication service providers currently expand their offered services to gain a higher market share and to increase their per user profit. Some of the added services are spam protection and anti-virus applications. Most of the companies are performing in more than just the pure communication service provider role; typically, they are also engaged in the application and hosting service provider roles.

2.1.3 Support Provider Roles

The support provider roles offer services that support and keep the ISP roles running. The services include layer 2 connectivity services, financial transactions and the supply, maintenance and service of technical equipment.

2.1.3.1 Carriers

Incumbent Local Exchange Carrier (ILEC)

The *Incumbent Local Exchange Carrier (ILEC)* closes the local loop to the end-user by offering layer 2 connectivity between the edge router of the end-user to a POP of the AISP.

The ILECs are telecommunication providers with their own circuit-switched local-loop networks. Originally, many ILECs did not have any IP infrastructure and therefore no possibility to connect directly to the Internet; they needed OLOs (see below) to do so (see Figure 2.4). The offered services are layer 2 and layer 1 services (see Figure 4.1). In most countries, ILECs are affected by telecommunication regulations, as they normally use voice lines that underlie additional regulation. An exception is modern broadband connections like ADSL, which are only used for data transportation. This has a big influence on the number of players in this segment. In the past, there was normally only one big state-owned monopolist per country. This was especially true in Europe. With a growing deregulation of the telecommunications market in Europe, this type of service is offered by a growing number of companies.

The ILECs used to play a very important role in the Internet service market, for without the local-loop infrastructure of the ILECs, Internet access would not have been possible for so many people, and the prices would have been higher. In the future, it will probably be difficult to find companies offering ILEC-only services, as they will tend to enhance their portfolio to act as a combined ILEC/OLO/AISP role, see Lakelin and Wood (2000).

The trend in the ILEC market goes towards broadband access, thus replacing the Plain Old Telephone Service as the access medium. However, the revenue is mostly gained by POTS and not by broadband service. This is not expected to change in the near future, thus making ILECs depend on POTS in the short to medium term, see Boardwatch (2004), (ii). Deutsche Telekom AG in Germany and the Regional Bell Companies in the United States provide the services of the ILEC role.

Long-Distance Carrier

Long-Distance Carriers provide INSPs with layer 1 and 2 connectivity, for example, leased lines, between two POPs.

This service is needed as not all INSPs can afford or want to buy the infrastructure for their own networks. In some countries, the telecommunication sector is still regulated and INSPs are not allowed to own the layer 2 infrastructure for their networks. Leased lines range from POTS telephone cables to optical lines. The players in this segment are generally the former telephone monopolists who also provide the ILEC service like Deutsche Telekom AG in Germany and the Regional Bell Companies in the United States as well as new players providing high-bandwidth infrastructure.

High-bandwidth services also include dark fibre services, see, for example, Dominion Telecom (2004). The dark fibre service offers customers fibre strands to which they can apply their own optronics to light the fibre. Today, dark fibre services are usually available in MAN markets, as only 5–10% of office buildings have fibre access. Nevertheless, the dark fibre market is expanding because of the dramatic price reductions and the increasing flexibility of dark fibre contracts.

2.1.3.2 Other Local Operator

Other Local Operators (OLOs) typically offer translation services from the ILEC's telecommunication layer 2 networks to the AISPs layer 3 (IP) networks.

The OLO service portfolio includes services like termination of calls, indirect access and number translation. The reason for the existence of the OLOs lies within the offered portfolio of the ILECs. At the beginning of the Internet, a lot of telecommunication-based ILECs were very slow in investing in IP technology. This opened a market segment for the OLOs that connected the ILECs to the Internet by providing the modem banks to translate between the circuit-switched Plain Old Telephone Systems (POTS) and the IP backbone infrastructure. With increasing competence in IP technology in the telecommunication companies, the importance of pure OLO services has decreased rapidly. The OLO role will instead be part of the ILEC or AISP service portfolio (like Deutsche Telekom AG, which provides switching services for small ISPs), see Lakelin and Wood (2000). The trend towards broadband access, in particular, makes OLOs pointless, as the medium itself is digital and therefore does not need any translation services.

2.1.3.3 Internet Exchange Points

The *Internet Exchange Point (IXP)* provides an exchange point in which ISPs can connect with one another in interconnection arrangements.

The structure of the exchange point can range from one exchange facility to several exchange facilities, connected with each other. The IXPs are a key component of the Internet backbone as they offer the possibility of global connectivity. Typically, an INSP is connected to a number of IXPs and these IXPs to a large number of peering partners. The IXP and interconnection topic will be discussed in more detail in Part III of this book.

2.1.3.4 Colocation Provider

A *Colocation Provider* provides carrier-neutral data centre services as well as management services.

The data centre, also called *colocation* facility, is a network-connected secure commercial facility for the housing of carrier and IT infrastructure. Colocation providers also offer services such as equipment housing, on-site engineering and maintenance. The carrier-neutral data centres enable ISPs to manage their own connectivity by negotiating directly with underlying carriers. Two players of this role are Telehouse Europe and Interaxion.

2.1.3.5 Financial Service Provider

A *Financial Service Provider* provides services around the money transfer between the provider and the customer.

The most typical form of the financial service provider role is that of a service provider who takes over the billing for his customers. An example of such a provider is the billing specialist Aurora UK Ltd, see Boardwatch (2003), (vi).

2.1.3.6 Consulting Service Provider

The *Consulting Service Provider* offers consultant services to their customers. Its services cover help on how to run an ISP through all of the company life phases.

Network-specific consulting services are especially appealing for small ISPs that often cannot afford to build up the specialised know-how needed to run their business.

There are two kinds of companies performing the consulting service provider role in the Internet market. The first are independent companies, like Accenture, that specialise in the consulting business. The others are companies that offer a product in a different segment. SAP, for example, has their own consulting department to implement their products.

2.1.3.7 Network Component Service Provider

The *Network Component Service Provider* offers and maintains hardware and software components that are necessary to operate the Internet infrastructure.

They can be differentiated by the type of components they sell to hardware component service providers (e.g. Cisco) and software component service providers (e.g. Oracle). Examples of the offered components are routers, line cards and web servers. Providers that offer installation and maintenance services for these components are also classified as network component service providers.

2.1.4 End-users

Another key part of the Internet market, besides support providers and ISPs, are the end-users who consume the offered Internet services. End-users are classified into *business end-users* and *private end-users*.

The **business end-users** are business entities that use Internet services to generate revenue. The offered services possess the character of investments. The **private end-users,** on the other hand, consume the offered Internet services while using the Internet for private purposes.

2.2 Classification of Selected Providers

Today, the top 10 AISPs in the United States account for almost three out of four Internet users. AOL and MSN, the top two ISPs, alone account for 41% of the Internet users, see Boardwatch (2004), (i). The stiff competition in the market, mainly on the DSL and flat rate markets, forces the AISPs to search for possibilities to stand out from the crowd. The trend goes to generating more revenue from e-commerce transactions and advertising. AOL, for example, offers among other services, a news portal, online gaming, online shopping, and Services For Mobile Phones (SMS), additionally to their access service, see America Online (2006). So AISP companies rarely act solely in the AISP role, they tend to add services to their business portfolio to gain a better market position, see Lakelin and Wood (2000). This expansion trend of AISPs tends to be towards the information provider roles at the moment and can be found throughout the market. Cooperations between AISPs and other ISPs are common as well, for example, AOL has Google's search service embedded within their homepage.

This is just an example from the AISP market and can be found in most of the other markets as well. The tense economic climate in the world forces the competition in almost all of the ISP markets to heat up, thus forcing the companies to find new revenues by differentiating themselves from others. A common possibility to accomplish this is by providing additional value-added services. As can be seen in the following Figure 2.5, most companies act in more than one role. To show that the described role model complies

| | Internet Service Provider | | | | | | | | Support Provider | | | | | | |
| | INSP | | SSP | | | IP | | | | | | | | | |
	End-User Netw. Op.	Access ISP	Backbone SP	Content Delivery Netw.	Hosting SP	Application SP	Internet Retailer	Content Prov.	Communication SP	Carrier	OLO	IXP	Colocation Prov.	Financial SP	Consulting SP	Network Comp. SP
Akamai				•	○											
Amazon, Ebay							•									
AOL		•			○	○	○	○	○							
Cable&Wireless		○	•		○										○	
T-Online		•	○		○	○	○	○	○							
DFN		○	•													
GMX		○				○		○	•							
Google									•							
HRZ	•															
MSN		•				○	○	○	○							
Oracle, SAP						•									○	○
Sprint			•	○	○					○	○		○			
WorldCom		○	•		○											
Aurora														•		
Cisco																•
T-Mobile		○														
T-Com			○							•	○					
Interaxion													•			
LINX, DE-CIX												•				

Abbr	Kind of Role
•	Core Role
○	Additional Role

Figure 2.5 Classification of Selected Providers

with reality, actual companies and their services in the Internet market are classified using the role model derived above.

Akamai

Akamai Technologies offers software and services to enable companies and government agencies to deliver Web content and applications (including video and other high bandwidth content). Through its network of more than 14,000 servers in 70 countries, Akamai services analyse and manage Web traffic, transmitting content from the server geographically closest to the end-user using Akamai's EdgeSuite product. The company also offers audio and video streaming services, content targeting applications and consulting services. According to the introduced role model, Akamai's main role is that of a content delivery network service provider.

America Online

America Online (AOL), the Internet division of Time Warner, is the world's largest AISP with more than 30 million subscribers using its services. AOL customers are mainly private end-users. The revenue mix of subscription, advertising, e-commerce services and Internet sales also reflects the ISP roles involved. According to the introduced role model, AOL is an AISP company, which incorporates the following roles.

- Access ISP: This is the core role. AOL offers dial-up service as well as broadband to end-users.
- Hosting Service Provider: AOL offers web space and hosting services on its servers for end-user homepages.
- Application Service Provider: Online gaming is one example for the application services offered by AOL.
- Internet Retailer: AOL offers on its homepage a marketplace for other Internet retailers to sell their products. AOL gets a commission on total sales revenue.
- Content Provider: Videos, news and various other content are offered mainly for AOL subscribers, but some content is offered for the public as well.
- Communication Service Provider: AOL offers email services as well as mobile services such as sending mobile messages.

Amazon and Ebay

Amazon offers millions of books, CDs, DVDs and videos, as well as toys, tools and electronics. It has a large market share, especially in the Internet book sales market. According to the introduced role model, Amazon is a classical Internet retailer.

Ebay offers a marketplace for all kinds of used and new products. As a marketplace provider, it is also classified as Internet retailer according to the introduced role model.

Deutsche Telekom and T-Online

Deutsche Telekom is the biggest telecom company in Europe and one of the largest in the world. It is divided into four subsidiaries. Its **T-Mobile International** division serves wireless phone customers. The **T-Com** unit is one of the largest carriers in Europe with about 58 Million connections. The company's **T-System** division is specialised in IT services. And finally, the **T-Online** subsidiary, with 13.1 million customers, is one of the leading ISPs in Europe:

T-Online has a diverse business model with both access and non-access businesses in its portfolio. According to the introduced role model, T-Online is, like AOL, an Access ISP that incorporates the following roles.

- Access ISP: This is the core role. T-Online offers dial-up service as well as broadband to end-users.
- Hosting Service Provider: T-Online offers web space and hosting services on its servers for end-users' homepages.
- Application Service Provider: The web-based organiser is one example for the application services offered by T-Online.
- Content Provider: T-Online offers videos, news and various other content to its subscribers. Part of the content is also publicly available.
- Communication Service Provider: Email services as well as international roaming access are offered by T-Online in its role as communication service provider.

DFN

DFN – Deutsches Forschungsnetz – is Germany's National Research and Education Network (it is similar to other nations' research networks). Its main tasks are to provide backbone infrastructure to the German research and education community and to create a testbed for science and development of new techniques. According to the introduced role model, DFN is a backbone service provider company that incorporates the following roles:

- Backbone Service Provider: This is the core role. DFN operates a backbone network, the *G-WiN*. Considering the size and geographical spread of the G-WiN, DFN can be considered as a national BSP.
- Access ISP: DFN offers the DFN@home service for students and scientists who want to access the network of their institution from their home computers.

GMX

GMX offers paid and free email services. To differentiate itself from other companies and to gain more revenues, it added new services like Internet access, online virus scans and an online organiser. GMX is one of the top five email service providers in Germany, see ECIN (2003). According to the introduced role model, GMX is a communication service provider company that incorporates the following roles.

- Communication Service Provider: This is the core role. GMX's main services are the different email services. Included in their ProMail service are additional Communication Service Provider services like fax and voice messages.
- Access ISP: GMX offers access to Internet, using different price schemes like flat rate, volume-based and time-based pricing.
- Application Service Provider: The applications provided by GMX are, for example, an online organiser.

- Content Provider: News and Information about sports, entertainment, lifestyle and much more are provided on the GMX portal.

Google

Google offers a targeted search engine that indexes and ranks Web sites according to the number of links leading to that site. Google is the most-used site in the world for Web searches, it serves more than 80 million users per month[6]. Other Google offerings include newsgroup sites and the web-based email service Googlemail; in addition, it licences its technology to other companies like America Online. According to the role model, Google is a content service provider but also active as communication service provider.

University Network Centres

University Network Centres like, for example, the "Hochschul–Rechenzentrum", of the Darmstadt University of Technology operate university networks and manage the network edge (connection) to the universities' AISPs. According to the role model, these network centres are end-user network operators.

Sprint

Sprint is a global communications company that operates a tier 1 Internet backbone. Sprint is one of the largest BSPs, serving 26 million business and end-users in more than 100 countries. According to the introduced role model, Sprint is a backbone service provider company that incorporates the following roles.

- Backbone Service Provider: The core role. Sprint operates a Tier 1 Internet backbone network.
- Content Delivery Network/Hosting Service Provider/Colocation Provider: Spring also provides global voice, video, data and Internet communications services, web hosting and colocation services.
- Carrier/ILEC: The company's telecommunications operation provides local telephone service through over eight million access lines in 18 states.

2.3 Summary and Conclusions

The Internet service provider (ISP) business is a complex, relatively new and quickly evolving business. This book deals with INSPs, which are ISPs that offer packet forwarding services. The different types of INSPs were discussed in this chapter, also supporting providers like Internet exchange points and carriers that are relevant, for example, for interconnections (see Part III of this book) were discussed here.

For the discussion, a role model for ISPs was introduced. The main advantage of this model is that it reflects the real world and can thus be used to classify and compare actual

---

[6] According to Nielsen/NetRatings 06/03 cited at www.google.com/corporate/facts.html.

ISP companies with each other. This was demonstrated towards the end of the chapter, in which several ISPs were classified using the model. At the same time, the individual roles the model contains describe exact sets of closely related technical services and can be used for analysis of individual services, for example, in technical and scientific works. We make reference of these roles in the rest of this book.

3

Performance Analysis Basics

In this chapter we present basic methods for analysing, predicting and optimising the performance of networks. We start with the standard theory for networks, the queueing theory. After that, we address the network calculus, a relatively new theory for deterministic and stochastic queueing networks. Network calculus is on its way to becoming a system theory for queueing networks but being a new discipline still lacks wide-scale recognition, and there are still many issues to be solved. Finally, we give a primer of optimisation techniques that come in handy for optimising networks.

3.1 Queueing Theory

3.1.1 Introduction

Queueing theory is the main method for the analysis of networks. There is a large amount of literature about queueing theory, also because it can be applied to more areas than just computer networks. The canonical reference for queueing theory is still Kleinrock (1975, 1976). An up to date overview can be found in Bolch *et al.* (1998).

Queueing theory deals with stochastic queueing systems. It tries to answer questions like, for example, the mean waiting time of a packet in a queue or the mean utilisation of a router. The behaviour of the system in its equilibrium state is in the foreground of the analysis, since results of systems in transient states are relatively hard to get. The stochastics of the system lie in the arrival and service processes. For analytical reasons, it is often assumed that distributions for arrival and/or service processes are memoryless, even though there are some results for general distributions. In the last couple of years, progress has been made in the direction of more realistic arrival processes. Nonetheless, the core results of queueing theory are still heavily based on the memorylessness of the underlying distributions – an assumption that is very much in question for Internet traffic, see Section 5.2. This means that results obtained from queueing theory should not be trusted blindly. Despite this, queueing theory is still the most important method for analysing networks. We now discuss the basic queueing systems that are important for analysing networks. They are used in various places throughout the book.

The Competitive Internet Service Provider: Network Architecture, Interconnection, Traffic Engineering and Network Design
Oliver Heckmann © 2006 John Wiley & Sons, Ltd

3.1.2 Kendall's Notation

The basic queueing model is shown in Figure 3.1. Packets arrive randomly and are stored in a queue until they are processed by the server (or node or router). The arrival of packets is characterised by the *arrival process*, the service time by the *service process*. The order in which packets are processed is determined by the *service discipline*. There can be *multiple servers* for the same queue and the *buffer size of the queue* can be limited. Depending on how these five parameters are chosen, we determine different queueing systems. The parameters allow the short description of a queueing system using Kendall's notation. In Kendall's notation, a queueing system is described with five abbreviations

$$A/S/m/B - S$$

where A is the arrival process and S the service process. The following abbreviations are common for the arrival and service processes:

M (Markovian) A Markovian process has a Poisson distributed arrival of events (packets). A Poisson distribution has exponentially distributed interarrival times – that is the time between two packet arrivals when talking of the arrival process or the time to process a packet when talking of the service process. The probability distribution $A(t)$ is therefore $A(t) = 1 - e^{-\lambda t}$ with λ as the only parameter. The expected interarrival time is $E(t) = 1/\lambda$ and the variance $\text{Var}(t) = 1/\lambda^2$. To describe arrival processes, λ is typically used as a parameter and for service processes μ is used accordingly.

D (Deterministic) A deterministic 'distribution' has a constant value. Constant bit rate traffic could be characterised by a deterministic arrival process.

G (General) The general 'distribution' in Kendall's notation stands for a distribution where nothing except (in most cases) the mean and the variance is known.

Parameter m stands for the number of parallel servers. When trying to model a single router, m will typically be one. The maximum size of the buffer space available for the queue is described by B (counted in packets). Often, an infinite buffer size is assumed. This is represented by dropping parameter B from the notation. In an infinite queue, no packets will ever get dropped and the dropping probability becomes zero. At the same time, the queueing theory will raise a stability condition that has to be met so that the system's average queue length does not become infinite.

The last parameter S stands for the service discipline. The most important service disciplines are the following:

FIFO (or FCFS) The First In, First Out (aka First Come First Serve) service discipline is the default discipline and is assumed if the parameter S is not explicitly specified. Packets are served in the order in which they arrive.

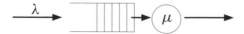

Figure 3.1   Queueing Model

PS (Processor Sharing) In a processor sharing service discipline, all packets in a queue receive service. The rate is split evenly among all waiting packets. A packet leaves the queue when it has acquired enough time slices.

PRIO (Priority) Each packet is assigned a priority and the server selects as the first packet that with the highest priority and serves it. If a pre-emptive priority discipline is assumed and a packet of a priority higher than the one currently being served arrives, then the service of the current packet stops and the higher priority packet is served. For a non-pre-emptive priority discipline, the service for the current packet is first fully finished until the high priority packet is being served. Non-pre-emptive priority scheduling is typically used for analysing communication networks.

A service discipline is called *work conserving* if it always serves a packet if there is at least one packet waiting in the queue. This means that as long as packets are waiting, the server is not idle. This can be assumed for the purpose of network analysis with queueing theory for all practical cases.

The most famous queueing system is the M/M/1 queueing system. It has a Markovian arrival process (interarrival times are exponentially distributed) and a Markovian service process (service times are exponentially distributed). There is only one server; it has infinite buffer for the queue and serves in First in, First out (FIFO) order. This system is discussed below. Afterwards the M/M/1/B system is discussed; it is similar to M/M/1 but only has a finite buffer size. This means that packets can get lost, and we can derive an average dropping probability. Finally, we look at a system where the service time is no longer Markovian. But first, we look at one of the most important laws in queueing theory, Little's law.

3.1.3 Little's Law

Little's law was first described in Little (1961). It is a very general law that holds even for G/G/1 queues and all service disciplines that are work conserving. Let $E(S)$ describe the *expected queueing delay* of a packet in the system, also called *sojourn time*. $1/\mu$ is the average service time; $1/\lambda$, the interarrival time; and $E(W)$, the waiting time. The *waiting time* describes the time in queue only (not in the complete system) so that $E(S) = E(W) + 1/\mu$. Let $E(L)$ be the expected number of packets waiting in the queue and $E(N)$ be the number of packets in the complete system (queue and server).

Little's law says: The average number of packets in the queue is equal to their average arrival rate, multiplied by their average waiting time

$$E(L) = \lambda E(W) \tag{3.1}$$

For systems without loss, this can be easily applied to the whole system as well[1].

$$E(N) = \lambda E(S) \tag{3.2}$$

---

[1] For lossy systems, the throughput and not the arrival rate would have to be used.

3.1.4 M/M/1 Queueing Systems

The M/M/1 queue has an infinite buffer space and uses a FIFO service discipline. The arrival process has exponentially distributed interarrival times with an average interarrival time of $1/\lambda$. The service process has exponentially distributed service times with an average service time of $1/\mu$. The arrival and service *rates* are accordingly λ and μ.

The state of the system can be described with the Markov chain depicted in Figure 3.2 (a). The nodes represent the number of packets k in the system, the arrows represent the state transitions.

We analyse the system in steady state; that means in a state where the system is stable and the probabilities do not change over time. The probability that there are k packets in the system is denoted by p_k. The sum of all state probabilities p_k has to be 1

$$\sum_{k=0}^{\infty} p_k = 1 \tag{3.3}$$

For the steady state, the flow in and out of state 0 has to be equal

$$\mu p_1 = \lambda p_0 \tag{3.4}$$

The same holds true for all other states k

$$\lambda p_{k-1} + \mu p_{k+1} = \lambda p_k + \mu p_k \tag{3.5}$$

Equation (3.5) can be reformulated as

$$p_k = \left(\frac{\lambda}{\mu}\right)^k p_0 \tag{3.6}$$

Combining (3.3), (3.4) and (3.6) leads to

$$p_0 = 1 - \frac{\lambda}{\mu} \tag{3.7}$$

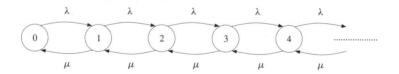

(a) Infinite Markov Chain of the M/M/1 Queue

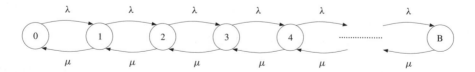

(b) Finite Markov Chain of the M/M/1/B Queue

Figure 3.2   Markov Chains

With (3.6) and (3.7), the system can be described. Some other metrics are useful as well. The *utilisation* ρ of the system is defined as the fraction of time when there is a packet in the system. It is

$$\rho = 1 - p_0 = \frac{\lambda}{\mu} \tag{3.8}$$

The expected number of packets in the system $E(N)$ is given by

$$E(N) = \sum_{k=0}^{\infty} k p_k = \frac{\rho}{1 - \rho} \tag{3.9}$$

if the condition $\rho < 1$ is met. This condition is the stability constraint for the M/M/1 queue. The expected queueing delay (sojourn time) $E(S)$ can be found with Little's law (3.2)

$$E(S) = E(N)/\lambda = \frac{1/\mu}{1 - \rho} \tag{3.10}$$

The delay is depicted in Figure 3.3 as a function of the utilisation ρ.

3.1.5 M/M/1/B Queueing Systems

The M/M/1/B queue is similar to M/M/1 except for the limited buffer size B. Its Markov chain is shown in Figure 3.2 (b). The state probabilities can be derived the same way as before which yields for $\rho = \lambda/\mu$

$$p_0 = \frac{1 - \rho}{1 - \rho^{B+1}} \tag{3.11}$$

$$p_k = p_0 \rho^k \qquad 1 \leq k \leq B \tag{3.12}$$

Please note that for the M/M/1/B queue the M/M/1 stability condition $\rho < 1$ is not necessary; the system is stable even for $\rho > 1$.

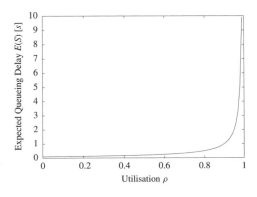

Figure 3.3 Delay of the M/M/1 Queue as Function of the Utilisation with $1/\mu = 0.1s$

The expected number of packets in the system $E(N)$ is

$$E(N) = \sum_{k=0}^{B} kp_k = \frac{\rho}{1-\rho} - \frac{B+1}{1-\rho^{B+1}}\rho^{B+1} \tag{3.13}$$

With Little's law (3.2) follows for the queueing delay (sojourn time)

$$E(S) = E(N)/\lambda = \frac{1}{\mu}\left(\frac{1}{1-\rho} - \frac{B+1}{1-\rho^{B+1}}\rho^{B}\right) \tag{3.14}$$

The loss probability p is the probability that an arriving packet finds the system full, which means that the system is in state p_B

$$p = p_B = \frac{(1-\rho)\rho^B}{1-\rho^{B+1}} \tag{3.15}$$

The loss probability and queue length of the M/M/1/20 queue are depicted in Figure 3.4.

3.1.6 M/G/1 Queueing Systems

The M/G/1 Queue is a generalisation of the M/M/1 queue where the service time follows an arbitrary distribution for which the mean value $E(x)$ and the standard deviation $\text{Var}(x)$ are known.

The exponential interarrival process with parameter λ of the queue is a memoryless process. From this follows an important property of the M/G/1 queue, called PASTA. This stands for **P**oisson **a**rrivals **s**ee **t**ime **a**verages. It means that a new packet arriving at the queue sees exactly the same statistics of the number of packets in the system as when looking at the system at any random time in steady state. This does not hold for other arrival processes where the system can have different properties for different random points.

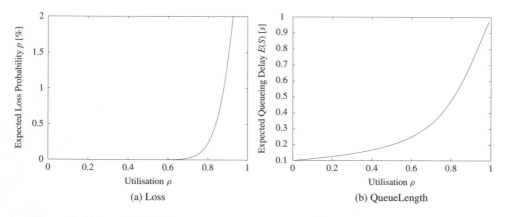

(a) Loss   (b) QueueLength

Figure 3.4 Loss Probability and Queueing Delay of an M/M/1/B Queue with $B = 20$

We define the utilisation ρ for this queue as $\rho = \min\{1, \lambda/E(x)\}$. The expected number of packets in the system $E(N)$ and the expected queueing delay (sojourn time) $E(S)$ for the M/G/1 queue are given by the so-called Pollaczek–Khinchin mean value formulas

$$E(S) = E(x)\left(1 + \frac{\rho(1 + C_v^2)}{2(1 - \rho)}\right) \tag{3.16}$$

$$E(N) = \rho + \frac{\rho^2(1 + C_v^2)}{2(1 - \rho)} \tag{3.17}$$

where C_v^2 is the squared coefficient of variation $C_v^2 = \frac{\mathrm{Var}(x)}{E(x)^2}$. One can immediately see that the M/G/1 queue gets unstable as the utilisation ρ approaches 1. Since $E(x) = 1/\lambda$ and $\mathrm{Var}(x) = 1/\lambda^2$ for the exponential distribution, the Pollaczek–Khinchin mean value formulas simplify to (3.9) and (3.10) for the M/M/1 queue.

From the Pollaczek–Khinchin mean value formulas it follows that the squared coefficient of variation C_v^2 of the service time distribution has a strong influence on the expected delay and queue length. This is visualised in Figure 3.5. The lowest delays can be achieved when C_v^2 is zero, that is, when the service time is deterministic (constant). As a side remark, this is a valid assumption for ATM networks because of ATM's fixed cell size. For IP routers, the service time can be modelled as a constant processing overhead plus the time to put the packet on the outgoing link which is proportional to the packet size. The packet size distribution shows a few clear peaks and can be used as an input, see for example, Appendix D.2.

3.1.7 Other Queueing Systems

There are countless works on more advanced queueing systems. Let us cite one interesting example in Boyer *et al.* (2003). Here, a heavy-tailed M/G/1-PS (processor sharing) queue is being analysed under certain given assumptions. Heavy tailed refers to the service times, which decline slower than exponential. The most important constraint here is that impatient users are considered as well. This is justified by the assumption that the rates

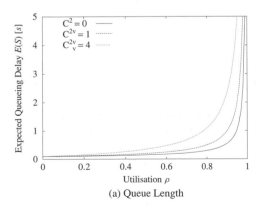

(a) Queue Length

Figure 3.5 Expected Queueing Delay of an M/G/1 Queue for Different Coefficients of Variation

of single users decrease when many users are active at the same time, and some will get impatient and cancel their transfer. The analysis of this queue is used to connect the access control of elastic data flows and the number of the impatient users.

3.1.8 Queueing Networks

The modelling of entire *networks* of queueing systems is very complex compared to the single queueing systems discussed above. On the basis of the highly simplified assumptions, such as the memorylessness of all underlying stochastic systems, a relatively simple form for the probability of the state of the whole systems arises. It results as the product of marginal distributions of the single queueing systems; hence, the outcome is the so-called *network in product form*. Yet, a loosening of the strict assumptions, for example, the introduction of priorities or not requiring memoryless service times, will quickly lead to systems that are analytically not manageable anymore. Then, only numerical methods can help calculating the probabilities for the states of the systems. Numerical methods provide only approximations, however, and quickly reach calculation limits because of their recursive nature. Here, we refer to Whitt (1995) and Kouvatsos and Denazis (1994).

An expansion of queueing networks results from the introduction of negative packets, see Gelenbe (1993). With the arrival of a negative packet at a queue, a positive packet is erased; hence both the negative and positive packets disappear from the system. This expansion is relevant for the Internet, as for example RED (Random Early Detection) can be modelled with it.

Besides the other more general references mentioned above, Robertazzi (2000) provides an overview of queueing networks.

3.1.9 Conclusions

In conclusion, queueing theory is an analytical approach and it has a more locally oriented scope. Although there are queueing systems with feedback, they are typically not used in the context of communication networks. Queueing theory is mainly a model for explanations and is less suitable for forecasts and decisions than the network calculus, because it usually works with mean values.

3.2 Network Calculus

For performance analysis based on worst-case assumptions, the *traditional network calculus* offers a mathematical framework. It is a theory for deterministic queueing models. The Intserv guaranteed service (see Section 6.2.2.4) is a service based on network calculus considerations.

The network calculus is based on min-plus algebra (see Baccelli *et al.* (1992)), i.e. an algebra in which the operations + and * of conventional algebra are substituted by the calculation of the minimum and +. The beginnings of this theory can be traced back to Cruz (1991). It was further developed and described in detail by Le Boudec and Chang, among others, in Le Boudec and Thiran (2001) and Chang (2000). Chang elaborates mainly on the filter theory.

3.2.1 Basics

In the analysis of network behaviour, the network calculus focuses on the contemplation of worst-case characteristics of a data flow and thus delivers deterministically applicable statements. This is usually based on the assumption that guarantees can be based on a traffic description and admission control and a bound for the system behaviour.

The details of the traffic description are abstracted with the help of the so-called *arrival curves:* Assume that the total amount of transmitted data of a traffic flow over time is described by function $R(t)$. $R(t)$ is obviously wide-sense increasing. An arrival curve $\alpha(t)$ indicates the maximum amount of traffic that is permitted *over all possible time intervals*. Examples for arrival curves are token buckets and TSpecs. A *token bucket* is an algorithm used to describe shaped traffic. Assume a bucket that is filled with tokens at a rate r. The bucket can store no more than b tokens; b is called the bucket size. For transmitting a packet a token is taken from the bucket. If the bucket is empty, no packet may be transmitted. A *TSpec* is an extended token bucket that is used to specify traffic in Intserv; see Section 6.2.2.4.

A token bucket arrival curve $\alpha(t) = rt + b$ is depicted in Figure 3.6 (a); it has to be interpreted in the following way: The y-axis shows the amount of traffic in bytes or packets. The x-axis represents the duration $t - s$ of every possible time interval $[s, t]$ since the beginning of the traffic flow. For every time interval, the maximum amount of traffic that is allowed according to the traffic description is given by the arrival curve. This means that if we select any point of time s of the lifetime of the traffic flow R, the amount of data that can be sent up to time t ($t \geq s$) is limited by $\alpha(t)$:

$$\alpha(t - s) \geq R(t) - R(s) \; \forall s \leq t \tag{3.18}$$

For the token bucket in Figure 3.6 (a), a traffic flow can send a traffic burst of maximum size b at a point in time s but in the long run the average rate will not exceed r.

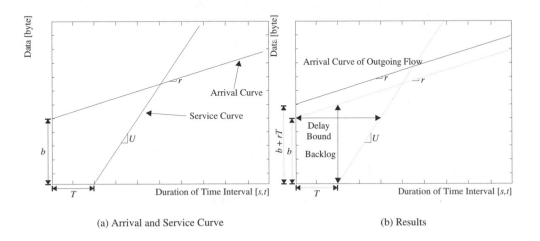

(a) Arrival and Service Curve   (b) Results

Figure 3.6 Network Calculus Example

The second basic element of network calculus are *service curves*. A service curve describes the scheduling behaviour of a node on an abstract level. In network calculus, it is assumed that a node (or system) S allocates capacity to flow $R(t)$ to offer guarantees. These guarantees are expressed by the service curve $\beta(t)$. An example service curve is depicted in Figure 3.6 (a). It is a rate latency service curve

$$\beta(t) = \begin{cases} 0 & t \leq T \\ U \cdot (t - T) & t > T \end{cases}$$

the system imposes a delay of up to T and serves the flow with a long-term average rate of at least U.

A crucial result of network calculus is that the output (outgoing flow) $R^0(t)$ of a system (node) S is given by

$$R^0 \geq R \otimes \beta \tag{3.19}$$

The $\otimes$ operator is the min-plus convolution, which is defined as

$$(R \otimes \beta)(t) = \inf_{s|0 \leq s \leq t}\{R(s) + \beta(t - s)\} \tag{3.20}$$

If R is bounded by its arrival curve α, the arrival curve α^0 of R^0 can be calculated with the min-plus deconvolution $\oplus$

$$\alpha^0 = \alpha \oplus \beta \tag{3.21}$$

The min-plus deconvolution is given by

$$(\alpha \oplus \beta)(t) = \sup_{s|s \geq 0}\{\alpha(t + s) - \beta(s)\}$$

Another crucial result of network calculus is that the service curve β_{series} of two nodes in series is the min-plus convolution of the two service curves β_1 and β_2 of the single nodes.

$$\beta_{series} = \beta_1 \otimes \beta_2 \tag{3.22}$$

3.2.2 Example

Let us assume that a traffic flow $R(t)$ is bounded by the token bucket arrival curve $\alpha(t)$ and served in a node with the rate latency service curve $\beta(t)$ of Figure 3.6 (a). Then the arrival curve of the outgoing flow $R^0(t)$ is

$$\alpha^0(t) = \sup_{s|s \geq 0}\{\alpha(t + s) - \beta(s)\}$$
$$= b + rt + rT$$

The result is depicted in Figure 3.6 (b).

The *backlog* describes the amount of buffer necessary in a system. The backlog is bounded by the vertical deviation between the arrival and service curves. The maximum backlog B for this example is shown in Figure 3.6 (b); in this case it is $b + rT$.

The network calculus also helps in determining a *delay bound* for traffic going through system S. This is very helpful for delay bounded services like the Intserv guaranteed service. The delay bound is given by the maximum horizontal deviation between the arrival and service curves. It is also depicted in Figure 3.6 (b). The delay bound is $T + b/U$ in this example.

Please note that the backlog and delay bound depend on the steepness of the arrival and service curves; they are not necessarily positioned as depicted in Figure 3.6 (b).

3.2.3 Conclusions

The traditional network calculus is an analytical approach that provides a model for the explanation of worst-case behaviour of data flows and can be used for forecasts and decisions. If all the calculations are done using the maximum traffic possible, one faces the risk of allocating the resources too conservatively and thereby causing a bad utilisation ratio. This effect amplifies with the aggregation of data flows. Therefore, one trend in network calculus research is the development of a statistical network calculus that could help at least partly in handling this problem; see the following text.

Since in network calculus it is assumed that the traffic meets a certain limit, feedback is normally not taken into account. Dynamics in terms of interaction with other data flows are also limited. In general, network calculus is only applicable to a limited number of use cases.

3.2.4 Outlook

The traditional network calculus is a system theory for deterministic queueing systems. The traditional network calculus is good for systems that give deterministic guarantees. Statistical guarantees, however, have a broader applicability and are more attractive for Internet Network Service Providers that wish to achieve a high utilisation of their network. Therefore, the traditional network calculus is currently being extended towards a statistical network calculus. Statistical network calculus (see e.g. Boorstyn *et al.* (2000)) is based on the assumption that an arrival curve – called effective envelope here – will be met only with a certain probability. Works on a statistical network calculus go back as far as Chang (1994) and Yaron and Sidi (1993). Lately, however, statistical network calculus is receiving much more attention. In Burchard *et al.* (2002), an effective service curve is introduced, which allows for the calculation of an output envelope with the convolution of effective envelope and effective service curve. The statistical network calculus is a promising field of research as it could develop a system theory for statistic queueing systems also. For the state of the art, we refer to Burchard *et al.* (2002); Ciucu *et al.* (2005); Fidler and Recker (2005); Fidler *et al.* (2005); Jiang and Emstad (2005a,b).

Another promising extension of the network calculus is that towards a calculus for sensor networks, see for example Schmitt and Roedig (2005).

3.3 Optimisation Techniques

3.3.1 Introduction

Queueing theory and network calculus allow us to analyse, understand, and predict the basic behaviour of nodes and smaller networks. In this section, we look at some basic optimisation techniques. Many of the problems INSPs face can be seen as *optimisation problems*: decisions have to be made, for example, about how and when the network capacity will be expanded. We call the parameters to decide upon as *decision variables* or in short just *variables*. The decision depends on many *input parameters* that are given or that have to be researched before the decision can be made. We call the input parameters shortly *parameters*. A certain function of these parameters and variables is to be maximised or minimised, for example, the costs of the network expansion. This function is called the *objective function*. Often there are certain restrictions that have to be observed, for example, that the traffic demand is sufficiently satisfied at all times. They are expressed as functions of the variables and parameters and are called *constraints*.

We place the focus on linear programming (LP) models and (mixed) integer programming models here. These models are used throughout this book to model various problems important for INSPs. There are many other optimisations problems and solution methods that we cannot present and discuss here owing to space limitations. We refer to the literature on operations research for more details. A good standard introductory book is Hillier and Lieberman (2001) that also forms the basis of this chapter.

In the next section, we look at how optimisation problems can be modelled mathematically. After that we discuss the standard methods used for solving these problems exactly.

3.3.2 Modelling Optimisation Problems

The standard approach to solving optimisation problems is to identify the exact problem, the variables, constraints, objective function, and parameters first. Then the objective function and constraints are formulated as a mathematical model. Certain parameters might have to be researched and often some simplifying assumptions are necessary to obtain the functions. Then the problem is solved, typically with the standard methods we discuss below. To increase confidence in the model and to find potential mistakes, often some small problems are solved first and the solutions are analysed carefully. If the confidence in the model is high, then the actual problem is solved. Because input parameters can be uncertain or incorrect, a sensitivity analysis is often performed after obtaining a solution to find how sensitive the solution is to changes of input parameters before the solution is applied.

Let us model a very simple optimisation problem now. The problem is very similar to the one in Hillier and Lieberman (2001), Chapter 3, in case the reader is interested in a more detailed discussion. It is an LP model:

- Assume that a company wants to produce two new products X_1 and X_2 without changing the already existing production line. Each batch of product X_1 makes a profit of \$3,000 and each batch of X_2 makes a profit of \$5,000. The company has three different machines, A, B and C that have different amounts of production time left per week. Producing the different products needs different production time per batch on all the three machines. Table 3.1 lists the production time for the different products and

Table 3.1 Production Time per Batch

Machine	Production Time per Batch		Total Available Production Time
	X_1	X_2	
A	1	0	4
B	0	2	12
C	3	2	18

the total available production time left on the machines. The goal is to maximise the total profit.

- To model this problem mathematically, we first introduce the decision variables. Variable x_1 describes the number of batches that is produced of product X_1 and variable x_2, the number of batches of product X_2. This helps us in formulating the objective function as a mathematical function: the total profit P (in thousand dollars) has to be maximised

$$\text{Maximise } P = 3 \cdot x_1 + 5 \cdot x_2 \tag{3.23}$$

Next the constraints have to be formulated. The constraints are given by the available production time of the different machines. As there are three machines, this results in the following three constraints

$$1 \cdot x_1 + 0 \cdot x_2 \leq 4 \tag{3.24}$$

$$0 \cdot x_1 + 2 \cdot x_2 \leq 12 \tag{3.25}$$

$$3 \cdot x_1 + 2 \cdot x_2 \leq 18 \tag{3.26}$$

Finally, we have to constrain the variables x and y to non-negative values to avoid the otherwise possible absurd results like produce -3 times product X_1 and 5 times X_2. The non-negativity constraints are

$$x_1, \ x_2 \geq 0 \tag{3.27}$$

The complete optimisation problem is now given by (3.23) to (3.27). If we abstract from the parameter values and replace the given number of products and machines with indices p and m, the general optimisation problem can be specified as in Model 3.1.

This optimisation problem is a *linear programming problem*, because all functions are linear and the variables are continuous. The word programming does not refer to computer programming but is used as a synonym for planning.

One potential problem of modelling the above-mentioned problem as an LP problem is that the variables are continuous. For example, the solution $x_1 = 3$ and $x_2 = 4.5$ is a valid solution of the problem mathematically. However, selling half a product might be impossible in reality. If that is the case, the optimisation problem has to be reformulated:

- Variables x_1 and x_2 have to be forced to integer values. This can be represented in the mathematical formulation by replacing (3.27) with

$$x_1, \ x_2 \in \{0, 1, 2, 3, \ldots\} \tag{3.28}$$

and (3.32) with

$$x_p \in \{0, 1, 2, 3, \ldots\} \quad \forall p \tag{3.29}$$

The resulting problem is no longer an LP problem but an *integer programming problem*. You will see next that integer programming problems are much harder to solve than LP problems. In some optimisation problems, integer variables and continuous variables are mixed. These problems are called *mixed integer programming problems* (MIP) and generally the same methods used for pure integer programming problems can be used for these.

Model 3.1 Example Optimisation Problem

Indices

$m = 1, \ldots, M$	Index for the different machines
$p = 1, \ldots, P$	Index for the different products

Parameters

M	Number of machines
P	Number of products
t_{pm}	Production time of product p on machine m
T_m	Available production time on machine m
w_p	Profit for one batch of product p

Variables

x_p	Number of batches produced of product p

$$\text{Maximise} \quad \sum_p w_p x_p \tag{3.30}$$

subject to

$$\sum_p t_{pm} x_p \leq T_m \quad \forall m \tag{3.31}$$

$$x_p \geq 0 \quad \forall p \tag{3.32}$$

3.3.3 Solving Optimisation Problems

3.3.3.1 Linear Programming Problems

Graphical Solution   The simple optimisation problem shown in the preceding text can be solved graphically as shown in Figure 3.7. The two decision variables x_1 and x_2 form the axes. Figure 3.7 (a) shows the constraints. The area marked by the constraints (see Figure 3.7 (b)) is the solution space. All valid solutions lie in that area. The objective function (3.23) is drawn in Figure 3.7 (b) for three different objective values (10, 20, 36). Obviously, they all have the same slope. All solutions that yield a certain objective value have to be in the solution space – otherwise they would break one or more constraints – and they have to touch the objective function for that objective value.

The goal is to find the objective function with the highest objective value that still touches the solution space and therefore contains at least one valid solution. This can be done graphically by shifting an objective function parallel towards the boundary of the solution space as indicated in Figure 3.7 (b). This solution – or these solutions as there can be multiple solutions that yield the same objective value – is the optimal solution. In the example shown previously, the optimal solution is $x_1 = 2$, $x_2 = 6$, which yields an objective value of 36 (which stands for a profit of \$36,000).

This graphical solution method works well with small problems of two and to some extent, three variables. We discuss solution techniques for much larger problems next.

Simplex Several well-researched methods for solving LP problems exist. The oldest algorithm is the *simplex* algorithm presented in Dantzig (1951). It is the most commonly used method. There are different versions of the simplex algorithm. The original algorithm is most commonly found in textbooks as it is a straightforward algebraic procedure. However, there are advanced versions of the algorithm that can be implemented more efficiently. A widely used efficient version of the algorithm, the so-called revised simplex algorithm, can be found in Hillier and Lieberman (2001), Chapter 5.2.

There are several open source implementations of the simplex algorithm available. A very useful free open source tool is lp_solve which is available at http://lpsolve.sourceforge.net.

Alternatives There are various alternatives to the simplex algorithm. The *Khachiyan ellipsoid* method described in Khachiyan (1979), for example, has polynomial running

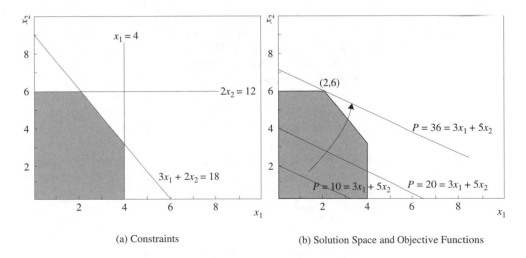

(a) Constraints   (b) Solution Space and Objective Functions

Figure 3.7   Graphical Solution of the Example

time, but is often found slow and inefficient in practice, see Korte and Vygen (2002). Also polynomial in time but faster in practice is the *interior point* method, see Karmarkar (1984).

3.3.3.2 Integer Programming Problems

Mixed integer programming (MIP) problems are much harder to solve than LP problems. For them, no general algorithm with polynomial time is known. A standard approach to solving MIPs is *branch & bound* with *LP relaxation*:

- An integer programming problem has only a finite number of feasible solutions owing to the fact that the variables can only take integer values. A full enumeration could be used to evaluate all feasible solutions and find the optimum one, but that would not be efficient as the number of feasible solutions is often extremely large. The branch & bound strategy is an enumeration method that tries to look at only a tiny faction of all feasible solutions while still finding the optimal solution. It consists of two steps, the branching and the bounding step. Both steps are executed many times, until the optimal solution is found.
- In each *branching* step, the set of feasible solutions is split subsequently into smaller and smaller subsets. This can be done in different ways. A standard approach is to pick one integer variable and force it to 0 in one subset, to 1 in the next, and so on. The algorithm has to keep track of all subsets, for example, by storing them in a tree-like data structure. All subsets are evaluated (one per bounding step) and are
  - either discarded,
  - split into further subsets (branching again) that then have to be evaluated,
  - or the best valid solution so far is found in the subset during the bounding phase.
  In the last case, that solution is stored. If later a better solution is found, the better solution replaces the old one. When all subsets have been evaluated, the best stored solution is the optimal solution of the whole optimisation problem. If no valid solution is found at all in that process, the whole problem is infeasible.
- In the *bounding* steps, the subsets are evaluated by solving a *relaxation* of the optimisation problem on that subset. The relaxation is chosen such that the relaxed problem can be solved relatively quickly. An (mixed) integer programming problem can always be relaxed towards an LP problem by making the variables continuous. This is called *LP relaxation*. LP relaxation increases the solution space because non-integer solutions become valid. The LP relaxation can be solved with simplex or the other methods discussed above. Looking at the optimal solution of the LP relaxation, two situations are possible:
  - The optimal solution of the relaxation might include only integer values for the variables. In that case, this solution is valid for the original integer programming problem. The objective value is compared with the best valid solution found so far and the better of the two solutions is retained.
  - Often, however, the optimal solution of the relaxation also includes non-integer values for one or more variables. In that case, the solution is valid for the LP relaxation but not for the original problem. Still, the objective value of the solution is a bound for the best valid solution for the original problem in this subset. A valid

integer solution can never be better than the best solution of the relaxation. This bound can be used to check whether this subset can be discarded or not:

- If the bound is not better than the best valid solution found so far, this subset can be safely discarded. It cannot contain an integer solution that could beat the currently best solution.
- If the bound is better than the best integer solution, this subset cannot be discarded and a branching step has to be performed on the current subset to split it into further subsets that are evaluated in later bounding steps.

When the branch & bound is finished, it returns the optimal solution. The running time is normally drastically smaller than a full enumeration, but that is not guaranteed and depends on the exact branching strategy and the order of bounding steps.

The branch & bound algorithm can be interrupted any time. The gap between the best bound of non-discarded subsets and the objective value of the currently best integer solution is called the *optimality gap*. The optimal solution of the problem cannot beat the currently best solution by more than this gap.

Besides branch & bound, the cutting plane method and the Lagrangian relaxation are other methods for solving MIP problems. For more information on MIPs we recommend Korte and Vygen (2002); Schrijver (2003); Wolsey and Nemhauser (1999).

There are commercial and free open source software packages available for solving MIPs. Ilog CPLEX (see http://www.ilog.com) is probably the most famous commercial package and lp_solve (see http://lpsolve.sourceforge.net) the most famous open source package. Both are easy to use as standalone products to solve an occasional problem but can also be integrated easily into other software.

While methods for finding the optimal solution for MIP problems are well researched, they can be computational and memory intensive for certain problem structures. In that case, the use of a MIP solver can still be useful. Typical MIP solution strategies such as branch & bound with LP relaxation as explained above give an upper bound on how much better the optimal solution can be, compared to the best solution found so far. These solution strategies can also be interrupted after some time while maintaining control over the quality of the solution, especially compared to meta-heuristic approaches such as simulated annealing, tabu search and genetic algorithms, see Rayward-Smith *et al.* (1996).

3.4 Summary and Conclusions

In this chapter, we presented some basic methods for analysing, predicting, and optimising the performance of networks. Queueing theory is the standard technique for stochastic queueing systems. It is employed in various ways for different types of networks and queues. It mainly provides results about the average properties of a queue or a network of queues.

Network calculus was originally developed for deterministic queueing systems and employed for the Intserv guaranteed service (see Section 6.2.2.4). Deterministic network calculus is conservative and based on worst-case assumption, which can lead to a waste of resources. Therefore, network calculus is currently being extended to stochastic queueing systems. In the third section of this chapter, optimisation models were discussed and methods to solve optimisation problems were presented.

4

Internet Protocols

The predecessor of today's Internet was the ARPANET (Advance Research Project Agency Network). In autumn 1969, the first computer was connected to a node at the University of California and by the end of the year, the ARPANET consisted of four connected computers with different operating systems. The ARPANET grew continuously and soon the TCP/IP suite was adopted as the official protocol suite. TCP/IP was used by other networks to link to ARPANET since 1977, which led to a rapid growth.

In 1989, the World Wide Web (WWW) was invented. Mosaic, the first graphical Web browser was released in 1993 and the first search engine – 'Yahoo' (Yet Another Hierarchical Officious Oracle) – went online in 1994. The WWW led to the Internet becoming 'attractive' for ordinary people, which led to an even higher growth rate. Today, more than 1 billion people are connected to the Internet and this number is still rapidly growing.

The Federal Networking Council passed a resolution defining the term Internet in 1995 (see Federal Networking Council (FNC)):

Internet refers to the global information system that

1. is logically linked together by a globally unique address space based on the Internet Protocol (IP) or its subsequent extensions/follow-ons;
2. is able to support communications using the Transmission Control Protocol/Internet Protocol (TCP/IP) suite or its subsequent extensions/follow-ons, and/or other IP-compatible protocols; and
3. provides, uses or makes accessible, either publicly or privately, high-level services layered on the communications and related infrastructure described herein.

In this chapter, we look at the TCP/IP suite. First, we discuss the Internet protocol stack and the layer model. After that we discuss the most important Internet protocols.

4.1 The Internet Protocol Stack

Throughout this book when protocol stack layers are mentioned, the five layer reference model of the Internet is used; it is shown in Figure 4.1. It is described for example by Tanenbaum (2002). It is a hybrid of the OSI reference model of Zimmermann (1980) and

The Competitive Internet Service Provider: Network Architecture, Interconnection, Traffic Engineering and Network Design
Oliver Heckmann © 2006 John Wiley & Sons, Ltd

5	Application layer
4	Transport layer
3	Network layer
2	Data link layer
1	Physical layer

Figure 4.1   Hybrid 5 Layer Reference Model of the Internet

the TCP/IP reference model of Clark (1988); Leiner *et al.* (1985). Each layer consists of a set of *protocols* for communication with another entity on the same layer and of *communication services* that are offered to the next higher layer. The layers can be distinguished as follows:

- The *physical layer* defines the mechanical, electrical and timing interfaces of the network.
- The *data link layer's* main task is to transform the raw layer 1 transmission facility into a line free of undetected transmission errors between two *directly connected systems*, typically by using the concept of data frames.
- The *network layer* is concerned with forwarding and routing of packets from sender to receiver *end systems*. The basic network layer protocol of the Internet is IP (Internet Protocol); it offers a connection-less datagram forwarding service.
- The *transport layer* uses the network layer to provide sender to receiver *application* communication. The most important transport layer protocols of the Internet are the connection-oriented virtual error-free TCP (Transmission Control Protocol) and the connection-less UDP (User Datagram Protocol).
- The *application layer* contains high-level protocols such as, for example, HTTP. It handles issues like network transparency, resource allocation and problem partitioning for an application. The application layer is not the application itself; it is a service layer that provides high-level services.

We now shortly address the basic network and transport layer protocols of the Internet (IP, TCP, UDP) and then the lower layer protocols that are relevant for ISPs. Some selected application layer protocols are discussed in Chapter 5 in the context of the applications that use them and the properties of the traffic they generate.

4.1.1 IP

4.1.1.1 IPv4

IP is the glue that holds the Internet together. It is a connection-less unreliable layer 3 protocol. 'Unreliable' in this context means that there are no guarantees that an IP datagram will be successfully delivered. IP is the least common denominator on top of the different layer 2 technologies that are used as infrastructure in the different autonomous systems that form the Internet. This explains why it is a relatively simple protocol. In fact, a basic design principle of the Internet is the *end-to-end principle* that is described by Saltzer *et al.* (1984). The authors argue that processing by intermediate systems can be made simpler, relying on the end-system processing to make the system work. This leads to the model of the Internet as a 'dumb network' that has smart end systems (terminals), a completely different model to the previous paradigm of the smart network with dumb terminals like the traditional telephony networks.

The transport layer takes data streams and breaks them up into datagrams that are transported with the IP to the target end system. Packets (datagrams) are transmitted independently from each other. IP datagrams can be as large as 64 kB, but in reality they are typically limited by a MTU (Maximum Transmission Unit) of 1500 byte[1]. The MTU determines the largest possible IP datagram size that can be transmitted by an IP router without it needing to be fragmented.

The IP version 4 (IPv4) header is depicted in Figure 4.2. It has a size of 20 bytes if no options are used. IPv4 options should be avoided because IP packets with options are often processed on the slow path of an IP router and this can lead to additional delay and performance problems. The *source* and *destination address* fields contain the sender's and receiver's IP address. Information about the sender/receiver port is transport layer information and is therefore not contained in the IP header but in the TCP/UDP header instead.

In practice, the *time to live* field is used as a hop counter used to limit packet lifetimes. It is decreased at each intermediate router, and the packet is discarded if it reaches zero. The default start value for the time to live is 64 according to RFC 1700. Some operating systems, for example, older versions of Microsoft Windows, use a value of 32, which is widely considered too small today.

The *protocol* field contains information about the transport layer protocol that generated the packet. For UDP, the field is 17 and for TCP, it is 6; see RFC 1700 for other protocols. The *header checksum* verifies the IP header only. It has to be recomputed at each hop because of the changing *time to live* field.

In Diffserv networks, the *type of service* field is redefined as the *Diffserv byte*, see Section 6.2.4. For best-effort networks, the *Type of Service (ToS)* field was intended to be used for selecting different types of service. However, as there was, for example, no way

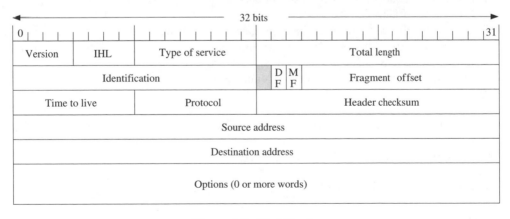

Figure 4.2   IPv4 Header

---

[1] This value comes from the maximum frame size of most Ethernet links on layer 2.

to stop end systems from always requesting the best possible service for all its packets, the *type of service field* is typically ignored by routers.

IP routers forward IP datagrams towards their destination. We discuss the different forwarding architectures in Section 6.3.

4.1.1.2 IPv6

IP version 6 (IPv6), also sometimes called IPng (IP next generation), is based on recommendations in RFC 1752. The core set of IPv6 protocols form an IETF Draft Standard since 1998. The protocol is described in Deering and Hinden (1998). Many other RFCs describe further details of IPv6 architectures and the transition from IPv4 to IPv6.

The main motivation behind the development of IPv6 was the predicted shortage of IPv4 addresses in the near future. IPv6 is therefore designed to 'never' run out of addresses. IPv6 addresses are 16-byte addresses which increase the address space drastically compared to the 4-byte IPv4 addresses. Multicasting is improved by a *scope* field as part of the multicast addresses and a new address type called *anycast* is introduced. Anycast addresses are group addresses where only one member of the group responds, for example, the member of the group that is closest to the source. Anycast addresses are potentially very interesting because the closest router or, for example, the closest name or time server can be accessed with that concept. For more details about the IPv6 addresses see Hinden and Deering (1998).

Besides this, IPv6 contains other improvements over IPv4. The most important ones are as follows:

- Simplification of the IP header. By comparing the IPv4 header in Figure 4.2 with the IPv6 header in Figure 4.3, one immediately recognises the streamlined header layout. Many IPv4 fields are dropped or made optional. This allows routers to process packets faster and can thus improve throughput.
- IPv6 header options are encoded differently compared to those of IPv4. This results in less stringent limits on the length of options and greater flexibility in introducing new options in the future.
- The IPv6 header contains a 20-bit *flow label* that can be used for marking packets belonging to certain flows to give them preferential treatment. The 8-bit *traffic class* field can be used as Diffserv byte similar to the *type of service* field in IPv4.
- Authentication, data integrity and data confidentiality are other important features of IPv6.

IPv6 maintains the good features of IPv4 and discards some of the bad ones. Owing to the increasing number of IP addresses needed in rapidly developing countries like China, internetworking with IPv6 networks will drastically increase in importance in the next years.

For more details on IPv6 see Loshin (2004) and Hinden and Deering (1998).

4.1.2 UDP

UDP (User Datagram Protocol) is the Internet's connection-less transport layer protocol. It is specified in RFC 768 (see Postel (1980)). UDP is a minimalistic protocol, its 8-byte

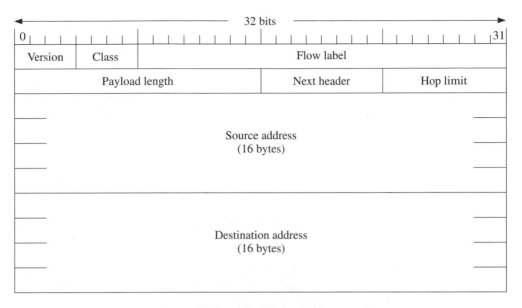

Figure 4.3   IPv6 Header

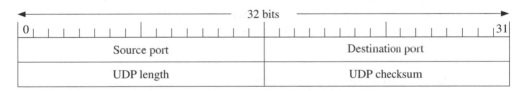

Figure 4.4   UDP Header

header is depicted in Figure 4.4. *Source* and *destination ports* identify the applications on the end systems (the source and destination IP addresses are in the IP and not the in UDP header). Calculating the *UDP checksum* is optional.

UDP does not support flow control or the reliable or even-ordered delivery of datagrams. UDP is mostly used for multimedia application protocols such as VoIP or video streaming. At the time of writing, a promising alternative to UDP for these applications is currently under development in the IETF. The Datagram Congestion Control Protocol (DCCP) (DCCP, see Kohler *et al.* (2005)) is a message-oriented transport layer protocol like UDP but has congestion control built in, like TCP, but without the TCP's in-order and retransmission features.

4.1.3 TCP

4.1.3.1 Introduction

TCP (Transmission Control Protocol) was designed to transmit a byte stream reliably using the unreliable IP datagram service. TCP is the most commonly used transport protocol today. It is best suited for application protocols such as SMTP, FTP or the HTTP that need a reliable connection-oriented service. It is less suited for real-time streaming applications that do not need the retransmission of lost packets and that prefer to have more influence on the transmission rate. The main features of TCP are the following:

- TCP is a connection-oriented protocol. A TCP connection is a byte stream, not a message stream. This means that the TCP stack and not the application splits the byte stream into packets (called *TCP segments*) that are transmitted through the network.[2]
- TCP connections are full duplex (traffic can go in both directions) and point to point (no multicast or broadcast).
- TCP takes care of the reliable in-sequence delivery of the TCP segments. Lost packets are retransmitted and out-of-sequence segments are reordered at the end system.
- TCP's window-based flow control mechanisms allow a slow receiver to slow down a fast sending sender.
- TCP has congestion control mechanism that tries to detect congestion in the network and adapt the window size accordingly.

TCP was first formally described in RFC 793 and then clarified, extended and changed in several other RFCs; see Table 4.1 for a selection of important TCP-related RFCs.

The 20-byte TCP header is depicted in Figure 4.5. Besides the *ports,* one notices the *sequence number* of the transmitted data (measured in bytes) and with the *acknowledgement number* the next expected byte in the opposite directions. The *SYN* bit is used for the three-way handshake at connection setup and the *FIN* bit for closing a connection. The *window size* contains the receiver window that tells the opposite end system how many bytes it may maximally send starting with the byte indicated by the

Table 4.1 Selected RFCs Related to TCP

Name	Title
RFC 793	Transmission Control Protocol
RFC 1122	Requirements for Internet hosts – communication layers (contain TCP clarifications and bug fixes)
RFC 3782	The NewReno modification to TCP's fast recovery algorithm
RFC 2018	TCP Selective Acknowledgement (SACK) Options
RFC 2883	An extension to the Selective Acknowledgement (SACK) Option for TCP
RFC 2581	TCP congestion control
RFC 2988	Computing TCP's retransmission timer
RFC 3042	Enhancing TCP's loss recovery using limited transmit
RFC 3390	Increasing TCP's initial window
RFC 2861	TCP congestion window validation
RFC 3168	The addition of Explicit Congestion Notification (ECN) to IP (this also influences TCP)
RFC 1323	TCP extensions for high performance
BCP 28	Enhancing TCP over satellite channels using standard mechanisms

---

[2] There are, however, means for an application programmer to influence how the byte stream is split into segments.

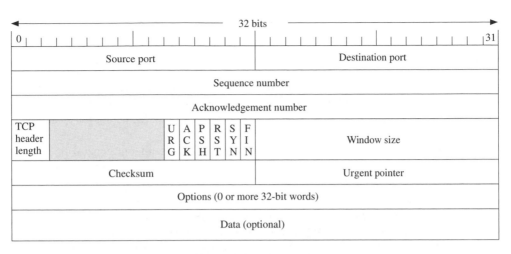

Figure 4.5 TCP Header

acknowledgement number. The window size is used for flow control. The 16-bit field for the window size is too small for high-bandwidth high-latency connections. Therefore, in RFC 1323 a window scale option was proposed, which is now widely supported by different TCP implementations. It allows shifting the window size field by up to 14 bits to the left, thus allowing windows of up to 2^{30} bytes. The original congestion control algorithm is also problematic for high-speed connections; there are several modifications for this under discussion, see for example, High-Speed TCP (HS-TCP) (see Floyd (2003)) or FAST TCP (see Jin *et al.* (2005)).

4.1.3.2 Flow and Congestion Control

TCP interprets packet loss as an indication for congestion in the network and reacts by decreasing its window size and, therefore, the number of packets it can have in the network at one point in time[3]. A TCP sender keeps track of two windows for sending data. The advertised window of the receiver (for flow control) and the congestion window (for congestion control). The maximum amount of data that can be sent unacknowledged at one point in time is given by the minimum of the receiver window and the congestion window.

TCP is a self-clocked algorithm. This means that the window size is adapted in intervals proportional to the round-trip time. TCP starts in a phase called *slow start*. The initial value of the congestion window *cwnd* is one maximum segment size (MSS). Each time an ACK is received while in slow start, the congestion window is increased by one segment size. Therefore, if the sender receives its full window's worth of ACKs per RTT, *cwnd* is doubled per RTT.

Another variable, the slow-start threshold *ssthresh*, is used at the sender to keep track of when to end the slow-start phase and enter the congestion avoidance phase. The names are misleading as the slow-start phase is actually the phase in which *cwnd* is increased

---

[3] This behaviour creates problems for wireless networks, where a packet drop is not necessarily a sign of congestion.

faster. The default value of *ssthresh* is 64 kB (which corresponds to 42–44 segments). Short-lived connections that only transmit little information might therefore never leave the slow-start phase.

When *cwnd* is less than or equal to *ssthresh*, TCP is in slow start. Otherwise, it enters congestion avoidance and *cwnd* is modified the following way: Each time an ACK is received, *cwnd* is increased by MSS $\cdot \frac{MSS}{cwnd}$. This means that if a full window's worth of ACKs are received per RTT, *cwnd* is increased linearly by one MSS and not exponentially as in slow start.

There are two indications of congestion: a retransmission timeout[4] occurring and the receipt of three duplicate ACKs in a row. TCP can generate an immediate acknowledgement (a duplicate ACK) when an out-of-order segment is received as that is a sign that the previous segment could be lost. When three duplicate ACKs are received in a row, TCP performs an immediate retransmission of the missing segment without waiting for the retransmission timer to expire. This mechanism is called *fast retransmit*.

When congestion occurs, *ssthresh* is set to one-half of the current window size[5]. Additionally, if the congestion is indicated by a timeout, *cwnd* is set to one MSS and slow start is re-entered. If the congestion is indicated by duplicate ACKs, no slow start is performed. This is called *fast recovery*. It is an improvement that allows high throughput under moderate congestion, especially for large windows. The reason for not performing slow start in this case is that the receipt of the duplicate ACKs tells TCP that there is only moderate congestion, as one packet got lost but a later packet (and the ACK) got through.

To summarise, slow start continues until the window is halfway to where it was when congestion occurred. The behaviour of TCP is visualised in Figure 4.6. When analysing long-lived TCP connections with losses of 5% and lower, the slow-start phase is often

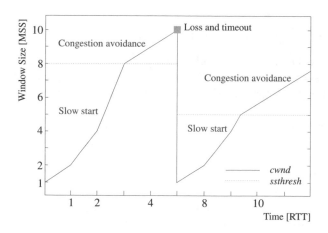

Figure 4.6   TCP Example

---

[4] The retransmission timeout is a function of the estimated round-trip time RTT. It is measured with Jacobson's and Karn's algorithm (Jacobson (1988); Karn and Partridge (1991)). In practice, the retransmission timeout is much greater than the actual RTT. Retransmissions based on the timer may therefore be slow. This is the reason for the duplicate ACKs as second congestion indicators.

[5] Precisely, that is, the minimum of *cwnd* and the receiver's advertised window, but at least two segments.

abstracted from and the TCP window assumed to move in a sawtooth behaviour (see Figure 4.7). Therefore, TCP congestion control is also called *additive increase/multiplicative decrease* (AIMD).

For an excellent source of more information on TCP, we recommend Stevens (1994).

4.1.3.3 TCP Flavours

There are different 'official' TCP versions plus countless slightly different implementations of TCP stacks in the different operation systems. The original TCP version is specified in RFC 1122. TCP *Tahoe* was developed later and distributed with the 4.3 BSD Unix in 1988. Its main advantage is that it includes the fast *retransmit* mechanism that was discussed above. In the 1990 BSD Unix, TCP *Reno* was developed and implemented. It includes all mechanisms of Tahoe plus the fast *recovery* mechanism explained above. TCP Reno has problems with multiple losses, because it does not see duplicate ACKs and times out. This led to the development of TCP NewReno in RFC 3782. It contains an improved fast recovery algorithm that deals better with multiple losses.

A basic problem of the all these TCP flavours is that they use cumulative ACK: one ACK acknowledges the correct reception of the indicated byte plus all previous bytes. If in a sequence of segments (1, 2, 3, 4) segment 2 gets lost but the other segments arrive, only segment 1 but not segments 3 and 4 can be acknowledged with cumulative ACKs. In the *SACK* TCP flavour of RFC 2018, selective acknowledgements are supported that allow acknowledging out-of-sequence segments. In combination with a selective retransmission policy, this can lead to considerable performance improvements.

A radically different TCP flavour is TCP *Vegas*. Vegas tries to use packet delay rather than packet loss as a congestion indicator. It uses the difference between the expected and the actual flow rate to estimate the available bandwidth in the network. When the network is not congested, the expected and the actual rates will be very similar. If the network is congested, however, the actually achieved rate will be significantly smaller than the expected rate. The difference between the rates can be expressed with the difference between the window size and the actual acknowledged packets. The sending rate in Vegas is adapted according to this measured difference. For details, we refer to Brakmo and Peterson (1995). A problem of TCP Vegas is that when competing with other TCP versions, it does receive less than a fair share of the bandwidth.

4.1.3.4 TCP Rate Estimation

For many practical purposes, it is important to get a feeling for the rate or the throughput of a single TCP connection. The TCP rate depends on several parameters. The most important ones are the round-trip time and the loss probability. We next discuss and present three different approaches to estimating the throughput or the rate of a TCP connection. We start with the famous square root formula, then give a better approximation formula and finally end with discussing a formula for short-lived flows.

The Square Root TCP Rate Formula   The so-called square root TCP rate formula can be easily derived, see Floyd (1991) and Lakshman and Madhow (1997). Assuming a long-lived TCP connection that is dominated by TCP's congestion avoidance phase, the

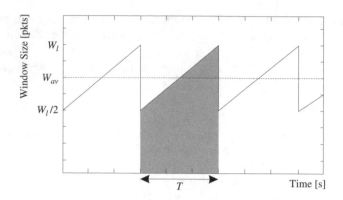

Figure 4.7   Square Root TCP Rate Formula

connection setup and the slow-start algorithm can be neglected. Assuming that losses occur with probability p in regular intervals every $1/p$ packets when the congestion window size has reached W_l, the window size over time will show the sawtooth behaviour depicted in Figure 4.7. If one ACK per packet is sent, the window size increases linearly by one per round-trip time RTT. Therefore, $T = \frac{W_l}{2} RTT$. The average window size W_{av} is $W_{av} = \frac{3}{4} \cdot W_l$ as follows from Figure 4.7. The average rate r_{av} in packets/second is therefore $r_{av} = W_{av}/RTT = \frac{3 \cdot W_l}{4 \cdot RTT}$. The area marked in Figure 4.7 equals the number of packets N sent in one cycle T. Therefore,

$$N = r_{av}T = 1/p \tag{4.1}$$

From this it follows that $\frac{3}{8}W_l^2 = 1/p$ and $W_l = \sqrt{\frac{8}{3p}}$. The average TCP rate is therefore $r_{av} = \frac{1}{RTT\sqrt{\frac{2}{3}p}}$ measured in packets/second or if expressed as r'_{av} in bytes/second $r'_{av} = \frac{MTU}{RTT\sqrt{\frac{2}{3}p}}$ with the packet size MTU. The MTU of TCP connections is typically 1500 bytes with and 1460 bytes without TCP/IP headers.

Measurements of packet traces have shown that the introduction of a proportional factor of 1.22 leads to a better approximation. This results in the well-known square root TCP formula

$$r = 1.22 \frac{MTU}{RTT\sqrt{\frac{2}{3}p}} \tag{4.2}$$

Measurements have also shown that the assumptions in this formula are not valid for loss rates of $p = 5\%$ and higher.

A Better Approximation   The square root formula (4.2) is a simple model of the congestion avoidance phase of a long-lived TCP connection. Padhye *et al.* (1998) derive a better steady-state model for the TCP rate which takes timeouts as well as duplicate ACKs as loss indication into account. A good approximation of their model is given with Model 4.1. It is based on the Reno flavour of TCP. The rate of a long-lived TCP connection can be limited by congestion avoidance algorithm and also by the maximal

---

Model 4.1 Advanced TCP Rate Approximation Model

Parameters

W_{max} Maximum receiver window size [pkts]

b Number of packets acknowledged by a received ACK

(typically $b = 2$)

p Loss probability

T_0 Retransmission timeout [s]

(initially $T_0 = 3s$, adapted according to RFC 793)

RTT Round-trip time [s]

r_{av} TCP rate [pkts/s]

Equation

$$r_{av} = \min\left(\frac{W_{max}}{RTT}, \frac{1}{RTT\sqrt{\frac{2bp}{3}} + T_0 \min\left(1, 3\sqrt{\frac{3bp}{8}}\right) p(1 + 32p^2)} \right) \quad (4.3)$$

---

congestion window size as advertised by the receiver. This maximal window size W_m is given by the buffer limit set aside for the connection on the receiver side. It is taken into account in the first part of the main *min* term in (4.3). The assumptions made for the derivation of the formula are that the effect of the fast recovery algorithm can be neglected and that the time spent in the slow-start phase is negligible. The latter holds true only for long-lived TCP connections. Among other assumptions, it is assumed that the round-trip time RTT is not affected by the window size. This assumption is acceptable for most connections except when the connections go through an extreme bottleneck such as a low-bandwidth modem line with a large buffer. The rate r_{av} is specified in packets/second. To obtain the rate in bytes/second, r_{av} has to be multiplied with the average packet size, which is typically around 1500 bytes.

For a derivation of the formula and more details, see Padhye *et al.* (1998).

Short-lived TCP Flows The aforementioned models predict the rate of long-lived TCP connections. Many HTTP transfers of web pages, however, are short-lived TCP connections that spend little or no time in the congestion avoidance phase of TCP. For these connections, the slow-start phase has to be taken into account. A detailed model for short-lived TCP flows is presented in Cardwell *et al.* (2000). Here we present a simplified version of that model assuming that no losses occur until the transfer is finished. The results are summarised in Model 4.2.

If only a few packets are transmitted, the initial handshake for TCP connection setup cannot be neglected. The total duration D of the short-lived TCP transfer consists of the

---

Model 4.2 TCP Latency Model

Parameters

D	Total duration of the TCP transfer [s]
r_{av}	Average TCP rate [pkts/s]
L	Duration of the TCP connection establishment [s]
T	Duration of the data transfer itself [s]
d	Number of data segments to be transferred [pkts]
RTT	Round-trip time [s]
W	Unconstrained window size at end of transfer [pkts]
γ	Slow-start growth rate
	($\gamma = 1.5$ if one ACK per two data segments is sent)
w_1	Initial congestion window size [pkts] (typically $w_1 = 2$)
W_{max}	Maximum receiver window size [pkts]

Equations

$$D = L + T \tag{4.4}$$

$$L = RTT \tag{4.5}$$

$$W = \frac{d(\gamma - 1)}{\gamma} + \frac{w_1}{\gamma} \tag{4.6}$$

$$T = \begin{cases} RTT \cdot \left(\log_\gamma \left(\frac{W_{max}}{w_1}\right) + 1 + \\ \frac{1}{W_{max}} \left(d - \frac{\gamma W_{max} - w_1}{\gamma - 1}\right)\right) & \text{when } W > W_{max} \\ RTT \cdot \left(\log_\gamma \left(\frac{d(\gamma-1)}{w_1}\right) + 1\right) & \text{otherwise} \end{cases} \tag{4.7}$$

$$r_{av} = \frac{d}{D} \tag{4.8}$$

---

connection setup time L and the time T for the data transfer itself. The connection setup consists of a three-way handshake; (4.5) shows the expected duration of the handshake, taking into account that the last ACK of the handshake already carries data (and is therefore part of T).

It is assumed that in total d segments are to be transmitted (d is approximately the number of bytes to be transmitted divided by the MSS which again is 1460 bytes in most cases). During slow start, there are two possibilities: If the maximum congestion window W_{max} is very large, the window size will be W at the end of the transfer. Duration T is then given by the second part of (4.7). Otherwise, the maximum congestion window W_{max} is reached during the transfer and T is expressed by the first part of (4.7).

4.1.3.5 TCP Root Cause Analysis

While the previous section described the theoretical throughput of a TCP connection as a function of the network parameter's loss and delay, TCP root cause analysis is concerned with determining reasons that limit the throughput of an actually measured TCP connection. This is important for end users as well as INSPs, as the network is not necessarily always the limiting factor for TCP connections. The possible rate-limiting factors as described in Zhang *et al.* (2002) are as follows:

- Opportunity (lifetime): Many short-lived flows do not transmit long enough to reach a high throughput.
- Bandwidth and Congestion: Packet loss limits the TCP throughput owing to TCP's congestion control as discussed above. Packet loss can be caused by the TCP rate approaching the total bottleneck bandwidth (e.g. the upload bandwidth of a ADSL connection) or by other flows competing for bandwidth in the network.
- Transport: The sender might be in congestion avoidance without experiencing loss.
- Receiver window: The sending rate can be limited by the advertised receiver window. This is called *flow control* and is used by slow receivers to throttle fast senders.
- Sender window: The sender window is constrained by the buffer space at the sender. It limits the amount of unacknowledged data outstanding at any time.
- Application: The rate with which an application produces data can of course also be an important limit of the actual throughput.

Zhang *et al.* (2002) introduce the tool T-RAT. On the basis of the backbone packet level traces and summary flow level statistics, they conclude that the dominant rate-limiting factors are congestion and receiver window limits. Siekkinen *et al.* (2005) discuss some drawbacks of the tool T-RAT and provide some new root cause analysis algorithms based on certain time series, for example, of the interarrival time of acknowledgements or of the number of unacknowledged bytes. These time series can be extracted from bidirectional packet header traces.

4.1.4 Lower Layer Protocols

INSPs connect their POPs (points-of-presence) typically by leasing lines from carriers. There are several layer 2 technologies available for carriers that can be employed by INSPs to run their IP overlay network over. In this section, we focus on the high-speed layer 2 technologies currently favoured by carriers.

4.1.4.1 Synchronous Optical Networking (SONET) / Synchronous Digital Hierarchy (SDH)

SONET (Synchronous Optical Networking) and SDH have replaced the Plesiochronous Digital Hierarchy (PDH) and are the most common link-layer technologies for today's high-speed wide area networks (WAN). SONET is a standard for optical communication, providing framing, as well as a rate hierarchy and optical parameters for interfaces ranging from 51 Mbps (OC-1) up to 9.8 Gbps (OC-192) and higher[6]; see ANSI T1.105 (1995); ANSI T1.119 (1995). SONET has been adopted as a standard for North America by the American National Standards Institute (ANSI), while a slightly advanced version – SDH, see ITU Recommendation G.707 (1996) – has been adopted by the International Telecommunication Union/Telecommunication Standardisation Sector (ITU-T) and is used in the other parts of the world.

SONET/SDH use time division multiplexing with a scan time interval of 125 μs, indicating the background of SONET/SDH, the telecommunication market where the standard sampling rate is 8000 samples/second for voice. SONET/SDH needs a tightly synchronised clocking environment for the synchronous transmission of the data streams.

Table 4.2 lists the line and payload rates of the optical SONET/SDH circuit hierarchy. The overhead carries information that provides 'Operations, Administration, Maintenance and Provisioning' capabilities such as framing, multiplexing, status, trace and performance monitoring. Higher-speed circuits are formed by successively time multiplexing multiples of slower circuits; for example, four OC-3 circuits can be aggregated to form a single OC-12 circuit.

For the transport of IP packets over SONET/SDH, the IP datagrams are typically encapsulated into Point-to-Point Protocol (PPP) packets. PPP provides link error control and initialisation. The PPP-encapsulated datagrams are then framed using high-level data link control (HDLC) and sent over a SONET/SDH circuit to the next hop. RFC 2615 describes this process, see Malis and Simpson (1999).

Simplified Data Link (SDL) is a very low overhead alternative to the HDLC-like encapsulation that avoids HDLC's byte-stuffing expansion and is designed for rates at OC-192 and above. It is described in RFC 2823, see Carlson et al. (2000).

4.1.4.2 10-Gigabit Ethernet (10GE)

Beginning with the 1-Gigabit Ethernet standard IEEE 802.3z, Ethernet is deployed not only in local area networks (LAN), the traditional domain of Ethernet, but also in metropolitan area networks (MAN). With the 10-Gigabit Ethernet (10GE) standard IEEE

Table 4.2 SONET/SDH Data Rates

SONET	SDH	Line Rate	Payload Rate	Overhead Rate
OC-1	–	51.840 Mbps	50.112	1.728
OC-3	STM-1	155.520 Mbps	150.336	5.184
OC-12	STM-4	622.080 Mbps	601.344	20.736
OC-48	STM-16	2488.320 Mbps	2405.376	82.944
OC-192	STM-64	9953.280 Mbps	9621.504	331.776
OC-768	STM-256	39813.120 Mbps	38486.016	1327.104

---

[6] SONET is also specified for non-optical digital circuits but because high data rates usually require fibre optic cable, we concentrate on OC (optical carrier) circuits here.

802.3ae, Ethernet can now be expanded also into wide area networks (WAN), the traditional domain of SONET/SDH. 10GE can work over SONET links and also without SONET as end-to-end Ethernet.

10GE is based entirely on the use of optical fibre and only full-duplex mode is supported. Two end systems can be connected directly; for more end systems, a switch has to be used. 10GE still shares the MAC (Media Access Control) protocol and the frame format with the slower Ethernet standards. But because there are only point-to-point connections rather than the multipoint connections that were used in the classic Ethernet networks (IEEE standard 802.3 before 802.3ae), the classic Ethernet collision detection mechanism CSMA/CD (Carrier Sense Multiple Access with Collision Detection) is no longer necessary.

10GE is, in contrast to SONET/SDH, an asynchronous protocol and differently clocked domains are interlinked by switches and bridges that buffer and re-synchronise the data. Therefore, 10GE requires less complexity and is generally cheaper than the SONET/SDH equipment. This was in fact one of the design goals of 10GE; see IEEE 802.3 High Speed Study Group (2002). Another cost-saving factor is that 10GE is capable of using lower-cost uncooled optics and multimode fibre for short-distance connections.

Over single-mode fibre, 10GE can bridge distances of 40 km and can therefore be used to build a pure Ethernet WAN. For compatibility with the existing SONET/SDH network, 10GE can also be operated on top of SONET OC-192/SDH STM-64 connections that only have a slightly slower transmission rate (see Table 4.2). This operation is described by the 10 GE WAN PHY (physical layer) specification. 1-Gigabit Ethernet (1GE) is already quite commonly found as the foundation for MAN[7].

Both Ethernet and SONET/SDH have their individual advantages, which are summarised in Table 4.3.

4.1.4.3 Wavelength-division Multiplexing (WDM)

Wavelength-division Multiplexing (WDM) is the generic name for frequency-division multiplexing in the optical domain. It is best understood as a fibre-multiplication technology: it allows multiple optical circuits to share a single physical fibre strand without interfering with each other as their signals use different carriers occupying non-overlapping parts of the frequency spectrum (virtual fibres).

The number of optical signals multiplexed within a window is limited only by the precision of the optical equipment. WDM can therefore increase the optical fibre bandwidth many folds without expensive re-cabling. However, electronic switching gear is commonly used at the ends of the optical circuits and forms the bottleneck in today's backbones. Optical switching technology promises to also remove this bottleneck and decrease the costs for bandwidth even further.

Because WDM operates at the photonic level, it allows different framing and transmission technologies to be used on each wavelength. Considering the vast investment carriers have made in SONET and SDH equipment and their experience with it, the integration of SONET/SDH and WDM seems a reasonable and likely step on the way to all optical

---

[7] See, for example, the services of Yipes (www.yipes.com), Cogent Communications (www.cogentco.com) and OnFibre (www.onfibre.com).

Table 4.3 Comparison of SONET and Long-distance Ethernet

	SONET/SDH	Ethernet
Historical traffic	Voice traffic	Data traffic
Historical network type	WAN	LAN
Standardisation gremium	ANSI/ITU-T	IEEE
Supported network types	MAN, WAN	LAN, MAN, WAN (10GE)
Link protocol	Synchronous	Asynchronous
Bandwidth scalability	52 Mbps to 40 Gbps	1 Mbps to 10 Gbps
Advantages	Survivable (50 ms restoration time with APS (automatic protection switching, 99.999% reliability; see Goralski (2002))	Lower equipment costs
	Optimised for voice traffic	Optimised for data traffic
	Widely deployed in WANs	Widely deployed in LANs; MANs/WANs can be connected to LANs without reframing
Annotations		Solutions exist for running Ethernet over existing SONET/SDH infrastructure

transport networks; see Cavendish (2000). Thus, one solution for IP over WDM is running IP over PPP/SDL over SONET/SDH over a WDM link.

It is possible to simplify the protocol stack by removing the complexity of SONET/SDH and send IP directly over WDM links using Multi-Protocol *Lambda* Switching (MPλS), a variation of the Multi-Protocol *Label* Switching (MPLS)(for MPLS, see Section 6.3.2) approach that uses wavelengths instead of labels. Packet-switching MPLS and wavelength-switching MPλS are subsumed[8] under the GMPLS (Generalised MPLS) framework that provides a generalised signalling control protocol standard for multiple types of switching. More information can be found in Banerjee *et al.* (2001); Berger (2003); Durresi *et al.* (2001).

With a set of tests over four testbeds in Finland, France, Sweden and Switzerland, Rodellar (2003) compares and evaluates three approaches for IP over WDM: (1) IP – Packet over SONET/SDH – WDM, (2) IP – native 1GE – WDM, (3) IP – DPT (Dynamic Packet Transport[9]) – WDM.

Among other things, the study shows that none of these solutions has a clear technical advantage over the other. The feasibility of real-time applications such as IP telephony or video across a 1GE link over a WDM network has been verified. The study, however,

---

[8] In the literature, MPλS and GMPLS are sometimes used as synonyms. This is not technically correct as GMPLS explicitly also addresses other kinds of switching besides wavelength switching, as for example switching in the time domain (time division multiplexing).

[9] DPT is a proprietary layer 2 switching solution from Cisco for transporting IP packets over ring networks.

also found that it is still necessary to separate the traffic of these real-time applications from other low-priority traffic. We investigate methods for doing this in Part II.

4.2 Summary and Conclusions

In this chapter, we discussed the basic network and transport layer protocols of the Internet: IP, TCP and UDP. TCP is the most commonly used transport protocol in the Internet and uses a complex congestion and flow control mechanism that was discussed in this chapter. Methods for estimating the throughput of a TCP connection for different network conditions were discussed as well. Finally, different lower layer technologies for ISPs and carriers were shortly discussed towards the end of this chapter.

5

Applications

Internet traffic is produced by many different applications. Table 5.1 shows the amount of traffic volume produced by different application protocols. It was measured on a backbone link connecting ADSL customers, see Azzouna and Guillemin (2003). We notice that *web* and *peer-to-peer* applications are responsible for the majority of today's Internet traffic. These two application types are discussed in this chapter. Traffic patterns in the Internet will change when new applications become successful. Two applications that will be very important in the future are also discussed in this chapter: *network games* and *Voice over IP (VoIP)* applications. Both have special quality-of-service requirements and their traffic models differ significantly from the web and P2P traffic models.

Depending on the general Quality-of-service (QoS) requirements, we distinguish elastic and inelastic applications. *Elastic* applications are flexible in their bandwidth requirement and adapt their rate to the network conditions. They typically use TCP as transport protocol and therefore TCP's congestion control mechanisms to react to packet losses and delay. We described in Section 4.1.3 how to estimate the throughput of these applications. Typical elastic applications are file transfer applications such as web applications that were not identified in the study browsers, P2P, mail or File Transfer Protocol (FTP) clients.

Inelastic applications are less flexible in their bandwidth requirements and typically need a certain minimum bandwidth to work properly. A typical example is a voice call. Almost all of today's Internet traffic is generated by elastic applications and most voice calls are still transported on dedicated infrastructure. At the time of writing, the total amount of data traffic is roughly 10 times as large as all voice traffic (traditional telephony and VoIP) and growing faster. Nevertheless, inelastic applications such as VoIP, videoconferencing, video streaming and (some) network games gain in importance and as they have special quality-of-service requirements and a high utility to the users, they need the attention of ISPs.

In order to differentiate between applications with different requirements in a network, the application a traffic flow belongs to has to be identified in real time in the network. Traffic classification is therefore important. It is discussed towards the end of this chapter.

5.1 World Wide Web

The World Wide Web (WWW) is based on the Hypertext Transfer Protocol (HTTP) to transport web documents from a web server to the web browser of the end-user. Web documents are typically HTML files and the pictures referenced in these files.

Table 5.1 Composition of Traffic by Application Type from Azzouna and Guillemin (2003)

Application type	Amount of traffic
World wide web (WWW)	14.6%
Peer-to-peer (P2P)	49.6%
File transfer protocol (FTP)	2.1%
Network news transfer protocol (NNTP)	1.9%
Other (it can safely be assumed that a large percentage is unidentified P2P traffic)	31.8%

5.1.1 QoS Requirements

Web browsing is an interactive application. The time from when the user clicks on a link until the web page is displayed in his web browser determines the perceived quality of service. This time largely depends on the throughput of the HTTP/TCP connection and this throughput again depends on the HTTP version and the network conditions (loss and delay) that determine the TCP throughput. The details are discussed in Section 4.1.3.

In Bhatti *et al.* (2000), user trials show that the tolerance of users for the time it takes until a web page is fully displayed in the browser depends on the task of the user, on whether a page is displayed progressively or not and on how long they have been interacting with the site. The study indicates that thresholds of acceptability change over time. Generally speaking, a web page should be fully displayed within very few seconds and the more interactive a user's task, the faster the transfer should be.

5.1.2 Traffic Model

A traffic model characterises traffic and is important for understanding how many resources are needed to support a certain traffic type, how to identify traffic by its behaviour and how to generate artificial yet realistic traffic, for example, for simulations.

The classic paper on modelling Internet traffic is Danzig and Jamin (1991). In this paper, a library of empirical traffic models for Internet applications that were common at the beginning of the 90s (FTP, SMTP, Telnet) is presented. Web traffic has not been included in it. Paxson (1994) later derived analytical models from traffic measurements by fitting probability distributions to the measured data. Today, there is a vast amount of work on traffic models.

For modelling *individual HTTP connections*, Mah (1997); Choi and Limb (1999) and Barford *et al.* (1999) are a good source, but also see the works cited therein.

- According to these studies, the average HTTP request from a client follows a lognormal distribution with a mean of 360 bytes per request.
- The request is answered by the server that sends back the requested web document. The size of web documents follows a heavy-tailed distribution. This means that most documents are relatively small. However, there is a small but significant chance that a random document is very large. To model the tail, typically a Pareto distribution

is used. A Pareto distribution has a heavy tail, while a Poisson distribution does not; this is visualised in Figure 5.1, where the PDF (Probability Distribution Function) of a Pareto distribution and a Poisson distribution are displayed. The probability of the Pareto distribution approaches zero much slower than the Poisson distribution for high-input values. This behaviour is called *heavy tailed*.

In Barford *et al.* (1999), a Pareto distribution for the tail is combined with a lognormal distribution of the body of the documents.

- A web document consists of one HTML page plus on average five to six objects. Most of these objects are pictures. According to empirical studies, a HTML page has an average size of 10 kB and the objects are around 8 kB.

- The viewing times of the user are around 40 s on average and can be modelled by a Weibull distribution. For more details, we refer to the cited papers.

If we assume that the complete transfer time for a web document including all pictures should not exceed 4 s, the minimum throughput for an average web document should be at least 100 kbps. In reality, owing to the TCP congestion control, a significantly higher amount of bandwidth should be calculated for web browsing.

As a side remark, the popularity of *individual* web documents in the Internet can be described by a Zipf distribution, see for example, Hubermann *et al.* (1998). This means that there will be a few documents that are extremely popular and are requested very often while most of the documents are not.

A widely used tool for generating web traffic is SURGE, see Barford and Crovella (1998). SURGE imitates a stream of HTTP requests from an assumed population of WWW users. Users follow an on–off process. If a user is on, it downloads web documents according to the aforementioned traffic models. Liu *et al.* (2001) describe a similar tool that follows a slightly higher-level traffic characterisation.

Aggregate web traffic shows self-similar behaviour. Self-similarity is a phenomenon observed often in the real world. Coastlines, for example, are statistically self-similar as parts of them show the same statistical properties at different scales. Typical network traffic has self-similar properties. This means that the traffic shows bursts not only on a small

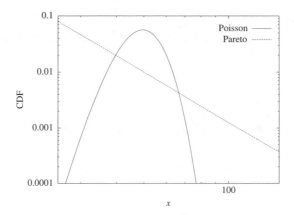

Figure 5.1 Pareto and Poisson Distributions (Logarithmic Scale)

timescale but also on a larger timescale (long-range dependency). An important conclusion from this is that simple traffic models using a Poisson distribution for packet arrival[1] are inaccurate, as for Poisson models the bursts disappear more quickly on larger timescales. This is visualised in Figure 5.2. Networks designed without considering self-similarity are likely to not have enough buffer space and to not work as expected.

Self-similarity has been shown for LAN traffic by Leland *et al.* (1994), for WAN traffic by Paxson and Floyd (1995) and for web traffic during busy hours by Crovella and Bestavros (1997). Self-similarity in web traffic can be explained by heavy-tailed file size distributions (see the preceding text) and by user reading times, see for example, Crovella and Bestavros (1997).

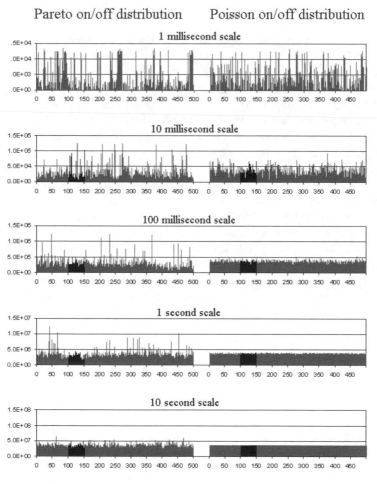

Figure 5.2   Self-similarity and Long-range Dependency (from Kramer (2004)). (Reproduced by permission of Glen Kramer)

---

[1] This does not necessarily imply that session arrivals cannot be Poisson distributed.

Aggregate web traffic can be well modelled as a superposition of a large number of individual on/off sources with heavy-tailed on/off period lengths. See Kramer (2004) for a simple traffic generator. For measuring self-similarity, tools such as SELFIS can be used, see Karagiannis and Faloutsos (2002).

For further information, we also recommend Beran (1994), Park and Willinger (2000), Mannersalo and Norros (2002), and Addie *et al.* (2002).

5.2 Peer-to-Peer Applications

As shown in Table 5.1, the bulk of today's Internet traffic is caused by P2P applications; see also Azzouna and Guillemin (2003); Fraleigh *et al.* (2003); Sandvine Incorporated (2003).

A *P2P system* is a self-organising system consisting of end systems (called 'peers') that form an overlay network. Peers offer and consume services and resources and have significant autonomy. The participating peers exchange services. Long-term connectivity of individual peers cannot be assumed in a P2P system. This means that a P2P system has to explicitly deal with dial-up users, variable IP addresses, firewalls, Network Address Translation (NAT) and that the system typically operates outside the domain name system. For general literature on P2P systems, see Oram (2001) and Steinmetz and Wehrle (2005).

Almost all of today's popular P2P applications are file sharing applications. They are used to exchange files between end-users. The large majority of the shared files are movies and music files, see for example, Heckmann *et al.* (2004). In 2005, despite increasing counter-measures of the music and movie industry, file sharing makes, to a large extent, illegitimate use of copyrighted material.

5.2.1 QoS Requirements

General file sharing applications are bulk transfer applications and have no real-time constraints and few requirements with respect to loss, delay or jitter. User satisfaction mainly depends on the duration a complete file transfer takes, which is a function of the long-term throughput. P2P traffic is typically treated as low priority or background traffic in most networks, if the network supports the differentiation of different traffic types. In order to do so, however, P2P traffic must be correctly identified in real time in a network. This is not trivial as port-based classification fails for a large part of the P2P traffic. This problem is discussed in Section 5.5.

5.2.2 Traffic Model

In the last years, the Internet has seen many different P2P file sharing applications emerging, becoming successful and then vanishing into insignificance for a variety of reasons. It is therefore hard to derive a general traffic model for P2P traffic. However, certain properties can be assumed:

- P2P applications are bandwidth greedy.

- Compared to WWW applications, they generate more long-lived and therefore reactive TCP connections over which the dominating part of traffic is exchanged. To support this claim, we did some measurements in the eDonkey network[2]:
  Our measurements in Heckmann *et al.* (2004) show that at the time of the study, an average eDonkey user was sharing 57.8 files with an average size of 217 MB, a large proportion of those files being movies. An average active TCP connection between two clients has a duration of almost 30 minutes, definitely long lived. During this time, on average 4 MB are transferred; this volume is mostly limited by the ADSL upload capacity that is typically almost fully used by the P2P application[3]. Few (around 1%), but extremely popular, files account for a very large part (>50%) of the generated traffic; this is also confirmed by Leibowitz *et al.* (2003) for the Kazaa file sharing network. In Kazaa as well as in eDonkey, files are either of medium (few megabytes) or of very large size (>600 MB). This is explained by most files being songs or movies. Measurements in Tutschku and Tran-Gia (2005) show that the flow size of eDonkey can be approximated well by a lognormal distribution. It seems that the heavy tail of the flow size distribution is reduced because eDonkey – like many other P2P file sharing applications – splits large files into smaller chunks.

If we look at the aggregate traffic, P2P traffic has some nice characteristics for ISPs, see Hasslinger (2005). It shows relatively little variability over time as the aggregate peak-to-mean rate over a day is usually smaller than 1.5. Web browsing and other applications typically have a factor of two and more, because of the fact that they are used mainly in the busy hours and less at night.

For web traffic, the popularity of Internet servers can change abruptly, for example, when a new service pack comes out that is downloaded by many systems within a short time or when a web page with previously little attention receives a great deal of attention within short notice, because it is referenced in a popular news magazine. The latter is also called the *Slashdot effect*. In P2P networks, new content quickly becomes more or less uniformly distributed in the network. Therefore, P2P applications lead to a more uniform distribution of traffic sources over the network, independent of sudden changes in the popularity. This makes it easier for ISPs to plan their capacity. P2P traffic is mostly symmetric traffic. Following the argument of Hasslinger (2005), aggregate P2P traffic approaches a Gaussian distribution.

5.2.3 The Future of P2P

The P2P communication paradigm is a powerful communication paradigm and is slowly adapted to other applications as well, because it promises scalability, cost savings, rapid deployment and more. Emerging P2P applications are the VoIP telephony application Skype (www.skype.com), groupware Groove (www.groove.net) or the P2P webcam network Camnet (Liebau *et al.* (2005)); for more applications see Steinmetz and Wehrle

---

[2] eDonkey was selected because according to Sandvine Incorporated (2003), the eDonkey/eMule network was with 52% of the generated file sharing traffic the most successful P2P file sharing network in Germany at the time of the studies.

[3] Keep in mind that a single client has multiple parallel TCP data transfers in progress at almost all times.

(2005). Therefore, future P2P applications can be expected to show much more variety than today's file sharing applications and their traffic can no longer be assumed 'low priority' or 'unwanted'.

5.3 Online Games

5.3.1 Computer Game Market

The computer game market and especially the online game market is a fast growing market with a tremendous amount of opportunities:

- According to ESA (2005), the computer and video game software sales reached 7.3 billion dollars in the United States of America and roughly 24.4 billion dollar worldwide in 2004.
- In 2005, IDC (www.idc.com) predicted an increase in turnover of 50% per year in the United States. For Asia, the turnover was 761 million dollars in 2003 with a prognosis of 1.84 billion dollars in 2008.
- Jupiter Research (www.jupiterresearch.com) forecasts a growth of the *online game* market in Europe from 96 million EUR in 2003 to 589 million EUR in 2007. Gamers are predicted to pay 79 EUR per month for games.

5.3.2 Classification of Computer Games

Figure 5.3 shows a classification of computer games by the type of game, the device the game is running on, the number of players, the interactivity between games and the network connectivity needed. The aspects of ISPs especially important with respect to QoS requirements are marked in grey: Online real-time games.

Online games can be persistent: If a player logs off for a while and logs on again later, he continues more or less from the previous state (e.g. with his previous character in a role-playing game), while for non-persistent games he typically starts a new gaming session, although certain information like the gamer's previous high scores might be kept.

The most important online games today are Massive Multiplayer Online Role Playing Games (MMORPG) such as Ultima Online, EverQuest and World of Warcraft; Real-time Strategy (RTS) games such as Starcraft and First Person Shooters (FPS, also called *ego shooters*) such as Counterstrike. For MMORPGs and some other online games, customers often pay a monthly subscription fee to be allowed to play online with/against other players. For World of Warcraft, the monthly fee at the time of writing was 14.99 dollars. MMORPGs can have multi-million subscribers.

5.3.3 Online Game Architectures

Some online computer games are played purely peer to peer with communication directly and exclusively between the participating parties; this is typical for most computer-based card and board games. However, most games use a client-server architecture where servers are used to distribute the information, and information exchange directly between the players is uncommon. Servers simplify the synchronisation between a larger number of

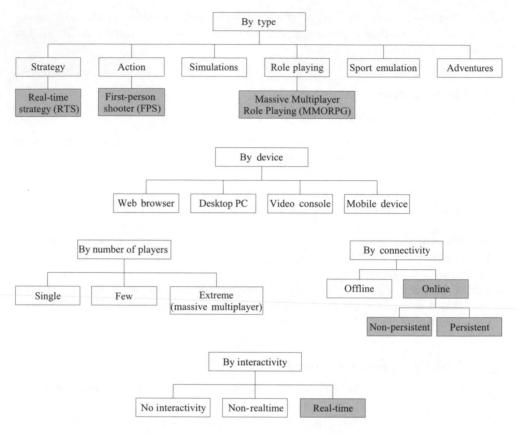

Figure 5.3   Classification of Computer Games

players and can be used to store game information persistently when players go offline. Servers do not necessarily have to be hosted by the producer of the game; in many games the server functionality is included in the game, allowing one of the players to start a session with his computer acting as a server for the duration of the session (Figure 5.4). For MMORPGs, large and widely distributed networks of servers are used to host the game and improve the quality of service by hosting the games with a server close to the gamers.

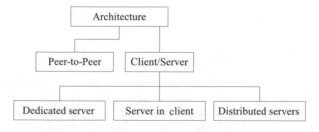

Figure 5.4   Computer Game Architectures

5.3.4 QoS Requirements

The QoS requirements of network games depend strongly on whether they are real-time games or not. Non-real-time games do not have any special QoS requirements. On the other hand, the QoS requirements of real-time computer games depend on the exact type of the game. Table 5.2 lists the results for the recommended upper limit of loss respectively latency from several different studies. The results are consistent with other studies of interactive applications such as those presented in Bailey (1989) that indicate an upper round-trip time of 200 ms for real-time interaction, MacKenzie and Ware (1993) that recommends less than 225 ms latency for interaction in virtual realities or Institute of Electrical and Electronic Engineers (1996) that recommends an upper bound of 300 ms on latency for military simulations; see also Henderson and Bhatti (2003).

As can be seen from Table 5.2, network games are generally more sensitive to delay than loss. The cited studies also show that better players are more affected by the delay in their performance and are generally more aware of QoS degradation. The most sensitive games are action games, especially first person shooters. Increased latency has the highest effect on shooting with precision weapons and only very little effect on actions like moving the game character, see Beigbeder *et al.* (2004). RTS games are relatively insensitive to loss and delay and online role playing games, even more.

5.3.5 Traffic Model

Internet real-time games with little tolerance for latency do not use TCP as transport protocol but use UDP instead to avoid the congestion control behaviour and the delay from retransmissions. This can be seen by comparing the transport protocol in Table 5.3 with the maximum latency in Table 5.2; only the very latency-tolerant MMORGPs use TCP. In addition to more control over the latency, using UDP gives the application full control of retransmitting a lost packet or not.

Table 5.3 lists traffic models for some common network games. Zander and Armitage (2004) list a number of references with advanced traffic models. As can be seen from the table, most of today's network games are designed to operate over dial-up Internet connections and therefore have a throughput of approximately 40–64 kbps. The traffic from the server to the client has significantly larger packet sizes than in the opposite way and has a packet interarrival time of 50 ms for most games.

To reduce the bandwidth requirements of online real-time computer games, game designers use mechanisms like dead reckoning and to reduce the effect of latency, they use mechanisms like buffering and artificial delays for actions on the local machine, time distortion and client predictions.

5.4 Voice over IP

5.4.1 QoS Requirements

Voice over IP (VoIP) applications use the standardised Session Initiation Protocol (SIP) or H.323 signalling protocols or proprietary protocols (like Skype). At the time of writing, SIP seems to be the protocol of choice although Skype also has a very large user base.

Table 5.2   QoS Requirements of Real-time Network Games

Game	Type	Source	Maximum Loss	Maximum Latency
Car racing	Action	Pantel and Wolf (2002)		100 ms
XBlast shooting game	Action	Schaefer et al. (2002)		140 ms
Quake 3	Action, FPS	Armitage (2001); Zander and Armitage (2004)	over 10%	150–180 ms
Half-life	Action, FPS	Henderson (2001)		225–250 ms
Halo	Action, FPS	Zander and Armitage (2004)	4%	200 ms
Unreal tournament 2003	Action, FPS	Beigbeder et al. (2004)	3%	150 ms
Madden NFL football	Sport emulation	Nichols and Claypool (2003)		500 ms
Warcraft III	Strategy, RTS	Sheldon et al. (2003)		800 ms
Everquest II	MMORPG	Fritsch et al. (2005)		1250 ms

The ITU G.114 standard recommends a one-way transmission delay up to 150 ms for voice communication, although a one-way delay of 400 ms is still considered acceptable (see International Telecommunication Union (2000)). Callers typically notice the delay if it is 250 ms round trip. The amount of tolerable jitter depends on the buffering strategy on the receiver side; if the jitter is high, more buffering is necessary, which adds to playback latency.

VoIP is not tolerant of packet loss for most codecs. For the 'standard' G.711 codec or the G.729 codec, 1% packet loss significantly degrades a call. Other more compressing codecs are even less robust. Packet Loss Concealment (PLC) or Packet Loss Recovery (PLR) algorithms can increase the acceptable packet loss rate to about 5%.

5.4.2 Traffic Model

The amount of required bandwidth depends on the used codec. Some of the standard voice codecs are G.711, G.729, G.726, G.723.1 and G.728. Bandwidth requirements and packet sizes depend on the codec and the configuration. Typically, a VoIP call will consume 25–100 kbps of bandwidth with 22–100 packets/second and a packet size of 60–200 bytes. G.711 has a 64 kbps voice bandwidth and if sampled every 20 ms the payload of each packet is 160 bytes. With 40 bytes IP/UDP/RTP header this leads to uniformly distributed constant bit-rate (CBR) flow with a bandwidth requirement of 80 kbps on IP layer.

Table 5.3 Traffic Models of Real-time Network Games

Game	Type	Source	Traffic Model
Half-life	Action FPS	Henderson (2001)	UDP, server to client 60–300 byte packet length (depends on the map used in the game) with interarrival times of 50 ms, client to server 60–90 bytes with regular interarrival times between 33 and 50 ms
Unreal tournament 2003	Action FPS	Beigbeder et al. (2004)	UDP, 63–70 kbps with a std. dev. of about 10 kbps. Median packet size around 70 bytes. Packet interarrival time server to client 50 ms, irregular in opposite direction (depends on user action)
Madden NFL football	Sport emulation	Nichols and Claypool (2003)	UDP, < 20 kbps/player, < 90 byte packet size (median 77 bytes). For high latencys, packets are aggregated and packet size increases accordingly
Counterstrike	Action FPS	Feng et al. (2005); Claypool et al. (2003)	UDP, 15–24 kbps/player, client to server average 40 byte packets, server to client average 130 bytes, large periodic bursts every 50 ms
Starcraft	Strategy RTS	Claypool et al. (2003)	UDP, 5.2–6 kbps/player, 120 byte median packet size, only small deviation from average packet size, packets sent uniformly over a range of 10–300 ms
Warcraft III	Strategy RTS	Sheldon et al. (2003)	UDP, mostly 46 or 49 byte packet size, interarrival rate 200 ms
Lineage II	MMORPG	Kim et al. (2005)	TCP, client to server average 59 bytes packet size, server to client average 358 bytes packet size
Everquest II	MMORPG	Fritsch et al. (2005)	TCP, client to server average 0.4 kbps (maximum 4.7 kbps), server to client 0.9 kbps (maximum 4.2 kbps)
ShenZhou online	MMORPG	Chen et al. (2005)	TCP, 7 kbps/player, 98% of packets smaller than 71 bytes, 30% are TCP acknowledgement

To calculate the necessary bandwidth for aggregate VoIP traffic, different traffic models are suited. Erlang B, extended Erlang B and Erlang C are the most commonly used ones; other models include Poisson and Neal–Wilkerson, see for example, Freeman (2004).

5.5 Traffic Classification

5.5.1 Port-based Traffic Classification

The standard way of identifying which application a data packet belongs to is by looking at the ports in the TCP/UDP header. The TCP/UDP ports can be distinguished into the so-called well-known ports from 0 to 1023, the registered ports from 1024 to 49151, and the dynamic/private ports from 49152 to 65535. A list of assigned ports is available at http://www.iana.org/assignments/port-numbers. The default ports of some important applications are listed in Table 5.4.

Compared to other traffic classification mechanisms, port-based classification is relatively cheap on a high bandwidth link in real time. However, there are ambiguities in the port registration; many applications are not listed in the port directories. In addition, nothing forces an application to use the assigned ports. In fact, many P2P applications allow their user to change the standard port or use random ports straightaway to avoid detection. In addition, many applications besides the web are tunnelled through HTTP (e.g. chat, streaming, P2P).

5.5.2 Advanced Mechanisms

The widespread usage of P2P file sharing applications and the problem of reliably identifying their traffic lead to the works on more advanced traffic classifications as discussed in the next section.

Table 5.4 Standard Ports of Some Applications

Application Protocol	(Main) Transport Protocol	(Main) Standard Ports
HTTP, HTTPS	TCP	80, 443
FTP	TCP	20, 21
Telnet	TCP	23
SSH	TCP	22
SMTP	TCP	25
POP, POPS	TCP	110, 995
IMAP, IMAPS	TCP	143, 993
DNS	UDP/TCP	53
Skype VoIP	TCP/UDP	Random and 80, 443
eDonkey P2P	TCP/UDP	4661–4665
Kazaa P2P	TCP/UDP	1214
BitTorrent P2P	TCP	6881–6889
Gnutella P2P	TCP/UDP	6346–6347

5.5.2.1 Signature Detection

Signature-based detection techniques are used in the context of network security and intrusion detection. Signatures can also be used for traffic classification. Sen *et al.* (2004), for example, present a traffic classification mechanism for P2P applications that uses application signatures. The signatures are application-specific bit patterns that occur in the payload of packets. A flow is classified depending on which signatures are identified in its packets.

Payload inspection has the drawback that it involves looking into the payload. This is costly and might involve legal considerations in some countries. Unknown applications cannot be classified and if the payload is encrypted, the method fails. Furthermore, some P2P protocols and other applications use HTTP requests and responses and can therefore not be distinguished from normal WWW traffic with this method. Despite all these drawbacks, payload inspection is the most common of the advanced techniques.

5.5.2.2 Traffic Statistics

Roughan *et al.* (2004) presents a statistical supervised learning approach for general traffic classification. It does not aim at identifying the exact application protocol; instead, it aims at identifying whether the application is interactive, a bulk transfer, streaming or transactional. The same application – for example a web browser – can be used for interactive transfers of web pages as well as for the bulk transfer of, for example, the latest Linux distribution CD image. In both cases, the same protocol (HTTP over TCP) is used; however, the QoS requirements differ. The approach of Roughan *et al.* (2004) promises to identify the QoS requirements of the flow more or less independent of the application protocol. To do so, traffic statistics such as the packet size, flow duration, bytes per flow, packets per flow, and so on, are used to classify a new flow into predetermined categories.

Moore and Zuev (2005) propose a Bayesian analysis that requires hand-classified network data as input. Zander *et al.* (2005) use machine learning techniques for self-learning traffic classification mechanisms.

Karagiannis *et al.* (2004) use two heuristics to identify P2P traffic in traffic traces. The first uses the fact that many P2P applications use TCP and UDP at the same time while few non-P2P applications do so. The second heuristic looks at the source/destination IP/port pairs. Web traffic has a higher ratio of the number of distinct ports versus the number of distinct IP addresses than P2P traffic. The mechanism is good for offline traffic characterisation. An extension of this concept is discussed in Karagiannis *et al.* (2005).

5.6 Summary and Conclusions

In this chapter, four important traffic types were discussed. Web traffic is based on the HTTP protocol and is mostly interactive traffic. P2P traffic is mainly caused by file sharing applications. It is mostly bulk transfer traffic and makes up the largest amount of traffic in today's Internet. As online games are becoming more and more important and as the important online games have special quality-of-service requirements, this traffic type will need increased attention of ISPs in the future. Voice over IP traffic is currently exploding and becoming one of the major traffic sources. It also has special QoS requirements. At the end of this chapter, methods for determining the application or application type of a traffic flow in real time were discussed. The drawbacks of port-based classification, which is mainly used today, were pointed out and advanced concepts were discussed.

Part II
Network Architecture

6

Network Architecture Overview

6.1 Introduction

Intelligent Network Service Providers (INSPs) offer layer 3 packet forwarding services by operating an IP network. The technical infrastructure of the network provides the core packet transport service that an INSP bases its business upon; for the quality and costs of the transport services it thus plays a vital role. Therefore, in this Part II of the book we discuss the network architecture which defines the characteristics of the network infrastructure:

With the term *Network Architecture* we describe the technology used for building the network of an INSP. The properties of a network depend on its network architecture and the configuration of that architecture. We distinguish the four sub-architectures that are depicted in Figure 6.1 as:

- **Quality of Service (QoS) Architecture**
  The QoS architecture describes the technical measures that provide quality of service. The nature of the QoS architecture has strict consequences for the forwarding and signalling architecture. For example, Intserv as QoS architecture makes the use of a QoS signalling protocol such as RSVP as part of the signalling architecture (see following text) very likely and works well with both a plain IP or a Multi-protocol Switching Label (MPLS) data forwarding architecture.
  We discuss QoS architectures in Section 6.2.

- **Data Forwarding Architecture**
  The data forwarding architecture describes the actual technical packet forwarding technology. INSPs can use plain IP packet forwarding where every hop in the path of the packet through the network is an IP router that looks up IP header information in its routing table to decide on how to forward the packet. An alternative data forwarding architecture is label switching packets using MPLS technology.
  Data Forwarding Architectures are discussed in Section 6.3.

- **Signalling Architecture**
  The signalling architecture encompasses the different signalling and control protocols to manage the network. This includes interior and exterior routing protocols, QoS

The Competitive Internet Service Provider: Network Architecture, Interconnection, Traffic Engineering and Network Design
Oliver Heckmann © 2006 John Wiley & Sons, Ltd

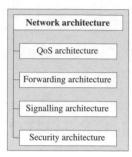

Figure 6.1   Network Architecture

signalling protocols and label distribution protocols. They are discussed in more detail
in Section 6.4.

- **Security Architecture**
  The security of an INSP's network depends on many factors, for example, the IP-level
  security architecture, the quality of its implementation, router operating system security
  and the physical security of the network. The IP-level security architecture of an INSP
  provides security at the IP packet level. Security issues encompass data encryption,
  authentication, confidentiality and network-level protection against denial-of-service
  attacks. The IP security architecture is discussed in Section 6.5.

In the remainder of this chapter, we give a detailed overview of the different QoS
architectures discussed in the community. Then, data forwarding, signalling and security
architectures are presented. Towards the end of the chapter, admission control mechanisms
are discussed.

In the next two chapters, we analytically and experimentally compare different QoS
architectures and admission control mechanisms with respect to several aspects to shed
light on the advantages and drawbacks of these architectures and mechanisms.

6.2 Quality of Service Architectures

We use the term *QoS architecture* to describe the general technology upon which actual
QoS systems are based. The range of technical forwarding services an INSP can offer
to his customers depends on his QoS system. The efficiency with which these services
are provided also depends on the QoS system. Therefore, the QoS architectures upon
which those QoS systems can be based are highly important for the purpose of this
book and are discussed in detail next. We will start by defining a *QoS system* and its
components (Section 6.2.1) and then we will discuss different QoS architectures for IP
networks.

- Integrated Services in Section 6.2.2,
- Stateless Core (SCORE) with Dynamic Packet State (DPS) in Section 6.2.3,
- Differentiated Services in Section 6.2.4,
- Several best-effort based approaches in Section 6.2.5 and
- Finally other more exotic approaches in Section 6.2.6.

We conclude the discussion of these architectures with a summarising classification in Section 6.2.7. Admission control plays an essential role in offering service guarantees and high quality services. In addition, many admission control works are relatively general and independent of specific QoS architectures. For these reasons, we present a separate overview and classification of admission control mechanisms in Section 6.6. In that context, we also present specific implementations of admission control mechanisms, for example, in the form of a Diffserv bandwidth broker which is used for the experiments later in this part.

6.2.1 Components of a Quality of Service System

The following definitions are based on Schmitt (2001); their inter-relation is shown in Figure 6.2.

A *QoS system* consists of the *QoS architecture* that describes the technical part of the QoS system and the *QoS strategy* that determines how an INSP exploits the technical features offered by the chosen architecture. The strategy involves the configuration of the architecture, policy decisions and tariffing. While there are only a low number of QoS architectures under discussion in the community, the number of QoS systems

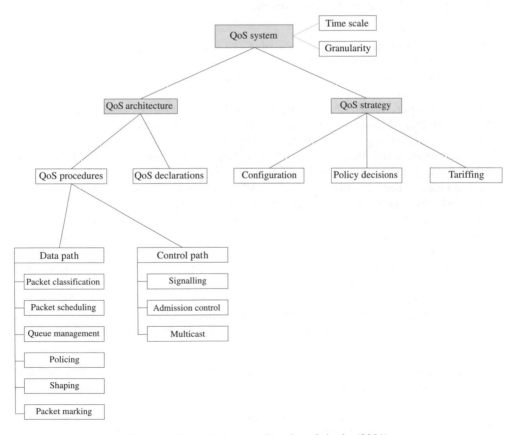

Figure 6.2   QoS System, Based on Schmitt (2001)

that can be built upon these architectures is much larger. This becomes visible for example, in Chapter 8 where more than 20 QoS systems based on three QoS architectures are evaluated.

A QoS architecture can be divided into *QoS declarations* and *procedures*. The QoS declarations form the static part of the architecture and contain properties like service classes, parameters and their specification units. QoS procedures constitute the dynamic part of the QoS architecture and consist of the data and control path mechanisms.

QoS procedures[1] on the control path are signalling, admission control and multicast. Some QoS architectures use a *QoS signalling protocol* as part of the signalling architecture to signal user demands, see Section 6.4.2.

QoS procedures on the data path are packet classification, packet scheduling, queue management, policing, shaping and packet marking. Packet classification is necessary to identify the service class, or flow, the packet belongs to, as that determines the service the packet receives.

If there are several packets competing for a link, then the **scheduling** algorithm decides the order in which these packets are sent. Sending packets on a First-come First-served (FCFS) basis is usually not enough to give delay guarantees or to split the bandwidth in a given proportion among flows or service classes. There are many different families of scheduling algorithms available, for example:

- *Priority* schedulers,
- the *EDF* (Earliest-deadline First) scheduler described in Liu and Layland (1973) and advanced EDF schedulers like *Rate-controlled EDF* as in Zhang and Ferrari (1994),
- Round Robin schedulers like *WRR* (Weighted Round Robin) and *DRR* (Deficit Round Robin, see Shreedhar and Varghese (1996)),
- the PGPS/WFQ family: *PGPS* (Packetised General Processor Sharing) as *WFQ* (Weighted Fair Queueing) (see Demers *et al.* (1989); Parekh (1992)), *SCFQ* (Self-clocked Fair Queueing, see Davin and Heybey (1994)), *FFQ* (Frame-based Fair Queueing, see Stiliadis and Varma (1996)), *SFQ* (Start-time Fair Queueing, see Goyal *et al.* (1997)), *WF2Q* (Worst-case Weighted Fair Queueing, see Bennett and Zhang (1996)),
- virtual-clock schedulers like the original *VC* (Virtual Clock, see Zhang (1990)) and *LFVC* (Leap Forward Virtual Clock, see Suri *et al.* (1997)),
- hierarchical schedulers like *CBQ* (Class Based Queueing, see Floyd and Jacobson (1995)), *HPFQ* (Hierarchical Packet Fair Queueing, see Bennett and Zhang (1997)), *HFSC* (Hierarchical Fair Service Curve, see Stoica *et al.* (1997)),
- and dynamic packet state (DPS) schedulers like *CSFQ* (Core-stateless Fair Queueing, see Stoica *et al.* (1998, 2002)),
- If the parameters of scheduling algorithms are not configured statically but instead adapted automatically based on current measurement information, we speak of *adaptive* variants of scheduling algorithms, see for example, Antila and Luoma (2003, 2004); Christin *et al.* (2002); Liao and Campbell (2001).

Queue management is typically closely connected to scheduling. While schedulers manage the access to an outgoing link's bandwidth, queue management controls the

---

[1] Contrary to Schmitt (2001) who counts traffic engineering and network design/engineering as (mid-term and long-term) QoS procedures we treat these procedures as part of a separate problem area (see Part IV of this book) because they affect the whole network, not only the QoS system.

buffer space inside a router that is used to store the packets that have not yet been served by the scheduler. If buffer space is running out, packets have to be dropped. Some queue management schemes drop only newly arriving packets (FIFO) while others can also drop already buffered packets, for example, from the head of the queue. Another decision with respect to buffer management is whether the buffer space is split up statically between the different queues.

Besides that, a router can employ *Active Queue Management* (AQM) strategies to actively keep the average queue length small. AQM promises to improve the end-to-end congestion control, to lower queueing delays, more fairness among the flows and buffer reserves for absorbing bursts of packets. This is done by *actively* signalling congestion early. Congestion is signalled by dropping packets or by marking packets if the sender supports *Explicit Congestion Notification (ECN)* as introduced by Ramakrishnan *et al.* (2001).

The classical AQM algorithm is *RED (Random Early Detection)*, see Floyd and Jacobson (1993). It maintains an exponentially weighted moving average of the queue length. When the average queue length exceeds a minimum threshold packets are randomly dropped/marked; if a maximum threshold is exceeded all packets are dropped/marked.

RED has been improved in a number of ways. The sensitivity of RED to parameter settings led to proposals like *Gentle RED* (see Rosolen *et al.* (1999)), *Adaptive RED* (see Floyd *et al.* (2001)), *Stabilised RED* (see Ott *et al.* (1999)) while other works like *Flow Random Early Drop* (see Lin and Morris (1997)), *RED with Preferential Dropping* (see Mahajan *et al.* (2001)) and *CHOKe* (see Pan *et al.* (2000)) aim more at improving the fairness of RED.

Virtual Queue (VQ) approaches like that of Kunniyur and Srikant (2004) maintain a VQ whose capacity is less than the actual link capacity. Packets arriving at the real queue are also accounted for in the VQ. If the VQ overflows, this is taken as congestion indication and packets arriving at the real queue are marked as dropped.

A control theoretic approach to AQM is the *Proportional Integrator (PI)* controller of Hollot *et al.* (2001). It is a based control theory applied to a linearised TCP/AQM model. PI regulates the queue length to a target value (queue reference) using instantaneous samples of the queue length contrary to the moving average of RED that can be influenced largely by past values of the queue length. An improved version of PI is presented by Heying *et al.* (2003).

The *Random Exponential Marking (REM)* AQM scheme of Athuraliya *et al.* (2001) uses a congestion measure labelled 'price'. This 'price' measures the mismatch between packet arrival (demand rate) and departure rates (service rate) and the mismatch between the actual and target queue lengths.

Another approach is called *Blue* by Feng *et al.* (2002); it is based on buffer overflow and link idle events contrary to the average queue length of RED. Several AQM mechanisms are compared and evaluated for example, in Bitorika *et al.* (2004); Le *et al.* (2003).

If a QoS architecture uses reservations or Service Level Agreements (SLA) that specify the amount of traffic a user is entitled to, a mechanism is necessary to control whether an arriving packet is conforming with the agreed traffic specification. The mechanism to detect non-conforming packets is called *policing*. The network can react to non-conforming packets by dropping these packets, by delaying these packets with a *shaper* until they conform or by downgrading the service these packets receive (the latter might require a *packet marker*). A shaper can also be used at an outgoing link, for example, at an interconnection to make the traffic conformant to a service level agreement with the next interconnection partner or just to smooth out bursts. Packet markers can also be necessary at ingress nodes for example, in Diffserv networks to write the Diffserv CodePoint (DSCP) into the IP header or in routers that use explicit congestion notification (ECN) to mark packets.

We now discuss different QoS architectures and then summarise them by the classification of Section 6.2.7.

6.2.2 The Integrated Services Architecture

6.2.2.1 Overview

The term *Integrated Services Network* was introduced by Scott Shenker. It describes one network for all kinds of applications, especially real-time multimedia traffic like voice, video conferencing and TV like applications. In the early 1990s, the Internet Engineering Task Force (IETF) realised that the Internet's egalitarian best-effort model is not suited for this kind of real-time multimedia traffic if the network is significantly loaded. The IETF's first answer to this problem was the Integrated Services architecture. Later, with the Differentiated Services architecture a second fundamentally different approach was pursued (see following text).

The general Integrated Services (Intserv) architecture is specified in RFC 1633 (see Braden *et al.* (1994)). It builds upon a QoS signalling protocol. The IETF proposed signalling protocol is RSVP. The IETF Intserv specifications can be broken into two parts, the signalling as RSVP part in RFC 2205 (see Braden *et al.* (1997)) and the integrated service specifications in RFCs 2211 and 2212 (see Shenker *et al.* (1997); Wroclawski (1997)); because of this the Intserv architecture is often described as 'RSVP/Intserv'.

Guarantees are given for individual flows, for each flow a path is reserved through the network. A flow is defined as a *distinguishable stream of related datagrams that result from a single user activity and require the same QoS*; it can be seen as a hybrid between the Virtual Circuit model of ATM and the pure datagram model of IP.

The Intserv service model is based on the distinction between real-time and elastic traffic. Elastic traffic is treated as the traditional best-effort traffic. Contrary to Diffserv, no differentiation of the elastic traffic flows is supported. The default service is best-effort; applications using it do not need any modifications.

The real-time traffic is further categorised by whether it is tolerant to loss and whether it is (rate/delay) adaptive. Multicast support was considered vital by the IETF during the development of Intserv and is widely supported by the architecture.

6.2.2.2 Intserv Control Path

Using RSVP, the applications on the end systems request a specific end-to-end QoS for one session from the network. A *session* in the context of RSVP/Intserv is defined by the triple destination IP address, protocol ID and optionally a destination port. As the destination address can be a multicast address, a session is a data flow from possibly

multiple senders to multiple receivers. The reservation process is described in RFC 1633
and 2205 (see Braden *et al.* (1994, 1997)) and depicted in Figure 6.3:

- A sender application announces itself by sending a PATH message to the destination
  unicast or multicast address. If multicast is used, each receiver must first join the associated multicast group using a multicast group management protocol like Internet Group
  Management Protocol (IGMP) (see Cain *et al.* (2002); Fenner (1997)) for IPv4 and
  Multicast Listener Discovery (MLD) (see Haberman (2003)) for IPv6. This, however,
  is not part of the QoS negotiation process and RSVP.
  The PATH message
  - contains a traffic specification (*TSpec*).
  - establishes path state in the intermediate routers.
    This path state is used for propagating back reservation requests on the reverse
    path. Unlike more traditional signalling protocols from telecommunication networks,
    RSVP does not set up an explicit route for the data transmission; this task is left to
    the routing protocols (see Section 6.4.1).
  - Optionally, the sender may include an advertisement specification (*AdSpec*) in its
    PATH message in order to advertise to receivers the characteristics of the commu-
    nication path. On their way downstream, the advertisements accumulate information
    about the hop count, minimum propagation latencies, minimal individual link band-
    width along the path, the path MTU, service-specific parameters, and whether all
    routers along the path support RSVP.
- Each receiver individually determines its QoS requirements. Therefore, the whole process of QoS negotiation is called *receiver-oriented*. The decision is obviously based
  upon the TSpec and the AdSpec of the PATH message but can be influenced by any
  knowledge about the locally available resources (e.g. maximum resolution of a video
  display), application requirements, service prices and so on.
- The receiver then initiates the actual reservation process by responding to the PATH
  message with a RESV that is routed along the previously set up path back to the sender.
  The RESV message contains:
  - A flow specification (*FlowSpec*) describing
    - the requested service class,

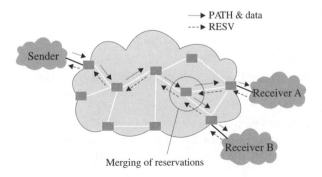

Figure 6.3   Intserv Control Path

- the specification of desired QoS (*RSpec*), and
- the description of the data flow (*TSpec*).

  ○ A filter specification (*FilterSpec*) that identifies the packet subset of the session that has these QoS requirements via the reservation style (see following text).

- Along their way to the sender, the RESV messages have to pass an *admission control* test in each router along the path.

  If admission has to be rejected in one of the intermediate systems, a reservation error is raised and signalled to the receiver with a *RESVERR* message. The receiver can try to initiate another reservation with a less demanding FlowSpec or give up. This QoS negotiation process is called *One-pass with Advertising*, see Shenker and Breslau (1995).

- In the multicast case, a distribution tree is created by *merging* reservations: Multiple receivers indicating a need to receive from the same sender do not install separate reservations. Rather, the largest reservation is granted and the rest are assumed to be using the same resources. Therefore, propagation of a RESV message ends as soon as the reservation encounters an existing distribution tree with sufficient resources.

  Besides having multiple receivers, a multicast group may also have multiple senders. For some applications, for example, video conferencing, where it can be expected that only one person is talking at a time, it is desirable that a resource allocation can be shared among multiple senders of a multicast group. This is supported by the RSVP reservation style. The *reservation style* specifies to which extent intermediate routers may merge the reservation requests from different receivers in the same multicast group. RFC 2205 (see Braden *et al.* (1997)) defines three reservations styles:

  ○ If the *wildcard filter* is used, all traffic from *all* senders directed to the receiver may be merged.

  ○ With the *shared explicit filter,* the receiver explicitly identifies the list of senders that share *one* reservation.

  ○ The *fixed filter* allows for a fixed set of simultaneously transmitting senders; the receiver can specify a set of sources and *for each* of them a certain amount of resources is reserved.

- The state in the intermediate routers is the *soft state*, that is, it times out after a certain period. Therefore, RSVP sends PATH and RESV messages periodically. The PATH refreshments will set up a new path in the case of node and link failures and RESV refreshments can also be used to adapt the resource allocations. Also, the soft state mechanism automatically times out and recovers orphaned reservations.

6.2.2.3 Intserv Data Path

Intserv uses a number of QoS procedures on the data path, see also Figure 6.4:

- Packet Classification
  For each incoming packet, the flow it belongs to and the reservation state associated with it have to be identified from the IP header information at line speeds; multiple fields (destination IP, port, etc.) are used in this classification.

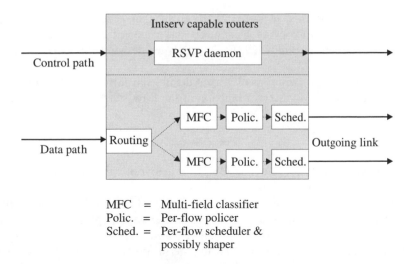

MFC   = Multi-field classifier
Polic. = Per-flow policer
Sched. = Per-flow scheduler &
          possibly shaper

Figure 6.4 Intserv Routers

- Policing
  At least at the edge of the network, policing is necessary to ensure that a host does not violate its promised traffic characteristics.
- Scheduling and Queue Management
  Different queues have to be managed and the packets waiting in the queues have to be scheduled so that the service guarantees are fulfilled. Intserv does not assume one specific scheduler. A wide variety of schedulers can be used, however, the error terms of the scheduling algorithm influence the amount of resources that have to be allocated to provide a certain QoS and thus the efficiency with which a certain service can be provided; see Section 6.2.2.4 for details.
- Shaping
  If multiple senders are used, reshaping is necessary at all points of the multicast distribution tree where traffic from two different sources that share the same reservation merge. Also, at points where the multicast distribution tree from a source branches to multiple distinct paths with differing TSpecs, reshaping is necessary on the outgoing links that have 'lower' TSpecs than the upstream link, see Shenker *et al.* (1997).

6.2.2.4 Intserv Guaranteed Service

Because of the importance of the Intserv guaranteed service (GS) as the 'strongest' service in today's QoS architectures, we now discuss GS in some more detail.

GS offers a deterministic service with zero-loss guarantees and delay bound guarantees: If every router in the flow's path supports guaranteed service (or adequately mimics GS), the flow experiences a delay-bounded service with no queueing loss for all conforming packets. Please note that it does not aim at minimising the jitter.

The Intserv GS is specified in RFC 2212 (see Shenker *et al.* (1997)). Other QoS architectures can be used to provide guaranteed service or at least similar services; see Section 6.2.3 or the central Bandwidth Broker (BB) we developed for the experiments of Chapter 8.

The flow's delay bound d consists of a fixed delay (transmission delay, etc.) d_t and the maximum queueing delay d_q that is a function of the flow's arrival curve and the service function allocated for the flow:

$$d = d_t + d_q \tag{6.1}$$

The mathematical foundation for the GS is the network calculus, see Section 3.2 for an introduction to network calculus.

The arrival curve is given with the TSpec that consists of a token bucket with rate r and buffer b that specifies the flow plus a peak rate p which specifies the maximal rate at which the source may inject bursts into the network and the maximum datagram size M and the minimum policed unit m. The long-term average rate of the flow does not exceed the token-rate r, the maximal burst sent into the network within a short period of length T does not exceed $M + pT$, see Figure 6.5. To assure that a flow conforms to these specifications, policing and reshaping are used.

The service a flow receives at a router is mathematically described by the service curve; it is specified by a service rate R and a latency L; see Figure 6.5. The latency L depends on the scheduling algorithm. The service rate R is specified in the RSpec by the receiver and represents the share of the link's bandwidth the flow is entitled to. Via the service rate R, the receiver can influence the delay bound. If there is a difference between the desired delay bound $\overline{d_q}$ and the bound $d_q = f(R)$ obtained by the chosen service rate R, this difference can be expressed with the slack term S of the RSpec that allows intermediate routers to reduce their resource reservations accordingly. The buffer size B represents the buffer space in the router that the flow may consume.

As the theoretical model behind GS is a fluid model, the rate dependent error term C_l and the rate independent error term D_l of a router (as outgoing link) l are used to express the difference between the fluid model and a real scheduling algorithm operating on packetised data. For WFQ, see Demers *et al.* (1989); Parekh (1992) and other non-preemptive scheduling algorithms, the rate independent error term is given by the delay

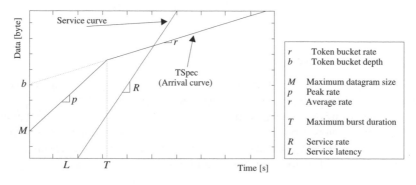

Figure 6.5   Guaranteed Service (Arrival and Service Curve)

caused by a maximum sized packet (size MTU_l) blocking the link with bandwidth bw_l for a conforming packet of the flow that arrives shortly after the maximum sized packet:

$$D_l = \frac{MTU_l}{bw_l} \tag{6.2}$$

The rate dependent error term C_l expresses the backlog of the queueing/scheduling algorithm against a fluid bit-by-bit service, for WFQ it is equal to the flow's maximal packet size M

$$C_l = M \tag{6.3}$$

The scheduler error terms are summed up (e.g. in the AdSpec) to the total error terms $\sum_{\forall l} C_l$ and $\sum_{\forall l} D_l$ for calculating the end-to-end delay. With the maximum burst duration

$$T = \frac{b - M}{p - r} \tag{6.4}$$

the end-to-end delay bound for a given R is given by

$$d_q = \begin{cases} T \cdot \frac{p-R}{R} + \frac{M+\sum_{\forall l} C_l}{R} + \sum_{\forall l} D_l & \text{for } p > R \geq r \\ \frac{M+\sum_{\forall l} C_l}{R} + \sum_{\forall l} D_l & \text{for } R \geq p \geq r \end{cases} \tag{6.5}$$

Besides allocating the rate R, a router l also has to allocate a buffer B_l to ensure no loss. This buffer is

$$B_l = \begin{cases} M + T \cdot (p - R) + \sum_\lambda C_\lambda + R \sum_\lambda D_\lambda & \text{for } p > R \geq r, \, L \leq T \\ M + p \left(\frac{\sum_\lambda C_\lambda}{R} + \sum_\lambda D_\lambda \right) & \text{for } R \geq p \geq r, \, L \leq T \\ b + r \left(\frac{\sum_\lambda C_\lambda}{R} + \sum_\lambda D_\lambda \right) & \text{for } L > T \end{cases} \tag{6.6}$$

with the overall scheduler latency L

$$L = \frac{\sum_\lambda C_\lambda}{R} + \sum_\lambda D_\lambda \tag{6.7}$$

where $\sum_\lambda C_\lambda$ as $\sum_\lambda D_\lambda$ are the error terms summed up from the first hop as the last reshaping point to link l.

If the peak rate p is unknown, it is assumed to be infinite; the arrival curve becomes a token bucket (r, b) and the end-to-end delay bound simplifies to

$$d_q = \frac{b}{R} + \frac{\sum_{\forall l} C_l}{R} + \sum_{\forall l} D_l \tag{6.8}$$

and the buffering required at a router l to

$$B_l = b + \sum_\lambda C_\lambda + \sum_\lambda D_\lambda \cdot R \tag{6.9}$$

6.2.2.5 Intserv Controlled Load Service

The Controlled Load (CL) service is specified in RFC 2211 (see Wroclawski (1997)). It provides flows with approximately the QoS that they would receive using the traditional best-effort service in an unloaded network. Though CL does not provide strict boundaries of QoS parameters like loss and delay, it ensures that a very high percentage of the delivered packets will not experience loss or delay higher than the basic packet error rate as the minimum transit delay along the path. Admission control and policing have to be used to control the amount of CL flows and – obviously – an estimate of the data traffic has to be given (in form of a TSpec).

The CL service allows much more freedom in its implementation than that of the GS. The idea is that the service allows for extremely simple implementations on one side as well as implementations with evolving scheduling and admission control algorithms for a highly efficient use of network resources on the other side.

6.2.2.6 Complexity and Scalability Discussion of Intserv

Two types of per-flow state are needed in Intserv networks:

- Forwarding state to pin the forwarding path of a flow, and
- the FlowSpec/FilterSpec state used by the admission control on the control plane as well as the packet classifier and scheduler of the data plane.

Therefore, each Intserv router has to process per-flow signalling messages, maintain the FlowSpec/FilterSpec tables per flow and perform per-flow packet classification and scheduling. It is obvious that this complexity has its costs, especially in backbone networks with a large number of flows where it can cause scalability issues.

Karsten (2000) and Karsten et al. (2001) discuss the scalability of the control path: the complexity of the RSVP daemon and the signalling. These works present and analyse an open-source RSVP implementation (see Karsten (2004)) that with some optimisations like fuzzy timer control can handle 50,000 flows on an off-the-shelf PC with a 450 MHZ Pentium III processor and 128 MB RAM. More importantly, they show that RSVP scales linearly with the number of flows.

There are also some works to reduce the scheduling complexity including proposals that require only constant time complexity, see for example Davin and Heybey (1994); Stephens et al. (1999); Wrege and Liebeherr (1997); Zhang and Ferrari (1993), although there is a natural trade-off between the complexity of a scheduler and its flexibility as discussed in Knightly et al. (1995).

Also, in packet classification there have been recent and very promising advances, see for example Gupta (2000); Singh et al. (2003); Srinivasan and Varghese (1999b) and the works therein[2].

There are proposals to reduce the amount of state via reducing the number of flows by aggregating micro-flows that follow the same path through the network into one macro-flow, see for example Baker et al. (2001).

---

[2] Packet Classification is also necessary for IP routing lookups, see Section 6.3.1.1.

A careful analysis shows that the scalability of an Intserv network is not as critical as often assumed but still has to be taken seriously. Therefore, we next investigate a number of proposals that have a focus on reducing the state complexity compared to the Intserv/RSVP QoS architecture.

6.2.3 Stateless Core Architectures

6.2.3.1 Overview

Because of the scalability concerns with Intserv especially in backbone networks, considerable research went into analysing stateless core (SCORE) QoS architectures. The general idea is to have a network where only edge routers have to perform per-flow management while core routers do not. The IETF QoS architecture Diffserv is an example of the SCORE idea. Besides Diffserv, there are some other proposals, the most famous one is based on Stoica (2000) and called *Dynamic Packet State* (DPS). Because of the importance of Diffserv, we discuss Diffserv in a separate section and focus here on DPS.

The basic idea of DPS as described in Stoica (2000) and Stoica and Zhang (1999) is that the ingress (edge) router inserts information into the IP header. This information is used and updated by the core routers to provide deterministic service guarantees like Intserv's GS. The core routers are using a special scheduling mechanism that only depends on the DPS and does not require per-flow state on the data path. In addition, the control path is made stateless in the core as the aggregate reserved rate needed for admission control at one link can be derived from the packet state, too. We now discuss the data path and the control path of DPS before addressing related works and applications.

6.2.3.2 SCORE Data Path

Stoica (2000) and Stoica and Zhang (1999) present the DPS technique and a *Core-jitter-virtual-clock* scheduling algorithm that can approximate Intserv GS without requiring per-flow state on the data path; it is a combination of a delay-jitter rate-controller and a VC scheduler. The algorithm works as follows:

- Each flow is assigned a rate r that is stamped into the packet header and thus does not have to be stored by the core routers.
- In a router, each packet is assigned an eligible time and a deadline upon arrival. A packet is not sent before its eligible time; the scheduler is thus not work-conserving. The next packet to serve is chosen among all eligible packets according to their deadlines (earliest deadline first).
  - One goal of the algorithm is to send packets close to their deadline but not after the deadline, thus incurring the maximal allowed delay and reducing jitter. The extent of time a packet is transmitted before its actual deadline in a node (the local fluctuation) is stamped into the packet header. In the next node the packet is not eligible unless that time has passed, thus the local fluctuation of node n is balanced out at node $n + 1$.

o The eligible time has to be at least as high as the deadline of the previous packet belonging to the same flow. Normally, a scheduler would have to keep per-flow state information to track the deadlines but by introducing a third variable that is stamped into the packet header at the edge (where per-flow state is allowed) and that is updated at each hop, this per-flow state can be eliminated, too. The eligible time at a node n can now be calculated as the sum of the arrival time, the local fluctuation of the previous node and the new slack variable. The slack variable effectively introduces an additional delay for a packet at each hop making sure that a packet is not sent before the deadline of the previous packet. Stoica and Zhang (1999) and Stoica (2000) show that this delay does not increase the overall delay compared to the non-SCORE version of the scheduling algorithm (which does not use the slack variable and instead keeps per-flow state).

o The deadline of a packet is the eligible time plus the time it takes to transmit the packet with rate r assigned to the packet's flow (as encoded in the packet header).

Stoica and Zhang (1999) and Stoica (2000) propose using the IP header to store the DPS but it is also imaginable to add a new header between layers 2 and 3 as in MPLS. These works also show that due to the traffic regulation of the scheduling algorithm, the number of packets in the server at any given time is significantly smaller than the number of flows, which further reduces scheduling complexity. The works further show that DPS can give the same guaranteed service as a network of routers with a WFQ scheduler – the 'typical' Intserv scheduler.

6.2.3.3 SCORE Control Path

The SCORE DPS approach assumes that RSVP messages are only processed by the edge nodes; inside the network a lightweight signalling protocol is used.

Before admitting a new flow, the admission control module at each node has to check whether the aggregated reserved rates and the new rate do not exceed the outgoing link's capacity. Just counting the aggregated reserved rates for each outgoing link without keeping track of the flows is no robust solution because flows can stop sending without notice, for example, because the sender crashed, without the system being able to free the resources again. So, normally the per-flow state would be needed to keep track of the rates of the already accepted flows and to be able to recover from the errors.

However, by introducing a fourth variable that is stamped into the packet header by the ingress node an upper bound of the aggregate reserved rate can be derived; thus, the control path per-flow state can also be eliminated. This variable contains the amount of data that the flow to which the packet belongs to was entitled to send according to its reserved rate since the previous packet. A node can add up these variables of *all* packets traversing a link for a certain period to get an estimate of the aggregate reserved rate on that link. The actual mechanism is more complex because it has to account for jitter, termination and other aspects; it is fully described in Stoica (2000); Stoica and Zhang (1999).

The DPS approach also depends on a route pinning mechanism like MPLS or the label-based one in Stoica (2000).

6.2.3.4 Related Works

The DPS approach can also be used to provide stateless flow protection or relative service differentiation in a network, for details see Stoica (2000); Stoica *et al.* (1998). Kaur and Vin (2003) describe a work-conserving core stateless scheduler for throughput guarantees.

A centralised admission control approach, contrary to the hop-by-hop approach above, is proposed in Zhang *et al.* (2000) for SCORE networks with DPS. It relieves core routers from the admission control functionality. A centralised BB is used instead that keeps track of the QoS reservation states. Besides that, the admission control is generalised with the virtual time reference system towards other types of schedulers and to class-based guarantees in Zhang *et al.* (2000). A methodology to transform any guaranteed rate per-flow scheduling algorithm into a SCORE version is presented in Kaur and Vin (2001).

In a SCORE architecture based on DPS, core routers have to trust the information carried in the packet headers; a single faulty router can disrupt the service in the entire core, therefore these solutions are not very robust. In Stoica *et al.* (2002), an enhancement of the SCORE fair queueing algorithm of Stoica *et al.* (1998) is presented. Core routers no longer blindly trust the incoming packet state (the rate estimates). Instead, they statistically verify and contain flows whose packets are incorrectly labelled.

To summarise, the SCORE approach with dynamic packet state (DPS) presents an interesting approach to provide guaranteed service or other services without having to keep per-flow state in core routers. This comes at the cost of additional fields used in the IP header or a shim header that has to be updated in each hop. The updates require relatively complex[3] operations, and expensive write access at high-speed routers.

6.2.4 *The Diffserv Architecture*

6.2.4.1 Overview

The Diffserv architecture is specified in RFC 2475 (see Black *et al.* (1998)), the Diffserv field in the IP header in RFC 2474 (see Nichols *et al.* (1998)). Diffserv can be seen as the IETF's response to the concerns about the complexity of Intserv/RSVP. Diffserv takes a more abstract and local view on resource allocation. It is a SCORE approach, the core nodes of a network do not have to keep per-flow state. Per-flow state is kept at edge nodes only where operations like policing and marking are also done exclusively.

On the data path, packets of different flows are aggregated into behaviour aggregates (BA) at the edge nodes. A BA is associated with a certain service class; it is identified by the six bit Diffserv CodePoint DSCP . The DSCP is contained in the Diffserv field[4] of the IPv4 IP header or the traffic class octet of the IPv6 IP header.

The heart of the Diffserv architecture is the Per-hop Behaviour (PHB) that specifies the forwarding behaviour of one router for packets of a DSCP that is locally mapped to that PHB. The edge-to-edge behaviour in a network of one service class – called *Per-domain Behaviour* (PDB) in the Diffserv terminology – results from the concatenation of PHBs. It is assumed that useful services can be constructed from the different PHBs in the standardisation process. The service construction process is mostly left to the INSPs.

---

[3] Compared to the standard write operations in IP routers: decreasing the hop count and updating the checksum, see also Section 6.3.1.

[4] The Diffserv field was called *type of service* byte before Diffserv was being standardised.

A Diffserv domain is a network over which a consistent set of differentiated service policies are administered in a coordinated fashion – typically, this equals the network of a single INSP. As a flow will typically pass several Diffserv domains for end-to-end QoS, a coordination of those is required. This coordination is done on the control path by the use of SLA and potentially BBs.

Figure 6.6 shows the main functionality of Diffserv edge and core routers. The ingress edge router of a Diffserv domain performs several operations on a packet arriving from outside the Diffserv domain:

- A micro-flow classification[5] is necessary to identify the flow, as flow aggregate to which the packet belongs, and to look up the associated traffic conditioning specification and the traffic profile (see following text).
- For most services, further processing by the traffic conditioning module is necessary. Depending on the service, packets are metered, marked, shaped and/or dropped.
  - A meter measures the traffic stream against the traffic profile and can influence the traffic conditioning actions that follow.

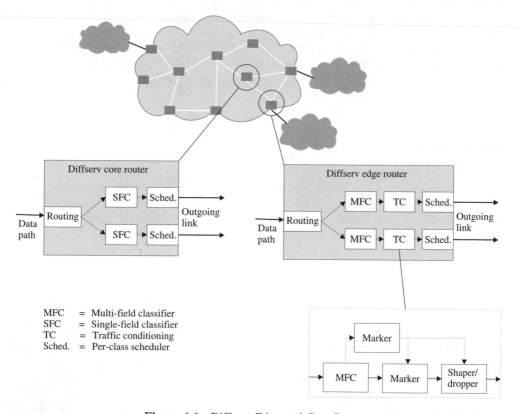

MFC   = Multi-field classifier
SFC   = Single-field classifier
TC   = Traffic conditioning
Sched. = Per-class scheduler

Figure 6.6   Diffserv Edge and Core Routers

---

[5] This is a multi-field classification based on the value of one or more IP header fields such as source address, destination address, and so on.

o The marker imprints a DSCP on the packet.

o Finally, traffic shaping may be applied to bring the micro-flow into compliance with a traffic profile. Alternatively, a dropper may be used to discard some or all out-of-profile packets.

The exact handling of in-profile and out-of-profile packets is described in the service level specification (SLS) that is part of the service level agreement (SLA), see following text.

• The buffer management and scheduling algorithm in the edge node treats the packet according to its DSCP and the PHB it has locally mapped to that DSCP.

When a packet arrives at a Diffserv core router, the operations are of less complexity:

• Only a single-field classifier is necessary; it reads the DSCP from the Diffserv field and determines the PHB that is locally mapped to that DSCP.

• The buffer management and scheduling algorithm treat the packet according to the PHB.

• Some optional functions like traffic shaping or packet remarking may be used at core routers.

6.2.4.2 Diffserv Services

As mentioned above, edge-to-edge services (Per-domain Behaviours, PDBs) are built by concatenating PHBs. As the number of the PHBs is limited, the queueing and scheduling complexity can be kept low in a Diffserv router. The packet classification in core routers is also relatively simple as the PHB is encoded in the DSCP that is stored in a single field (Diffserv field) in the IP header.

Per-hop Behaviours (PHBs) Besides a default PHB that corresponds to traditional best-effort forwarding, the following PHBs have been specified so far by the IETF:

Class Selector (CS) The CS PHB group is specified in RFC 2474 (see Nichols *et al.* (1998)) and consists of eight classes. CS is mainly intended for backward compatibility with the old IPv4 precedence bits contained in the type of service octet that is now used as the Diffserv field. Contrary to the assured forwarding PHBs (see following text), the CS precedence classes have an ordering with respect to timely forwarding: CS codepoints with a higher relative order have an equal or higher probability of timely forwarding than CS codepoints with a lower relative order.

The CS PHB can be used for relative service differentiation as it is discussed in Dovrolis and Ramanathan (1999). Contrary to an absolute service differentiation scheme where admission control is imposed on users and an admitted user receives absolute performance levels in a relative service differentiation scheme no admission control is necessary; performance guarantees are only given relative to the performance of other classes: a higher class will receive the same or better service than a lower class.

In the proportional differentiation model of Dovrolis and Ramanathan (1999), the INSP assigns a quality differentiation parameter c_i to each class i of his network. While no absolute performance levels are given for short-term performance measures p_i like the loss rate or the queueing delay, the ratio between all classes i and j is controlled by $p_i/p_j = c_j/c_i$. To achieve this ratio, special scheduling and queue management algorithms are necessary, see Dovrolis and Ramanathan (1999).

Expedited Forwarding (EF)   The EF PHB was originally specified in RFC 2598 (see Jacobson *et al.* (1999)) and later refined to a more rigorous definition in RFC 3246 (see Davie *et al.* (2002) and for more information also Armitage *et al.* (2002); Charny *et al.* (2002)). It can be used to build a low loss, low delay, low jitter, assured bandwidth service that is called *virtual leased line service* or *premium service* (see Nichols *et al.* (1999)). To provide this service, it is necessary that the aggregate traffic experiences no or at least only very small queues. To achieve this, it is necessary to ensure that the service rate R of EF packets on a given output interface exceeds their aggregate arrival rate A at that interface over long and short time intervals, independent of the amount of other (non-EF) traffic at that interface. It is difficult to define the appropriate timescale at which to measure the service rate R because too small timescales may introduce sampling errors and too large timescales may allow excessive jitter. Also, if there are not enough packets arriving at a queue in a certain interval – externally this might not be obvious – the service rate R cannot (obviously) be obtained. Because of these reasons the formal definition of EF calculates the ideal departure time of an arriving packet by assuming that it is served with rate R either immediately upon arrival or upon the departure time[6] of the previous packet. The deviation of the real departure time of a packet from the ideal departure time is bounded by an error term E. The scheduling requirements of the EF PHB are stricter than the service curve of a rate-latency scheduler[7]. A number of scheduling algorithms satisfy the EF requirements but differ in their error terms, for example:

- a strict non-preemptive priority scheduler where EF has priority over the other classes;
- worst-case fair weighted fair queueing (WF2Q), SFQ and SCFQ;
- DRR.

The EF PHB is intended for low loss services, RFC 3246 (see Davie *et al.* (2002)) leaves it optional to specify a region of operation for an EF node where no losses occur. If this is not used it means that in an RFC conformant operation of an EF node a limited number of EF packets can be dropped due to limited buffers.

For deterministic service guarantees, the worst-case aggregate arrival rate has to be bounded. However, the aggregate arrival rate depends on the topology and the routing through the Diffserv domain. Charny and Le Boudec (2000) derive a delay bound for general topologies that is a function of the maximal link utilisation α and the maximal number of hops h of a flow (as the network diameter); it is named *Charny Bound* after the first author.

For the general assumptions, the delay experienced by a single packet depends not only on the behaviour of the flows sharing at least one queue with the packet, but also on the behaviour of flows in the other parts of the network and potentially on past flows as shown in Charny and Le Boudec (2000).

It is assumed that the incoming flows are characterised by a leaky bucket and on each link l of capacity C_l, the aggregate rates are bounded by αC_l and the aggregate bucket depths by τC_l. The maximum packet size is MTU. For link l, a bound on the peak rate of

---

[6] The real or the ideal departure time of the previous packet, whichever is later.

[7] If a scheduler satisfies the EF requirements it also satisfies the rate-latency curve but not necessarily vice versa.

all incoming flows is given by Γ_l. If no further assumptions about the routing are made, Γ_l is given by the bit rates of all incoming links or if they are unknown $\Gamma_l = \infty$.

For $\alpha < \min_l \frac{\Gamma_l}{(\Gamma_l - C_l)(h-1) + C_l}$, a bound on the worst-case end-to-end queueing delay for EF traffic is

$$d_q = \frac{h}{1 - (h-1)u\alpha}(\Delta + u\tau) \tag{6.10}$$

with

$$u = \max_l \frac{\Gamma_l - C_l}{\Gamma_l - \alpha C_l} \tag{6.11}$$

$$\Delta = \max_l \frac{MTU}{C_l} \tag{6.12}$$

This bound is depicted in Figure 6.7 for different maximal hop counts h assuming a topology with a maximum in-degree of 5, a capacity C_l=155 Mbps, MTU=1500 bytes, EF flows with an average rate of 64 kbps, and a maximum burst size (bucket depth) of 600 bytes.

As can be seen, the delay bound explodes if α approaches $\min_l \frac{\Gamma_l}{(\Gamma_l - C_l)(h-1) + C_l}$; this does not mean that the delay is necessarily unbounded for these cases. It is possible that a better (lower) delay bound can be derived, yet at the time of writing no other delay bound with the same general, assumptions as the Charny bound is known to the author. Besides that, it has been shown in Charny and Le Boudec (2000) that for larger α there exists a large enough network such that the worst-case delay of some packet can exceed any D, even if the maximum hop count never exceeds h.

On the basis of this delay bound, a medium-sized network could only be utilised little more than 10% with EF traffic.

Improvements of the Charny bounds can be obtained if additional mechanisms are added to the Diffserv network; e.g. traffic shaping, see for example Cruz (1998); Fidler (2003); Ossipov and Karlsson (2003).

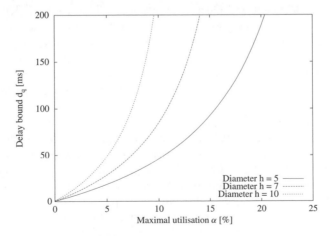

Figure 6.7   Charny Bound

Assured Forwarding (AF)   The assured forwarding PHB group is specified in RFC 2597 (see Heinanen *et al.* (1999)). It consists of four independently forwarded AF classes. Within each class, packets are marked with one of three different levels of drop precedences; they are often called *green*, *yellow* and *red*. The drop precedence determines the relative importance of the packet within the AF class. More AF classes or levels of drop precedence may be defined by INSPs for local use. There are no delay or delay variation requirements associated with the forwarding of AF packets. An active queue management algorithm such as RED (Random Early Detection, see Floyd and Jacobson (1993)) is required, but no details about the algorithm are prescribed by the RFC except that flows with different short-term burst shapes but identical longer-term packet rates should have packets discarded with essentially equal probability.

There is no specific order between the AF classes. At each node, a certain amount of forwarding resources (bandwidth and buffer) is assigned to each class. Typical implementations could use admission control to limit the load in the different classes to different levels, for example, by overbooking the classes with different overbooking factors. This, however, is not detailed by the RFC. An overbooking factor is the ratio of the maximal admitted traffic of one class to the resources assigned to that class.

Within a single class, packets are not reordered. A typical implementation could assign each class a different queue and use different RED weights for the different drop precedences within each class.

RFC 2597 (see Heinanen *et al.* (1999)) describes as an example an Olympic service that can be built with the AF PHB group in the following way: The Olympic service consists of the three service classes, bronze, silver and gold that are assigned to the AF classes 1, 2 and 3. Packets in the gold class experience lighter load and thus have greater probability for timely forwarding than packets assigned to the silver class. The same kind of relationship holds true for the silver and the bronze class.

Within each class, all three drop precedences could be used. Packets are marked at the ingress by the traffic conditioner module. Many different marking algorithms could be used, the most common ones designed with the AF PHB in mind :

- RFC 2697 (see Heinanen and Guérin (1999a)) describes with the *Single Rate Three Colour Marker* a way to mark packets according to three traffic parameters: *Committed Information Rate, Committed Burst Size, and Excess Burst Size*. A packet is marked green if it does not exceed the committed burst size, yellow if it does exceed the committed but not the excess burst size, and red otherwise. It is useful for ingress policing of a service, where only the length, not the peak rate, of the burst determines service eligibility.
- The *Two Rate Three Colour Marker* of RFC 2698 (see Heinanen and Guérin (1999b)) describes a way to mark packets based on two rates, the *Peak Information Rate* and the *Committed Information Rate*. A packet is marked red if it exceeds the peak information rate. Otherwise, it is marked either yellow or green depending on whether it exceeds or does not exceed committed information rate. It is useful for ingress policing of a service, where a peak rate needs to be enforced separately from a committed rate.
- Packet marking based on the running average bandwidth of the traffic stream compared to the *Committed Target Rate* and the *Peak Target Rate* is described by RFC 2859 (see Fang *et al.* (2000)), and called *Time Sliding Window Three Colour Marker*. Packets

contributing to the sending rate below or equal to the committed target rate are marked green, those contributing to the rate between committed and peak target rates are marked yellow, the others red. Because of the sliding window rate estimator, burstiness is taken into account and smoothed out to approximate the longer-term measured sending rate of the traffic stream.

Achieving service differentiation with profile-based marking in edge routers is not a straightforward task, especially for a mix of responsive (TCP) and non-responsive (e.g. UDP) flows. Several studies investigate these effects, for example Clark and Fang (1998); Feng *et al.* (1999); Nandy *et al.* (1999); Sahu *et al.* (2000); Stoica and Zhang (1998); Yeom and Reddy (1999).

Per-domain Behaviours (PDBs) Building services and PDBs out of the PHBs is mainly left to the INSPs. There are several Internet drafts describing PDBs but few reached RFC status. The *virtual wire PDB* (see Jacobson *et al.* (2000)) can be constructed with the EF PHB plus appropriate domain ingress policing. As the name says, it is intended to provide a service that behaves like a dedicated circuit by providing an assured peak rate and bounded jitter. The EF PHB efficiency concerns discussed above lead to efficiency concerns about his PDB, too.

The *assured rate PDB* (see Seddigh *et al.* (2000)) is intended to provide a rate assurance but no delay or jitter bounds. It is built with the AF PHBs and suitable policers at the ingress.

A different approach is taken with the *bulk handling PDB* as *lower-effort PDB* of Bless *et al.* (2003); Carpenter and Nichols (2001) that provides a less-than-best-effort service. This service may be 'starved' by other services (including the standard best-effort service) in times of congestion as high load and is intended for low value traffic. The effect that the low value traffic has on other traffic is limited. The CS or AF PHBs can be used to implement the service; policing at the ingress is not required.

Service Level Agreements (SLAs) Between a customer (INSP or end-user) and an INSP operating a Diffserv domain, *Service Level Agreements* (SLAs) are used to specify the service the customer receives. For more information and a general (non-Diffserv) example see Section 9.2.4.

SLAs contain a *Service Level Specification* (SLS). A SLS is a set of parameters and their values that together define the service offered to the traffic by a Diffserv domain as long as it adheres to the *Traffic Conditioning Specification* (TCS) which is an integral part of the SLS. The TCS is a set of parameters and their values which together specify a set of classifier rules and a traffic profile.

SLAs can be dynamic and static, see Figure 6.8. Static SLAs remain in existence and constant on a medium to long timescale; typically, they are set up and maintained manually. Dynamic SLAs change more frequently, and typically, they are negotiated and set up automatically by BBs (in the Diffserv environment), see Nichols *et al.* (1999). With dynamic SLAs, networks can be used more efficiently unless traffic patterns are very stable and constant.

Possible SLS formats for Diffserv Premium service can be found in Bouras *et al.* (2002); Hashmani *et al.* (2001) and the works therein. For SLA trading, we refer to Fankhauser (2000).

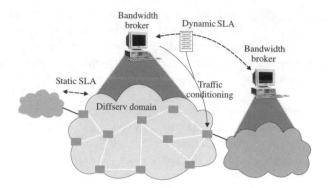

Figure 6.8   Diffserv Service Level Agreements

Preferential Treatment of Acknowledgment Packets   TCP flows are bidirectional with data packets being transported on one path and Acknowledgement (ACK) packets being transported back on the opposite path. The TCP throughput suffers when either of the two paths is congested. Because the packets take different paths and enter a Diffserv domain at different ingress nodes, they can receive completely different services. Papagiannaki *et al.* (2001) analyse in which ways ACKs have to be marked so that connections can achieve their performance goal despite congestion on one or both of the paths. In that study, the throughput of AF flows increased by 20% when the ACKs were sent with the highest service class over the throughput achieved when ACKs are marked as best-effort. The recommendations of that study in a nutshell are that ACKs should receive the same class of service as their data packets.

6.2.4.3 Bandwidth Broker

BBs for Diffserv were introduced in Nichols *et al.* (1999). A BB is *a software agent that manages the network resources of a Diffserv domain and makes the admission control decision for it.*

A BB needs inter-domain communication with other BBs to negotiate SLAs and intra-domain communication to allocate resources and configure ingress routers according to the TCS of a new SLA. For intra-domain communication, network management protocols like SNMP can be used. For inter-domain BB communication (and SLA trading), several protocols are under discussion: Border Gateway Protocol (BGP) extensions, RSVP extensions, the Internet Open Trading Protocol (IOTP), the BB transfer protocol, DIAMETER, Common Open Policy Service (COPS), and SNMP.

Schelen *et al.* (1999) describe a BB implementation that obtains a topological database through the Open Shortest Path First (OSPF) routing protocol, obtains link bandwidths through SNMP. The performance of the admission decision speed is evaluated. Other BB implementations are presented in Pop *et al.* (2001); Stattenberger and Braun (2003); Terzis *et al.* (1999a).

The Internet2 QBone group was working on a BB for an EF-based premium service that was later dropped in favour of a lower than best-effort service, largely due to deployment and other problems; see Teitelbaum *et al.* (1999), Teitelbaum and Shalunov (2002), and also Section 6.2.5.3.

In Fidler (2003), a BB is proposed that influences the routing of the Diffserv domain to give deterministic delay bounds.

In Chapter 8, we present and evaluate the design of centralised and decentralised BBs with focus on the admission control and efficiency.

6.2.5 Tuned Best-effort Architectures

In this section, we describe several approaches that we call "tuned" best-effort because they use the best-effort architecture as a foundation and try to improve QoS or offer service differentiation with relatively little changes to it.

6.2.5.1 Overprovisioned Best-effort

The currently dominating approach to improving QoS is adding bandwidth and buffer to a best-effort network. This approach is called the *overprovisioned best-effort* and is based on the fact that packets travelling through a relatively lightly loaded network experience little to no loss and little queueing delay – therefore relatively good QoS. For many applications, this can be enough.

The advantage of this solution is that the basic QoS architecture does not have to be changed. A disadvantage is that in the absence of admission control and service differentiation, an INSP cannot give (absolute or relative) service guarantees. Also, an INSP cannot offer technically different services with that approach. The latter can be necessary, for example, to use price discrimination as a means of increasing profits.

An important question to ask in the context of overprovisioning is *how much overprovisioning is needed?* Except for the work of Breslau and Shenker (1998), this question is not well answered in the scientific literature on an architectural abstraction level. Therefore, we address this question in the next two chapters of this book by comparing an overprovisioned best-effort architecture with other QoS architectures.

6.2.5.2 Price-controlled Best-effort

The Price-controlled Best-effort (PCBE) approach goes back to the works of Frank Kelly and others; see Kelly (2000, 2001a); Kelly *et al.* (1998). PCBE is one possibility to realise congestion pricing, see Henderson *et al.* (2001). The basic idea behind **congestion pricing** is that if congested network resources are priced, there is an incentive for users to back off in the case of congestion and thus reduce the congestion.

Smart Market   An *economically* efficient method to implement congestion pricing is described in MacKie-Mason and Varian (1995) and is called *smart market:*

Sending packets in an uncongested network is free of charge, but packets sent in a congested network are charged on a per-packet basis. Thus, the price to send a packet can vary minute-by-minute (or on even much shorter timescales) to reflect the current degree of network congestion.

MacKie-Mason and Varian propose an auction mechanism to realize this smart market:

- The sender puts the amount of money he is willing to pay (willingness-to-pay) for the transmission of the packet in the header of the packet.

- The network then admits all packets with a willingness-to-pay higher than the current cut-off amount, which is determined by the marginal congestion cost imposed by the next additional packet. Rejected packets could be bounced back to the user, or be routed to a slower network. Users then pay the market-clearing price for all transmitted packets, not their bidding price (second price auction).

The outcome of this mechanism is the classic supply-equals-demand level of service of economic theory that maximises social welfare. Social welfare is the total utility of all end-users and providers[8]. MacKie-Mason and Varian (1995) also show that the congestion revenues equal the optimal investment in capacity expansion.

Unfortunately, it is *technically impossible* to hold auctions on a per-packet basis:

- First, the Internet is not and cannot be managed centrally, therefore there cannot be a central auctioneer realising the auctions.
- A packet normally has to cross several boundaries between providers. If each provider is implementing an auction mechanism (e.g. to avoid the point above), there is another problem: If the auctions along the way are held in serial order (one after the other), the end-user might end up winning and paying most auctions along the way but loosing the auction at the last provider. This could be avoided if the auctions are held in parallel order but this again needs a central instance for coordination. Further discussions on these problems and possible solutions can be found in Courcoubetis and Weber (2003); Courcoubetis *et al.* (2001).
- Hard work is undertaken to make networks faster and faster. Collecting the bids, calculating the clearing price, selecting the packets to be forwarded etc. all add additional delays to the treatment of each packet in each router and requires additional buffer memory in the routers. This can be unacceptable for time critical real-time applications and very expensive for high-bandwidth links.
- The charging and accounting effort of this approach is also extremely high.

Despite these strong concerns about the technical feasibility of auction mechanisms, there is a long line of works about auctions for packets, micro-flows or higher aggregates in the scientific community, see for example, Courcoubetis and Weber (2003); Courcoubetis *et al.* (2001) and the works therein.

PCBE   The smart-market approach above has one appealing property: Resources are allocated according to the willingness-to-pay of the customers, leading to proportional fairness weighted by the willingness-to-pay. It has been shown in Kelly *et al.* (1998) that the additive increase-multiplicative decrease rate control of TCP also leads to proportional fairness. If the TCP congestion control is modified by introducing a weightage resembling the willingness-to-pay, weighted proportional fairness could be achieved, too. Such a proposal is made and studied in Crowcroft and Oechslin (1998), using a TCP implementation called *MulTCP*; unfortunately, it leads to difficulties in charging, accounting and policing. A quite similar approach without some of these difficulties is the price-controlled best-effort approach based on Kelly (2000, 2001a); Kelly *et al.*

---

[8] Costs of providers are counted as negative utility.

(1998). These works describe and model a QoS architecture that leads to the same results as the smart-market approach but that can be implemented technically more easily and in a distributed way. We call this QoS architecture *price-controlled best-effort* (PCBE); it is sometimes also called *ECN charging*, see for example Briscoe *et al.* (2003).

Architecture   Technically, the PCBE-architecture is based on a plain best-effort architecture with the ECN mechanism. If congestion is building up in a router, it notifies the end systems by randomly marking packets. Marking is done by setting a single bit[9] in the IP header. The receiver can notify the sender of the fact that it received a marked packet for example, via the TCP ACK. Upon notification, the "normal" reaction of a TCP sender would be to reduce the TCP congestion window in the same way as if the packet was dropped. If PCBE is used, the sender and/or receiver instead have to pay a certain (very small) amount of money for each ECN mark generated as received. Thus, a dynamic price for a stream of packets has to be paid. If that price exceeds the willingness-to-pay of a user, he (or an agent acting on his behalf) will back off by reducing his sending rate. Vice versa, if the willingness-to-pay exceeds the current price, a user will continue to send or even increase his rate. Obviously, this leads to an economically more efficient resource distribution than when all users would be forced to back-off, independent of their willingness-to-pay.

Modelling   The PCBE approach has been mathematically modelled in Kelly (2000, 2001a); Kelly *et al.* (1998). The system can be modelled as an optimisation problem and it can be shown that under certain assumptions – for example, increasing concave and differentiable utility functions – it maximises social welfare. Hansen and Naevdal (2000) shows that the resulting system can be treated well with standard economic theory; the work also analyses the resulting loss in social welfare if a (realistic) monopolistic control of network resources is assumed.

Discussion   Contrary to the 'typical' QoS architecture (e.g. Intserv and Diffserv) that focus on managing the available resources and thus the 'supply side of QoS', PCBE influences the 'demand side' by giving incentives to impose self-admission control. PCBE follows the end-to-end principle of the Internet which is keeping the network simple by putting the intelligence into the edges of the network.

PCBE maximises social welfare under the assumptions made. However, a competitive INSP is more interested in maximising his medium to long-term profits and not social welfare. This approach might not be ideal under that assumption.

The above described single-bit marking algorithm is a minimalist approach to QoS that does not require many changes in routers – most routers support ECN marking anyway. Significant changes, however, are necessary at the end systems and the INSP's charging and accounting systems. On the end systems, agents (called *dynamic price handlers*) are needed to react to the fluctuating prices on the users' behalf. Such an agent was developed in Briscoe *et al.* (2003). In the same work, the feasibility of a per-marked-packet charging and accounting system was shown.

---

[9] the ECN congestion experienced bit.

While PCBE is an innovative lightweight approach to QoS, it also has some significant drawbacks:

- Under the more realistic assumptions of monopolistic control of routers, social welfare is no longer maximised, see Hansen and Naevdal (2000). Also, if utility functions are not concave this goal is not met.
- The scheme needs a per-marked-packet accounting system that is potentially expensive.
- The price is dynamic and not known in advance to the user. User trials in Edell and Varaiya (1999) show that end-users will probably not like this situation although other later trials in the context of Briscoe *et al.* (2003) indicate a certain interest in dynamic pricing.
  To avoid dynamic prices for end-users, so-called guaranteed stream providers as brokers have been proposed to offer a kind of insurance service against dynamic prices, see Briscoe *et al.* (2003); Key (1999).
- It is not clear whether the sender or receiver of a marked packet should pay. On one hand, typically the receiver has the benefit of the data transfer. Therefore, it makes sense that he should pay if the transfer causes congestion. But this gives denial-of-service attacks a chance of inflicting direct economic damage on a receiver.
- An INSP is earning money with congested routers. The theory behind PCBE assumes that strong competition between INSPs and a transparent market forces them to upgrade their equipment where necessary and not to cheat on customers. This is not fully convincing, especially as there is hardly a possibility for a customer to control whether a packet was marked correctly.

6.2.5.3 Lower than Best-effort Service

While most QoS architectures and service models aim at introducing additional services that offer a better service than the traditional best-effort service, an interesting approach is to try to do the opposite: Offering a lower than best-effort service.

Carlberg *et al.* (2001) describe an implementation and experiments of this approach realised on a per-flow basis. Contrary to other approaches, such as the above-mentioned Diffserv bulk handling / lower-effort PDBs (see Bless *et al.* (2003); Carpenter and Nichols (2001)) or the QBone Scavenger Service we discuss below, that introduce a service class below the (default) best-effort class, the approach of Carlberg *et al.* (2001) aims at *actively degrading* the QoS of certain flows and to deny them resources even if those resources could *not* be used by other flows. This service is thus intended at punishing certain flows and discouraging certain behaviour: It is suited for punishing non-TCP-friendly flows, to reduce the QoS of certain applications – for example, peer-to-peer applications – or to punish flows suspected to be part of a denial-of-service attack. In Carlberg *et al.* (2001), a modified CBQ scheduling algorithm and some penalty algorithms are used to degrade the quality of flows by increasing their dropping probability. Flows that exceed a certain packet count or service rate are punished. Their experiments show that while it is hard to penalise an individual TCP flow – especially a short flow – and still maintain a minimum throughput, the concept works well for UDP flows and aggregates of TCP flows.

The QBone Scavenger Service (QBSS) follows a different approach. It creates a service class with a lower priority than the best-effort class. Strict priority queueing is not recommended; instead a small amount of network capacity is allocated to the QBSS class to avoid starvation of TCP flows in the QBSS service class during times when the best-effort service class is significantly loaded. Capacity not used by higher services is fully available to the QBSS service class – contrary to the approach of Carlberg *et al.* (2001). QBSS is thus designed not to punish certain behaviour but for bulk transfers that are currently run voluntarily during periods of low-utilisation (e.g. large nightly transfers of scientific data-sets, network backups, Content Delivery Network (CDN) content pushing). In addition, it is suited to downgrade the performance of non-critical traffic, such as peer-to-peer file sharing traffic at universities.

6.2.5.4 Alternative Best-effort

Alternative Best-effort (ABE), presented in Hurley (2001); Hurley *et al.* (2001), is an enhancement of IP best-effort. The idea is to have two service classes that provide a delay against throughput *trade-off*.

Each IP packet is marked green or blue. The green packets receive a lower delay but via a possibly higher loss probability lower throughput than the blue packets. Hurley *et al.* (2000) describe how the ABE colour can be encoded in the IP packet header.

One important property of the ABE service is that *green packets do not hurt blue packets*; if an application marks some or all of its packets green, the service – that is delay and throughput – received by applications that mark all their packets blue is not degraded. Therefore, if the ABE service model would be introduced in the Internet, unmarked packets would be considered to be blue packets and no harm would be done to applications unaware of the ABE service.

Applications aware of the ABE service would mark their packets blue or green or even mix both colours by marking some blue and others green. Packet sequence is preserved within the blue and within the green queues only, therefore when mixing the colours, packet reordering can be induced.

As green packets receive lower delay but higher loss, they are not strictly 'better' than blue packets and no incentive mechanism like pricing is needed to keep users from sending all their packets with the 'best' service. In addition, no policing mechanism is needed. All this has the additional advantage that the control and data path can be kept almost as simple as a traditional single-class best-effort network. The only additional complexity needed is that the scheduling mechanism has to make sure that green packets do not hurt blue ones. This requirement can be split into two parts:

1. The first part of the requirement is called *local transparency to blue*. It addresses the case of non-TCP-friendly sources.
   Local transparency to blue is defined over a plain best-effort scenario in which a node would treat all packets equally regardless of their colour. Local transparency to blue requires that every blue packet in an ABE node:
   - does not receive a larger delay than in the fictive plain best-effort scenario.
   - is not dropped unless it would also be dropped in the fictive plain best-effort scenario.
   A scheduling algorithm called *duplicate scheduling with deadlines (DSD)* is described in Hurley (2001); Hurley *et al.* (2001). DSD has elements from earliest-deadline-first

(EDF) and first-come-first-service (FCFS) schedulers. The packets are tagged with deadlines like in EDF, but this deadline is used only to determine which of the queues is to be served. Within each queue, FCFS is used.

Duplicates of all incoming packets are sent through a VQ which is served first-come-first-service (FCFS) with the rate of the plain best-effort scenario that is emulated by the VQ. The VQ is used to assign each blue packet a tag that indicates the time at which it would be served in the VQ. This acts as the deadline for the blue packet; a blue packet is always served at the latest moment the deadline permits, subject to work conservation. This ensures the local transparency to blue requirement; a blue packet that arrives when the VQ is full, is dropped.

Green packets are served in the meantime unless they have been in the queue for more than d sec, in the latter case they are dropped. For optimisation purposes a green packet also has to pass an acceptance test upon arrival to be put into the tail of the green queue, otherwise it is dropped. The acceptance test checks whether the queueing delay for the green packet exceeds d sec – imagine d to be in the order of magnitude of 20 ms. The scheduler events are summarised in Table 6.1.

It can happen that the deadlines of both the blue and the green packets at the head of their queues permit them to wait for the other packet to be served first. In that case, the packet to be served first is selected randomly, the probability that the green packet is selected is controlled by an additional parameter called the *green bias* g. The reason for this parameter becomes clear when looking at the second part of the 'green packets do not hurt blue packets' property.

2. If a TCP-friendly and greedy source is sending, its rate depends on the Round-trip Time (RTT) and the loss probability. If that source is sending green packets, it is possible that because the RTT decreases for green packets, that source would receive a higher throughput than when it would send blue packets only. This also carries the risk of hurting blue sources because of the increased rate. This leads to the second part of the requirement: *throughput transparency to blue* – a green flow shall receive a less or equal throughput than if it were blue.

This requirement is much harder to implement than local transparency to blue because an exact implementation would have to keep track of the exact end-to-end RTTs for every flow. The authors of ABE propose to use a controller in each node that adapts

Table 6.1   Duplicate Scheduling with Deadlines Events

Event	What is served?
Both queues empty	Nothing
Green queue empty, blue queue not empty	Head of blue queue
Head of blue queue cannot wait	Head of blue queue
Blue queue empty, green queue not empty	Head of green queue
Head of blue queue can wait, head of green queue cannot	Head of green queue
Head of both queues can wait	Randomly

the above-mentioned green bias parameter g of the DSD scheduler so that the through-put transparency (estimated via a TCP throughput formula) is ensured. For this, the controller assumes that all flows are greedy and have a total RTT equal to the queueing delay at this node plus a fixed virtual base value (e.g. 20 ms). This tends to underesti-mate the green RTT but that is no problem as the underestimation is conservative for the blues.

To conclude, the ABE service is a good example for a low overhead tuned best-effort service that can offer some advantages like lower delay without introducing much overhead. One of its drawbacks is that it depends on a special scheduling algorithm and has received little IETF support so far. As blue packets are not harmed, green packets do not receive the same high QoS premium that packets would receive in other QoS systems like Intserv / Diffserv – but that is actually a desired property of ABE because it removes the need of policing and pricing.

6.2.6 Other Architectures

Besides the above discussed QoS architectures, there are a number of other interesting approaches to the question of how and what QoS to provide; we now discuss some of these works.

The above discussed SCORE approaches with dynamic packet state (Section 6.2.3) and Diffserv (Section 6.2.4) can be used to give absolute QoS guarantees without keeping per-flow state on the data or network path. However, they need per-flow state in the edge routers; this can be problematic, too, because also edge routers can be performance bottlenecks. For example, if many interconnection partners are connected via that router (see Part III of this book for interconnections). They typically have important additional tasks like running BGP, counting traffic volumes to transit partners, etc. Also, it is a common policy with providers when core routers are replaced with more modern routers because they can no longer handle the ever-growing traffic, that they are moved 'towards the edge' and used as edge routers. If it was considered problematic to keep per-flow states with them in the old core network, then the same is true in their new role as edge routers.

Approaches like PCBE (Section 6.2.5.2) and ABE (Section 6.2.5.4) do not need per-flow state at the edges but can only give soft relative QoS guarantees. This train of thought leads to the question whether it is possible to give strong absolute QoS guarantees to flows, without the need for per-flow state at the edge *and* core (**stateless edge and core**). Machiraju *et al.* (2002) propose such a solution that can be used to offer a service like the Diffserv premium service (see Section 6.2.4.2):

- The *data plane* is simple: For the reserved (premium) traffic, a single queue is maintained.
- For the *admission control,* a soft-state protocol is used to reserve a peak per-flow bandwidth in the intermediate routers. The reservation is refreshed in a fixed well-known interval $T_{refresh}$. Routers only keep track of the aggregate reserved rate by adding the reservation refresh messages in one interval $T_{refresh}$[10].

---

[10] The actual mechanism is slightly more complicated to take care of refresh messages that arrive delayed due to jitter, see Machiraju *et al.* (2002) for details.

- On the *control plane,* misbehaving flows could send refresh messages without being admitted. This is countered with router-specific lightweight certificates generated by each router along the path. They are attached to the admission request message if a flow is admitted by the router. Refresh messages are accepted only with a valid certificate. Misbehaving flows could stop sending and resume sending later, without undergoing the admission control test again using their old certificate. To avoid this, routers change the keys for their certificates at regular intervals.
- Another problem arises if flows send more than one refresh message per $T_{refresh}$. This would allow them to send more than what is actually admitted by the admission control. This can be avoided by random sampling. Random packets are chosen and the flows they belong to are monitored for some time to detect extra refreshes.
- Similarly, flows sending more than what is allowed could be detected by monitoring a (limited) number of randomly selected flows. Machiraju *et al.* (2002) present an alternative called *recursive monitoring* that turned out to be superior in simulations. The basic idea is to monitor aggregates of flows. If the aggregate misbehaves, it is recursively split into smaller aggregates until the misbehaving flow is detected.

Another interesting approach called *Paris Metro Pricing* (PMP) is introduced in Odlyzko (1999). PMP is a minimalist relative service differentiation scheme using only the price as differentiation mechanism. The idea is to split the bandwidth among several channels, for example, with (WRR) scheduling. The channels differ only in their price. The QoS of each channel depends only on the load of that channel. Users choose a channel depending on the expected QoS and their willingness-to-pay. PMP relies on self-regulation: If an expensive class is too congested, users can be expected to back off because for them it is not worth the price. The opposite can be expected if the expensive class is not congested while the cheaper class is. An economical analysis of PMP with a single monopolistic provider in Jain *et al.* (2001) indicates that there are profit incentives for monopolistic providers to adapt PMP compared to offering a single channel only.

6.2.7 Classification of Quality of Service Architectures

As a summary, we classify and describe the most relevant QoS architectures in the Tables 6.2 and 6.3 of this chapter with respect to the following points:

- **Shortest Timescale of Control**
  The timescale of a QoS architecture describes the smallest possible timescale that a QoS system based on that architecture can work and react upon to influence the QoS. It ranges
  - from the per-packet timescale which implies manipulation of individual packets for a per-packet QoS
  - over the RTT timescale (round-trip time) that implies a reactiveness of the system with a delay that is in the RTT order of magnitude
  - up to network engineering and capacity expansion timescales; they imply that a reaction is possible only by extending the network capacity.
- **Reactiveness**
  Reactive architectures react to QoS relevant events like congestion while proactive systems actively try to avoid these events before they occur.

Table 6.2 Classification of QoS Architectures Part 1

	Intserv	SCORE with DPS	Diffserv	ABE	PCBE	Over-provisioning
Shortest Timescale of Control	Packet	Packet	Packet	Packet	RTT	Network engineering
Reactiveness	Proactive	Proactive	Proactive	Proactive	Reactive	Reactive
Services	Guaranteed service Controlled load service Best-effort service Open for other services.	Guaranteed service Best-effort service Other services possible.	Extreme variety of possible services.	Two alternative types of best-effort services	Price-controlled best-effort service	Plain best-effort service
Guarantees	Absolute, end-to-end, per-flow, deterministic, loss and delay guarantees	As Intserv	Absolute, per Hop, per-aggregate, deterministic, loss and delay guarantees	Relative guarantees	No guarantees	No guarantees
Data Path Procedures	Edge & Core: Packet classification, scheduling & queue management, policing	Edge: Packet classification, scheduling & queue management, policing, marking Edge & Core: Scheduling & queue management, packet remarking, core-stateless scheduler	Edge: Packet classification, scheduling & queue management, policing, marking Edge & Core: Scheduling & queue management	Edge & Core: Scheduling & queue management Special scheduler necessary	Edge & Core: Queue management	

Table 6.3 Classification of QoS Architectures Part 2

	Intserv	SCORE with DPS	Diffserv	ABE	PCBE	Over-provisioning
Data Path Complexity	Per-flow	Edge: Per-flow Core: Independent of number of flows	Edge: Per-flow Core: Per-class complexity	Low	Minimal	Minimal
QoS Signalling	Required	Required, lightweight signalling protocol in Core	Not required	Not necessary	Not necessary	Not necessary
Admission Control	Required, per-flow and per-hop admission control mechanism tied to the signalling protocol	See Intserv	Not required, Per-SLA admission control by central or decentral bandwidth brokers	None	Endpoint admission control mechanism	None
Control Path Complexity	Per-flow state and processing in every node	See Intserv, reduced complexity in core due to lightweight signalling protocol	Strongly depending on signalling and admission control mechanism, low complexity possible	Minimal	Minimal	Minimal
Implications for the QoS Strategy	Differentiated pricing necessary	Differentiated pricing necessary	Differentiated pricing necessary	No differentiated pricing necessary	Per-packet pricing and accounting	No restrictions

- **Services**
  Different architectures allow different types of services to be offered.
- **Type of Guarantees**
  One of the key questions of a QoS system is what kind of guarantees it can give. If architectures allow different types of guarantees, we focus on the strongest possible guarantees.
  Service guarantees can be absolute or relative, deterministic or statistical. QoS architectures also differ in the granularity of the guarantees, that is, whether guarantees are given per-flow or per-aggregate.
- **Data Path Procedures**
  The QoS architectures differ in the QoS procedures like marking and policing they need on the data path. For some architectures, it is important to split this aspect into two: the procedures applied at edge routers and in the core network. Only the data path procedures needed *in addition* to those available in best-effort routers are listed.
- **Data Path Complexity**
  The data path complexity describes the parameters on which the complexity of the data path procedures mainly depends.
- **QoS Signalling**
  Some architectures depend on the use of a QoS signalling protocol like RSVP.
- **Admission Control**
  Admission control is required in some QoS architectures, see Section 6.6.
- **Control Path Complexity**
  The control path complexity describes on which parameters the complexity of the control path procedures mainly depends.
- **Implications for the QoS Strategy**
  Most QoS architectures have some important implications for the QoS strategy, mainly for pricing and tariff services. If different services are offered and one service is strictly better than another, pricing or similar mechanisms have to be used to keep all users from requesting only for the better service.

6.3 Data Forwarding Architecture

With the term data forwarding architecture, we describe the actual technology used for forwarding packets at a node. In the core of a network, packets can be routed or label switched.

If a packet is *routed*, the router evaluates information from the packet's IP header, mainly the 'time-to-live' and 'destination address' field, and its routing table to decide locally and upon arrival of the packet with which outgoing interface the packet is forwarded to its next hop.

If a packet is *label switched*, it receives a label at the edge of the network and the path that packets with a certain label take is set up beforehand through the core of the network. A label switching router forwards arriving packets solely on the information contained in the label; it does not have to look into the IP or higher-layer headers.

6.3.1 IP Routing

6.3.1.1 Routing Lookup

IP routing occurs at layer 3 of the 5-layer model (Figure 4.1). A router uses a routing protocol (Section 6.4.1) to maintain a routing table that contains the information – which next hop lies on the shortest[11] path to which destination. The routing lookup has to be made upon the arrival of a packet, and on the basis of the result the packet is put into the outgoing queue of the interface connected to the next hop router. It is obvious that the routing lookup is a time critical operation.

Because Classless Inter-domain Routing (CIDR, see Fuller *et al.* (1993); Rekhter and Li (1993)) routing is used in todays IPv4 based Internet, routing tables contain variable-length address prefixes. For a routing lookup, the prefix that matches best with the destination IP address of the packet has to be found. It is the longest match; therefore, the lookup problem is called 'Longest-prefix Matching'.

As an example, a part of a routing table is shown in Table 6.4. The IP prefix is stored in dotted-decimal notation, the number after the slash indicates the length of the prefix in bits. The best match for destination 130.83.198.178 in the routing table would be interface #3, because the first 18 bits match with the destination address. The best match for destination 130.83.64.130 would be interface #2 with 16 matching bits.

Longest-prefix matching algorithms try to optimise the average and worst-case number of memory accesses for routing lookup (and thereby the lookup time) and the memory requirement of the routing table. The routing table size in typical routers has grown exponentially in the last years from 30,000 to 120,000 entries; see Narayan *et al.* (2003). According to Bu *et al.* (2002), the reason for the rapid increase lies mainly in address fragmentation, that is, the fact that an autonomous systems (AS) has several prefixes that cannot be aggregated. Multihoming and load balancing also contribute to this trend and their contribution is growing faster than that of the address fragmentation. The prefixes of a multihomed AS cannot be aggregated by all of its providers and for load balancing, an AS can announce different prefixes via different AS paths. Complementary, Narayan *et al.* (2003) analyse the structure of the routing table and the impact of that structure on routing lookup methods.

The classical solutions for longest-prefix matching algorithms are *trie-based schemes*. A trie is a tree-like structure that exploits the fact that various entries share prefixes of each other and store the shared parts in the same location; see Fredkin (1960). The bits

Table 6.4 Section of a Routing Table (Example)

Destination Address IP Prefix	Next Hop	Output Interface
130/8	145.253.4	#1
130.83/16	145.253.81	#2
130.83.192/18	145.253.183	#3
130.83.192/24	145.253.12	#4
. . .	. . .	. . .

---

[11] with respect to the used routing distance metric.

of prefixes are used to direct the branching, see Figure 6.9. Nodes that correspond to a routing table entry are marked. Finding the longest prefix in a trie is straightforward; the bits of the destination address are inspected in sequential order, every time a marked node is passed, it is stored as the longest-prefix match found so far, until the end of the trie is reached. Obviously, this type of search can lead to a lot of memory accesses. The binary tree algorithm can be improved in a number of ways:

- *Path compression* as in Gwehenberger (1968) as a Patricia tree as in Morrison (1968), for example, removes internal nodes with only one child and thus the size of the data structure. A skip count has to be stored instead that indicates how many bits have been skipped. The IP lookup implementation for the Backbone Service Provider (BSD) Unix kernel by Sklower (1993) is based on a similar mechanism.
- *Level compression* as another example replaces n complete levels of a binary trie with a single node of degree 2^n, leaving the number of nodes unchanged but shortening the search path.
  Nilsson and Karlsson (1999) present a longest-prefix matching algorithm that combines path and level compression.
- Some router vendors do IP lookups based on compressed *multibit tries/tree bitmaps*, see for example, Degermark *et al.* (1997); Srinivasan and Varghese (1999b). These works are based on inspecting multiple bits simultaneously; therefore, a multibit trie is used – a multibit trie node has 2^k children. A comparison of Degermark *et al.* (1997) with Nilsson and Karlsson (1999) can be found in Kencl (1998). Eatherton *et al.* (2002) present a similar work optimised for implementation in hardware.

Many other routing lookup schemes exist. An overview, taxonomy and complexity evaluation is given in Ruiz-Sanchez *et al.* (2001). Gupta (2000); Srinivasan and Varghese (1999a); Waldvogel (2000) also give an overview.

As a more novel approach, Dharmapurikar *et al.* (2003) propose the use of bloom filters for longest-prefix matching.

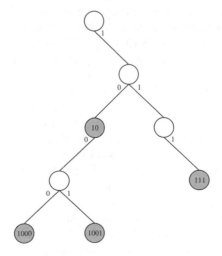

Figure 6.9 Trie Structure

Special hardware solutions like *Ternary Content Addressable Memories* (TCAMs, see McAuley and Francis (1993); Shah and Gupta (2001)) use parallelism to gain lookup speed. They can store the values 0, 1 and X; X is a 'don't care' value. TCAMs can compare a given destination address to all stored prefixes in parallel and return the longest match in a single memory access.

6.3.1.2 Other Routing Tasks

Besides the forwarding decision, a router has to perform some other tasks, see for example, Baker (1995):

- Decrementing the Time-to-Live (TTL) field.
  A router decreases the TTL field of the IP header. If it reaches zero, it is assumed that the packets loops in the network; the packet is discarded and an Internet Control Message Protocol (ICMP) error message is generated.
- Verification and update of the IPv4 header checksum.
  The checksum verification is often omitted for performance reasons because packets hardly ever are corrupted in transit and end systems will recognise the rare cases of corruption anyway. Therefore, IPv6 no longer has an IP header checksum (see Deering and Hinden (1998)).
  If the TTL field of an IPv4 packet is decreased, the checksum has to be updated. An efficient mechanism is described in Mallory and Kullberg (1990); Rijsinghani (1994).
- Fragmentation.
  IPv4 packets that are too large for a subnet are fragmented. However, IP fragmentation rarely occurs on high-speed links because these are designed to handle large enough packets.

6.3.2 Label Switching

As we have shown in the last section, the routing lookup is a time critical operation. It can be replaced with a simple index label lookup if a label switching mechanism like MPLS is used. Multi-protocol Label Switching (MPLS) is the IETF's standardised label switching packet forwarding architecture and has largely replaced prestandard and proprietary solutions like Ipsilon's IP Switching, IBM's Aggregate Route-based IP Switching (ARIS) (Aggregate Route-based IP Switching), Cisco's early Tag Switching and Toshiba's Cell Switch Router technology; see Armitage (2000). With and without explicit traffic engineering, it is growing in popularity for provisioning and managing core networks. Practically every modern router is able to do plain IP packet forwarding and MPLS. The early evolution of MPLS is summarised in Viswanathan *et al.* (1998).

In traditional IP routing, each router analyses the header of the arriving packet and independently chooses the next hop based on the distributed routing algorithm that uses a routing protocol (see Section 6.4.1) and the information of the IP header. Using the MPLS terminology of Rosen *et al.* (2001), an IP routing lookup partitions the IP packets destined to addresses with the same IP address prefix in the routing tables into one forwarding equivalence class (FEC). Each FEC is mapped to the next hop in a routing table; therefore, different packets in the same FEC are treated equally with respect to the forwarding decision. As the IP packet traverses the network, each hop re-examines the packet and assigns it to a FEC.

Contrary to that, in a network using MPLS as data forwarding architecture, the FEC assignment is done at the MPLS ingress node just once when the packet enters the MPLS domain. An MPLS domain is a contiguous set of nodes using MPLS as the forwarding architecture. The FEC is encoded into a 4 byte label (so-called shim header) that is attached to the packet between the layer-2 and layer-3 headers. Subsequent hops no longer have to examine the IP header of the packet, the label is used as an index into a forwarding table that specifies the next hop as outgoing interface and a new label that replaces the old one (label swapping). Each MPLS node (router/switch) is called a *Label Switching Router* (LSR). The path for one FEC through one or more LSRs is called Label Switched Path (LSP).

MPLS offers some advantages over the conventional IP forwarding architecture:

- MPLS forwarding could be done by switches that do not have to be capable of analysing the IP headers. MPLS forwarding is a simpler operation than IP routing and less expensive to implement for operations at state-of-the-art line speeds.
- The ingress router assigning the label can use any information to assign a label to a packet. Apart from analysing the destination address of the IP header, the transport layer ports could be evaluated, or the DSCP of a packet in a Diffserv domain.
- Additionally, the process of determining the label can become more and more sophisticated without any impact at all on the core routers.
- As information about the ingress router does not travel with an IP packet, traditional IP routing does not allow differentiation between packets from two different ingress routers in the core. With MPLS, this can easily be done if each ingress routers assigns a different label.
- For traffic engineering or policy reasons, it may be desirable to force packets to follow a path different to the standard shortest path as it is determined by the routing protocol algorithm. With MPLS traffic engineering, the path set-up can be controlled centrally and any path through the network can be used.

These advantages make it obvious that a network with MPLS-based forwarding architecture is well-suited for traffic engineering. We discuss and evaluate traffic engineering in Part IV of this book; works related to MPLS in the context of traffic engineering are discussed in Chapter 12.

For one LSP, the direction of the traffic flow is called *downstream*. The assignment of a particular label to an FEC is done by the downstream LSR and has to be signalled opposite to the traffic flow direction of the upstream LSR. The protocols used for signalling the label bindings and setting up an LSP are called *Label Distribution Protocols* (LDPs)[12]. The MPLS architecture of Rosen *et al.* (2001) does not assume one specific protocol; moreover, it does not even assume that there is only a single protocol used. LDPs are part of the signalling architecture of a network and are thus discussed in that context (Section 6.4.3).

---

[12] Unfortunately, one of the IETF LDPs is called exactly like the general term: *LDP*. It is discussed in the following text.

6.4 Signalling Architecture

This signalling architecture includes the different signalling/control protocols used to manage the network. We distinguish between routing, QoS signalling, and LDPs.

6.4.1 Routing Protocols

Routing protocols can be distinguished as interior and exterior routing protocols.

6.4.1.1 Interior Routing Protocols

Interior routing protocols are used to exchange routing information inside an INSP's network, based on that information the routers are enabled to fill their routing table by calculating the shortest path through an IP network with respect to a certain composite distance metric to a destination. The distance metric can be based on hop count, delay, link bandwidth, utilisation and so on.

Existing routing protocols can be classified as distance-vector or link-state protocols. **Distance-vector** protocols are based on the distributed Bellman-Ford algorithm and exchange their distances to all destinations with their neighbours; each node's calculation of the shortest path depends on the calculation of the other nodes. With *link-state* protocols, a node distributes its connectivity with its direct neighbours to all routers in the network, which can then reconstruct the complete topology and calculate their routing table by constructing the shortest-path tree. Generally, link-state protocols are more stable and converge faster.

The Routing Information Protocol (RIP) (see Hedrick (1988)) was the most widely deployed interior routing protocol for the early Internet. It is a distance-vector protocol. In the mid-1980s Cisco introduced with IGRP a distance-vector protocol that also supports multipath routing and avoids some performance problems of RIP in large heterogeneous networks; see Rutgers (1991). IGRP was widely replaced by its enhanced version EIGRP in the early 1990s; see Cisco (2003). EIGRP is still a distance-vector protocol but uses some features of link-state protocols to overcome some of the disadvantages of distance-vector protocols.

The OSPF (Open Shortest Path First, see Moy (1998)) routing protocol was the IETF's approach to overcome the limitations of RIP. OSPF is a link-state routing protocol using the shortest path first as Dijkstra algorithm (see Dijkstra (1959)) to derive the shortest path to each node. Like IGRP, it supports multipath routing. For scalability reasons, OSPF is a hierarchical protocol and allows a larger network to be split into subnetworks; all nodes of a subnetwork have identical topological databases but limited knowledge of the topology of the other subnetworks.

Another link-state protocol that can be used with TCP/IP networks is OSI's IS-IS routing protocol, see Callon (1990); ISO DP 10589 (1990).

6.4.1.2 Exterior Routing Protocols

For the route advertisement (see also Section 9.2) between two AS as INSP networks, exterior routing protocols like BGP (BGP, see Rekhter and Li (1995)) are used. Contrary to the interior routing protocols, exterior routing protocols are used between two INSPs to exchange reachability information, enforce policy decisions and hide the details of the internal topology from the interconnection partners. Contrary to the interior routing protocols, routes advertised by BGP consist of AS hops and not individual router hops.

BGP neighbours exchange full routing information after they establish their BGP connection that uses TCP as transport protocol. When changes to the routing table are detected, the BGP routers exchange only those routes that have changed; they do not send periodic routing updates and advertise only the optimum and not all possible paths to a destination network.

In order to support policy decisions, BGP associates certain properties with the learned routes. They are used to determine the best route when multiple routes exist to a particular destination. These properties are referred to as *BGP attributes*. For example, the *local preference* attribute is used to select the exit point for a specific route if there are multiple exit points from the AS. Related to that, the *multi-exit discriminator attribute* is used as a suggestion to an external AS regarding the preferred route into the AS that is advertising the attribute. The *origin attribute* indicates, for example, whether BGP learned about a particular route was via an exterior routing protocol or whether it was injected into BGP based on information from an interior routing protocol (then the route is local to the originating AS). To simplify administration, the *community attribute* allows group destinations – called *communities* – to which routing decisions (such as acceptance, preference and redistribution) are applied. Finally, the *next hop attribute* is the IP address that is used to reach the advertising router.

6.4.2 Quality of Service Signalling Protocols

Some QoS architectures depend on the use of a QoS signalling protocol. QoS signalling protocols can be receiver or sender oriented, based on which side initiated the process of requesting QoS from the network. The most famous signalling protocol for the Internet is the resource reservation protocol RSVP that is proposed for use with Intserv but can also be used for the label distribution in networks with a MPLS forwarding architecture; see Braden *et al.* (1997) and Awduche *et al.* (2001). The latter functionality is discussed in Section 6.4.3.1.

6.4.2.1 RSVP

The operation of RSVP in conjunction with the Intserv architecture was described in Section 6.2.2.2. The functional specification of RSVP is given in RFC 2205 (see Braden *et al.* (1997)), extensions to the QoS signalling functionality are given in RFC 2961 (Refresh Reduction, see Berger *et al.* (2001)) and RFC 3175 (RSVP Aggregation, see Baker *et al.* (2001)). For a scalability discussion of RSVP see Section 6.2.2.6.

The key functionality of RSVP can be summarised as follows:

- RSVP uses IP datagrams and alternatively UDP encapsulation for the message exchange.
- It supports heterogeneous receivers in large multicast groups by using a *receiver-oriented* reservation style based on the argument that the receivers know best about their QoS requirements.

- Dynamic membership in these large groups is supported with the receiver-oriented approach, too, and by the fact that the data transfer is handled separately from the control by RSVP. Receivers can join and leave the distribution tree installed by RSVP at any time during the data transmission.
- Multiple receivers are supported by the concept of multicast groups. At the same time, RSVP supports multiple senders sharing resources too. For this, the *reservation styles* are used (see Section 6.2.2.2).
- RSVP is independent of and does not interfere with the multicast group management protocol, the data path procedures or the routing protocols of the network.
- For the case of routing changes or network failures, a recovery mechanism is necessary to establish new and release old reservations. Because of the *soft-state* principle of RSVP reservations are frequently refreshed. In the case of a routing change or network failure, new reservations are set up when the refreshing takes place and old reservations are released automatically after some time.

6.4.2.2 Other Protocols

While RSVP is the dominant QoS signalling protocol, there are a number of other QoS signalling protocols for IP networks:

- The Internet Stream Protocol Version 2 (ST-2+) was an experimental IETF QoS signalling protocol. It is specified in RFC 1819 (see Delgrossi and Berger (1995)) and differs from RSVP in many aspects. It is a connection-oriented hard-state protocol. ST-2+ is operating parallel to IP and not compatible with the datagram service of IP. For a comparison of ST-2/ST-2+ and RSVP we refer to Delgrossi *et al.* (1993); Mitzel *et al.* (1994).
- YESSIR (YEt another Sender Session Internet Reservations), see Pang and Schulzrinne (1999, 2000)) is a QoS protocol based on RTP that was developed to avoid complexity and scalability issues that RSVP was believed to have. RTP itself is specified in Schulzrinne *et al.* (1996). YESSIR avoids message processing overhead in the end systems and routers and reduces the bandwidth consumption of the refresh messages. Reservations are triggered by the sender; the protocol is a soft-state protocol.
- The Boomerang protocol of Fehér *et al.* (2002, 1999) aims at reducing part of the RSVP overhead by using a sender-oriented approach. The sender generates a reservation message. Once this reaches the receiver, the reservation is already in place.

6.4.3 Label Distribution Protocols

An MPLS data forwarding architecture implies the use of a label distribution protocol to set up LSPs unless each switch would be configured statically by hand. Within the IETF, two label distribution protocols that also allow the set-up of explicit paths for traffic engineering are under discussion: RSVP-TE and Constraint-based Routing Support For LDP (CR-LDP).

6.4.3.1 RSVP-TE

RSVP-TE stands for RSVP with traffic engineering support. RSVP-TE (described in Braden *et al.* (1997)) is a set of extensions to the basic RSVP protocol (see Section 6.4.2). It is specified in RFC 3209, see Awduche *et al.* (2001).

RSVP messages are exchanged directly via raw IP datagrams. The protocol uses soft-state and refresh reduction allowing it to recover automatically from failure. RFC 3209 (see Awduche *et al.* (2001)) also describes a means of rapid node failure detection via a new HELLO message.

The label distribution method is downstream-on-demand: If an ingress LSR determines that a new LSP has to be set up to a certain egress LSR, a PATH message is sent containing a specified explicit route. That route can be different from the standard hop-by-hop route. The message also contains the traffic parameters for the new route. Each router along the path receiving the message builds up state. The egress router selects a label and answers with a RESV message that is routed back towards the ingress, finishing the set-up of the new LSP. Intermediate routers allocate resources and select a label when the RESV message reaches them. They update the message and forward it to the ingress routers via the interface by which the according PATH message was received.

6.4.3.2 CR-LDP

CR-LDP is a set of extensions to the LDP protocol of Andersson *et al.* (2001) that are specified in RFC 3212 (see Jamoussi *et al.* (2002)). CR-LDP stands for constraint-based routing support for LDP.

CR-LDP uses TCP connections for a reliable message exchange and is a hard-state protocol. It does not need to refresh the set-up of an LSP. With respect to failure recovery, it is not as well placed as RSVP-TE. The loss of the according TCP control connections also results in a failure of all associated LSPs.

The label distribution method is downstream-on-demand as in RSVP-TE. A label request message is sent by the ingress LSR towards the egress. It contains an explicit route. Contrary to RSVP-TE, intermediate routers reserve resources immediately when the label request reaches them. The egress-router responds to the label request with a label mapping message that contains the label and information about the final resource reservation. It is routed back to the ingress nodes.

While RSVP-TE and CR-LDP are quite different as pure protocols, they offer similar functions to the user. For a more detailed comparison of both protocols, we refer to Brittain and Farrel (2000).

6.5 Security Architecture

The *IPsec* security architecture is the security architecture of the IETF[13] for the Internet Protocol (IP); it is specified in Kent and Seo (1998). It offers cryptography-based security services at the IP layer and enables applications like virtual private networks (VPN).

---

[13] IETF IPSEC working group, http://www.ietf.org/html.charters/ipsec-charter.html

The architecture consists of a set of protocols, mainly the *Authentication Header* (AH, see Kent (1998a)) and *Encapsulating Security Payload* (ESP, see Kent (1998b)) protocols and the *Internet Key Exchange* protocol (IKE, see Kaufman (2004)):

Authentication Header (AH) and ESP can operate in two modes. The transport mode provides protection primarily for upper-layer protocols and the tunnel mode to tunneled IP packets. *AH* provides data integrity and data origin authentication and optional replay protection by embedding an additional header (the AH) that contains an authentication field into the IP datagram. The authentication field contains an integrity check value that is calculated over those IP header fields that do not change in transit, the AH header except for the authentication field and the entire upper-layer protocol data, and protects them against tampering. Replay protection is provided by an additional sequence-numbering mechanism.

The *ESP* protocol additionally offers confidentiality by encapsulating and encrypting the data to be protected. In transport mode, it encrypts and optionally authenticates the IP payload but not the IP header. In transport mode, AH authenticates the IP payload plus selected parts of the IP header. In tunnel mode, the entire IP packet is encapsulated within a new IP packet to ensure that no part of the original packet is changed.

The IPsec encryption mechanisms need a key exchange mechanism. This mechanism can be manual or automated. For automated key exchange, the IKE protocol is proposed in Kaufman (2004).

While security aspects are important for Internet service providers, they are not in the scope of this book and therefore not further discussed here. For more information about security architectures, we refer to Schneier (1995) for a general overview and to Frankel (2001) for more details.

6.6 Admission Control

Admission control is an optional aspect of the control path of the network. Admission control is used to keep the network load within certain bounds on a small timescale. There is a vast amount of general work on admission control and the different proposed admission control systems and schemes vary enormously. Most of them are independent of a specific network architecture. We give an overview and classification of admission control next.

An actual admission control system is characterised by a number of properties, the most important ones are shown in Figure 6.10. It is important to stress that the individual properties influence each other significantly. For example, the type of guarantees a system can support strongly depends on the flow and network behaviour assumptions and the location of the system.

The individual objects for which admission control decisions are made are called *flow* throughout this chapter; this shall neither imply that they are necessarily micro-flows such as individual TCP flows nor that they are unidirectional. They could also be macro-flows consisting of an aggregate of micro-flows, such as the complete traffic of one customer or a group of customers.

We now present a structured overview of different admission control systems.

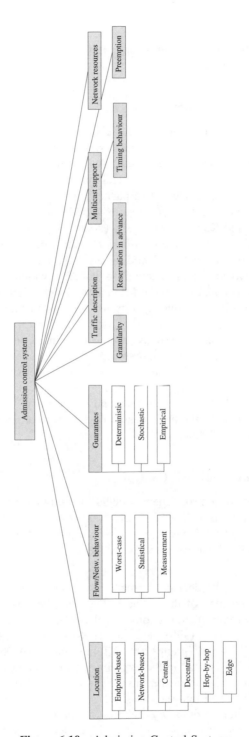

Figure 6.10 Admission Control Systems

6.6.1 Location

One important property of each admission control system is the location where the admission control decisions are made.

6.6.1.1 Endpoint Admission Control

In endpoint admission control schemes, the end-to-end admission control decision is made at the end systems themselves. No admission control instance and, therefore, less 'intelligence' is required in the network itself. As the endpoints have no control and no further information about the traffic of other endpoints, the decision is typically based on probing and measurement information like packet marking.

Pioneering work in the direction of endpoint-based distributed admission control has been done by Gibbens and Kelly (1999); Kelly (2000); Kelly *et al.* (1998). Their analysis shows the basic stability of distributed admission control based on marking at resources even in the case of feedback delays. Building on these results, some works shed light on the influence of delayed system reaction on stability, which presents bounds for the reaction delay, see Johari and Tan (2001); Massoulie (2000). Kelly *et al.* (2000) present a model for an Internet exclusively managed by the end systems and analyses the stability of this system.

As endpoint admission control systems assume no admission control instances in the network that could actually hinder non-admitted flows from sending, a mechanism is needed that forces or gives incentives to the end systems to perform the admission control algorithm and to behave according to its decision. In the works of Kelly (2000); Kelly *et al.* (2000), pricing per ECN mark is used as an incentive mechanism.

A simulative comparison of the basic design options for endpoint admission control is presented in Breslau *et al.* (2000b).

6.6.1.2 Network-based Admission Control

Network-based admission control schemes decide to admit or reject flows to one network. As a flow can pass through several networks, several sequential admission control decisions might be necessary. Network-based admission control schemes can be further distinguished as centralised and decentralised systems:

- In *centralised* systems, the decision is made at a central instance of the network that can have global knowledge of the network's current state. BBs (see Section 6.2.4.3) typically include a centralised admission control system. The bandwidth broker concept goes back to Nichols *et al.* (1999); Schelen (1998). Examples are given for example, in Khalil (2003); Khalil and Braun (2000); Terzis *et al.* (1999b); Zhang *et al.* (2001, 2000). The centralised BB used in the experiments of Chapter 8 is also of this type.
- *Decentralised* systems can be further divided into whether each link/hop or only the network edges are involved into the decision:
  - A typical example for a *hop-by-hop* admission control decision is the Intserv/RSVP admission control mechanism for GS and CL service as described in RFC 2212 (see Shenker *et al.* (1997)) as RFC 2211 (see Wroclawski (1997)). Each router along the

path of the flow through the network checks its resource availability before a new flow is actually admitted to the network, see Section 6.2.2 for more details.

o For *edge-based* admission control, the admission control decision is based on information locally available at the edge node of the network.
The admission control decision can be made

- exclusively at the ingress node (*ingress-based*) (see the decentral BB of Chapter 8),
- exclusively at the egress node (*egress-based*) (e.g. Cetinkaya and Knightly (2000)) or
- at both nodes (*ingress-egress-based*) (e.g. Bhatnagar and Nath (2003); Bhatnagar and Vickers (2001)).

Edge-based admission control mechanisms can also be distinguished by the nature of the local information they are using.

- The information can be *local traffic measurements* that can be constantly updated. Cetinkaya and Knightly (2000) present, for example, an egress-based admission control architecture. It treats the core network as a black box and is based on monitoring the aggregate traffic characteristics of one service class per path at the egress nodes. One-way per-packet delay measurements are used; these are, however, all but trivial to make. On the basis of these measurements, statistical traffic envelopes are derived and used as decision basis for admitting new flows. In Karsten and Schmitt (2002), ECN marks are counted and constantly updated at the egress at a per-ingress basis and used as estimation for the congestion level of the network.
- The information can also be the *status information of the whole network* that is stored in a distributed database at the edge nodes. This allows each edge to base its decisions on the same type of information that is available to a centralised admission control system. However, contrary to the centralised system, for the decentralised one a synchronisation and update mechanism is needed for the distributed database. The distributed database has to be updated on a relatively small timescale or the system will not work efficiently.
Systems implementing a decentral admission control algorithm based on a distributed database are described in Bhatnagar and Nath (2003); Bhatnagar and Vickers (2001). They use token passing.
Bhatnagar and Vickers (2001) specify a mechanism to provide bandwidth guarantees that requires only edge routers to implement the admission control scheme. No assumptions about the behaviour of the core routers on the data or control path are made, especially core routers do not have to be able to differentiate between different flows and not even between best-effort and reserved flows. The approach further assumes that the edge nodes have up to date information about the topology of the network and that there is a route-pinning mechanism for the network; this can, for example, be MPLS or IP source routing. RSVP is used as signalling mechanism, but only interacts with the ingress- and egress-router of a network. On the basis of the RSVP message exchange, the route between ingress and egress nodes is pinned.
The admission control mechanism uses a distributed database. Each ingress router has knowledge of the network's topology and a more or less up-to-date knowledge about the reservation state of the network. For synchronisation, a token passing

mechanism is used. A router is only allowed to change the reservation state of the network if it possesses a token that is passed around the edge routers. A reservation for a new flow does not become effective until the ingress router has fully circulated the updated token once among all edge routers. This prevents several edge routers from over-allocating bandwidth by simultaneously reserving bandwidth on a single link and it gives edge routers the opportunity to reduce the rates of best-effort flows sharing the same links as the new reserved flow. The latter is necessary because core routers cannot differentiate between reserved and best-effort flows.

Bhatnagar and Nath (2003) adapt the mechanism of Bhatnagar and Vickers (2001) to support GS in core-stateless networks (see Section 6.2.3) and improve the efficiency in several ways, for example, by marking potentially congested links and only requires a full token circulation before admitting a new flow when marked links are involved.

The drawback of these approaches is that compared to a central mechanism, they introduce additional delay (token circulation time) that can become long for large networks before admitting a new flow. In addition, they require each edge router to have the computational resources for managing and updating the database and add additional complexity to protect against lost tokens etc. Therefore, in most cases a specialised centralised system would seem the better choice.

- Finally, the local information used as a basis for an edge-based admission control decision can be a *contingent* as *resource budget* that is assigned off-line to the edge node. While measurement information as the distributed database is updated in rather small intervals, the contingents are updated only on much larger timescales and typically by a central instance based on the past performance of the system.

We call the latter mechanism *contingent-based admission control*. Among other things it is investigated in Chapter 8. Contingent-based admission control systems can be further distinguished by the contingent assignment.

- The contingent is assigned to the edge *node;* all flows entering the network through this node share this contingent.
- The contingent can also be assigned to each ingress and respective egress *link* of the edge node. Only flows with the same first and respective last hop through the network share a contingent.

  The problem with the first two contingent assignments is that they are very inefficient for deterministic guarantees as all – also pathological – traffic patterns through the network have to be taken into account when assigning the contingents. This results in very low contingents for deterministic services. A famous example is the Charny bound, see Charny and Le Boudec (2000) and Figure 6.7.

- The contingents can also be assigned to *tunnels* or MPLS label switched paths through the network if information about the traffic patterns is available. More state has to be kept in this case and it might be necessary to update the contingents more regularly than for the first two cases for this mechanism to be efficient. However, it promises higher possible contingents for deterministic guarantees as the information about the traffic patterns can be exploited in the contingent assignment process.

- A further alternative of contingent-based admission control schemes is assigning each ingress as egress node contingents for all links of the complete network.

This approach decentrally controls flows from the edge but can take the whole path through the network into account. The admission control test for a new flow predicts the path of that flow through the network – not a trivial task. Along the links of that path, the node checks for every link whether there are still contingents that were assigned to it available for that link. This approach necessarily loses efficiency compared to a centralised admission control, as an edge node cannot use the contingents assigned to another edge node for the same link. For this approach to be efficient, the contingent assignment process is of great importance. This approach is discussed in more detail in Menth (2004).

A comparative study of different contingent-based admission control schemes (and other admission control schemes) and a discussion of contingent assignment algorithms are presented in Menth (2004); Menth *et al.* (2003).

6.6.2 Flow and Network Behaviour

For the admission control test, certain assumptions about the flow and network behaviour have to be made.

6.6.2.1 Worst-case Assumptions

If worst-case behaviour of the flow and network elements is assumed, a conservative admission control test is performed. For decisions based on worst-case assumptions, the traditional *network calculus* offers a suitable mathematical framework, see Section 3.2.

6.6.2.2 Statistically Relaxed Assumptions

The admission control test can also be performed with statistically relaxed assumptions. They promise a higher resource usage at the cost of an increased but controlled risk of wrong decisions that will manifest themselves in violations of the (loss and delay) guarantees.

Among other methods, the stochastic network calculus and the queueing theory offer mathematical foundations for statistically relaxed flow and network behaviour assumptions; see Section 3.1 and 3.2.

6.6.2.3 Measurements

Admission control systems that use the two above-mentioned assumptions (worst-case and statistically relaxed) are based on mathematical models for predicting flow and network behaviour and maintain state information about the currently active flows. Contrary to that, predictions about the flow and network behaviour can also be based on measurements. In this case, we speak of *measurement-based admission control*. There are vast amounts of works on that topic; the works can be divided into admission control schemes with *active* and *passive* measurements. Active measurements actively probe the network by sending special probe packets while the passive measurements passively monitor the performance of normal data packets.

In addition, measurement-based admission control schemes can be classified by whether they measure the properties of individual flows (as e.g. in Karlsson (1998); Más and Karlsson (2001); Más et al. (2003)) or that of traffic aggregates (as e.g. in Jamin et al. (1997a); Qiu and Knightly (2001)).

The probe-based admission control scheme developed in Karlsson (1998); Más and Karlsson (2001); Más et al. (2003) for a loss-predictive unicast service and in Más et al. (2002) for multicast service is a typical example for an active measurement-based admission control scheme. A controlled load (see Section 6.2.2.5) type of service is offered. Before a new flow is accepted, a loss measurement is done by actively sending constant bit-rate probe packets at the maximum rate of the new flow for a sufficient time from the network ingress to the egress node. At the egress, the loss can be measured and reported back. The core network differentiates data packets from already admitted active flows and the probe packets so that probes do not disturb the active flows. A 0.5 to 2 s probing interval is recommended. In Más and Karlsson (2001) a simulative study of this scheme is contained; Más et al. (2003) use queueing theory to analytically evaluate the scheme for a single link and a single probing process. The comparison study of several endpoint and measurement-based admission control schemes also reports probing durations in the order of several seconds in Breslau et al. (2000b), whereas recent simulative work in Kelly (2001b) argues for much lower values for the initial probing phase.

Instead of using the measured packet loss as basis for the decision, measured delay such as delay variations are used in Bianchi et al. (2000, 2002). Kelly et al. (2000), Kelly (2001b) propose using ECN marks for a distributed measurement-based admission control system. Karsten and Schmitt (2002) also use ECN marks as congestion indication. So-called load control gateways running at a backbone network edge use the amount of measured ECN marks that data packets experience to estimate the current congestion level of the network.

The works of Gibbens and Kelly (1997); Gibbens et al. (1995) explicitly maximise the expected profit of an admission control that is defined by the reward of utilisation minus the penalty of packet-losses by calculating acceptance bounds for a specific set of flow types.

Jamin et al. (1997a) present an algorithm that uses the measured queueing delay of individual packets and the utilisation of the different service classes as inputs to derive an aggregate token bucket descriptor for each class. This measured token bucket is typically much smaller than the sum of the individual worst-case token buckets that describe the individual flows. Before admitting a new flow, the available bandwidth and the delay bounds are checked on the basis of these measured aggregate token bucket descriptors.

In Benameur et al. (2002), a measurement-based admission control mechanism is evaluated that explicitly considers elastic (TCP) flows besides real-time multimedia flows. In most other admission control schemes, the special characteristics of elastic flows are ignored or they are assumed to be in a low-priority best-effort upon which no admission control is applied.

The rate a new elastic flow would acquire is estimated either with a TCP phantom connection (an emulated TCP connection over the considered path) or by measuring the loss rate and applying the TCP formula. A new flow (elastic or not) is accepted only if it does not reduce the throughput of ongoing elastic flows below a certain threshold.

6.6.3 Guarantees

Three types of guarantees can be given to newly admitted flows such that their transmission or QoS requirements will be fulfilled by the network. The guarantees very strongly depend on the flow and network behaviour assumptions.

6.6.3.1 Deterministic Guarantees

Deterministic guarantees are based on worst-case assumptions. The Intserv GS of RFC 2212 (see Shenker *et al.* (1997)) is the typical example for a service with deterministic loss and delay-bound guarantees, see Section 6.2.2.4. Other works with deterministic service guarantees are, for example, Choi *et al.* (2000); Elwalid *et al.* (1995b); Knightly *et al.* (1995); Rajagopal *et al.* (1998). For example, Choi *et al.* (2000) offer delay guarantees by an offline worst-case delay calculation for QoS systems like Diffserv with aggregate scheduling.

6.6.3.2 Statistical Guarantees

Statistical guarantees are based on the statistically relaxed assumptions of flow and networking behaviour. There is a broad set of admission control algorithms for stochastic service guarantees; most of them fit into one of the following five classes according to Knightly and Shroff (1999):

- Average and Peak Rate Combinatorics
  Lee *et al.* (1996) use the peak rate and long-term average rate to predict the loss probability assuming a bufferless multiplexer. The loss rate is used as basis for the admission control decision. Ferrari and Verma (1990) use the delay-bound violation probability as basis for the decision and use the peak rates and worst-case average rates of the flows as inputs.
- Additive Effective Bandwidths
  The effective bandwidth is the bandwidth bw_f that has to be provided for a flow f to fulfill its service guarantees. It is a function of the flow's required loss probability and stochastic properties like the peak- and average rate or the mean burst duration. Overviews of the effective bandwidth concept can be found in Bodamer and Charzinski (2000); Gibbens and Teh (1999); Kelly (1996). There are different ways of computing the effective bandwidth, see for example, Courcoubetis and Weber (1995); Elwalid and Mitra (1993); Guérin *et al.* (1991); Kesidis *et al.* (1993). A simple admission control decision based on effective bandwidths makes sure that the added effective bandwidths bw_f do not exceed the link's capacity C: $\sum_f bw_f \leq C$.
- Refined Effective Bandwidths
  The above additive effective bandwidth approach has two shortcomings. First, the result is not applicable to traffic sources that show long-range dependency. Second, the economies of scale such as the multiplexing gain from adding a large number of sources are not exploited by adding the effective bandwidths, resulting in an inefficient admission control mechanism (see Knightly and Shroff (1999)). More advanced effective bandwidth approaches are not additive and incorporate the interdependences of the

traffic flows on each other when calculating the effective bandwidth, see for example, Courcoubetis *et al.* (1998); Duffield and O'Connell (1995); Kelly (1996).

- Loss Curve Engineering
  A loss curve models the loss probability as a function of the buffer size. It can be used as the basis for an admission control scheme. Assuming additive effective bandwidths (see preceding text), the loss curve is an exponential function of the buffer size. For the reasons mentioned above, the additive effective bandwidths and therefore the exponential loss curves are inefficient. Various techniques have been proposed, which seek to engineer the shape of the loss curve to better reflect empirical relationships, see for example, Baiocchi *et al.* (1991); Choudhury *et al.* (1996); Elwalid *et al.* (1995a); Shroff and Schwartz (1998).

- Maximum Variance Approaches
  Maximum variance approaches are based on estimating the loss probability via the tail probability of an infinite queue based on a Gaussian aggregate arrival process. The Gaussian characterisation of the traffic allows for different correlation structures as any function can be a valid autocovariance function, hence it can capture the temporal correlation of the traffic. Some maximum variation-based admission control schemes are Choe and Shroff (1998); Kim and Shroff (2001); Knightly (1997).
  Knightly and Shroff (1999) evaluate typical admission control schemes from these five categories in a set of experiments using Motion Pictures Expert Group (MPEG) video traces and Markov modulated on-off traffic sources. Among other things, they show that the assumption of bufferless network elements significantly reduces the admission control efficiency and network utilisation and that the accuracy of an admission control algorithm for one type of traffic does not assure accuracy for another type of traffic.

6.6.3.3 Empirical Guarantees

If a measurement-based admission control scheme is used, only *empirical guarantees* based on the past networking behaviour can be given. In Jamin *et al.* (1997b) and Breslau *et al.* (2000a), an extensive comparison of measurement-based admission control schemes finally results in the conclusion that all schemes perform fairly similar with respect to the utilisation they yield.

6.6.4 Other Properties

Most admission control systems need an explicit *traffic description* for new flows. At least for deterministic guarantees, usually a policer or shaper is used to force flows to comply with their traffic description. A wide variety of traffic descriptors can be imagined, for example,:

- Peak rate
- Average rate and maximum burst size, for example, a token bucket or if extended by peak rate and maximum packet size a TSpec, see Shenker *et al.* (1997)
- Effective bandwidth
- General arrival curve for network calculus
- Elastic flows could be characterised by their transfer volume alone, see for example, Benameur *et al.* (2002)

- Peak rate and long-term average rate as for example, in Lee *et al.* (1996)
- Peak rate and short-term average rate as for example, in Ferrari and Verma (1990).

Another characteristic is whether *multicast* flows are supported as for example, in Más *et al.* (2002); Shenker *et al.* (1997).

The above-mentioned criteria are in most cases sufficient to roughly classify the vast amount of works on admission control. However, real admission control systems can also be distinguished by a number of further criteria, for example, by whether they are *pre-emptive*.

- *Non-preemptive* admission control systems do not interrupt flows once they have been admitted while
- *pre-emptive* systems can interrupt an admitted flow in order to free resources for another flow, see for example, Yavatkar *et al.* (2000).

Access to different *network resources* can be managed by the admission control system.

- Link *bandwidth* is practically always used as the central resource.
- Additionally, some systems also check the availability of *buffer space*, for example, Shenker *et al.* (1997).

The *granularity* of the system describes which type of flows form the decision objects of the system, ranging from

- individual micro-flows (specified by the source and sink IP address, port and the protocol number) over
- sessions that can consist of multiple flows, senders and/or receivers (e.g. Intserv/RSVP)
- up to large aggregated macro-flows identified by other means.

The *timing behaviour* of the system describes whether the flows also specify their (expected) duration and whether this information is used for the admission control test. This is especially important if the system also supports *reservation in advance* (see e.g. Karsten *et al.* (1999)). Reservation in advance allows customers to request resources long before the actual transmission is started.

After this overview and classification of admission control mechanisms, it is also important to stress that besides testing the availability of resources before admitting a new flow – which the above-mentioned works do in a wide variety of different ways – it is also important for an INSP to apply certain *policies* to the admission control decision. With *policy,* we describe all kinds of non-technical rules that are applied besides technical rules to a certain decision. For the admission control decision, the technical rules are the ones that check the resource availability (see preceding text) while non-technical rules – policies – in that context can, for example, check the identity of the user, his contract and his solvency. On the basis of the policies a flow might be rejected despite resources being available. We do not further investigate the support of policies here but instead refer to Durham *et al.* (2000); Herzog (2000); Yavatkar *et al.* (2000).

6.7 Summary and Conclusions

In this chapter, network architectures were discussed. A network architecture consists of the QoS architecture, the data forwarding architecture, the signalling and the security architecture. The most commonly used QoS architecture is the plain best-effort architecture although Diffserv is becoming more and more popular as QoS architecture with the increased importance of QoS-sensitive applications like for example, VoIP (see also Chapter 5). We also discussed alternative approaches to QoS architectures that are not supported by the IETF but use interesting and innovative concepts.

With respect to the data forwarding architecture, label switching as provided by MPLS routers is an alternative to the standard approach of plain IP routing and is gaining importance. The signalling architecture encompasses the routing protocols, the QoS signalling protocols (if used), and the LDPs (if used). The security architecture adds cryptography-based security services at the IP layer. At the end of this chapter, we discussed the broad spectrum of admission control mechanisms. Admission control can be used to control the network load on a small timescale by not admitting certain traffic flows or customers to the network or at least to certain (high quality) traffic classes.

In the following two chapters of Part II of this book, different QoS systems are evaluated. Chapter 7 does so on an abstract level using two analytical approaches while Chapter 8 uses implementations of systems based on the IETF QoS architectures in an experimental study.

7

Analytical Comparison of Quality of Service Systems*

In this chapter, we use two analytical approaches to compare different Quality of Service (QoS) systems. We compare two QoS systems:

1. A QoS system using admission control and a reservation mechanism that can guarantee bandwidth for flows (Section 7.1) offers service differentiation based on priority queueing for the two service classes (Section 7.2)
2. and a system with no admission control and a single best-effort service class.

We call the second model *Best-effort (BE) model/system* and the first one *QoS model/system*.

Important for the evaluation in this chapter is the type of traffic application assumed. We use different application and traffic models. *Inelastic traffic* represents multimedia applications that require a certain rate. We speak of *strictly inelastic traffic* if no loss or delay bound violations are tolerated. Most multimedia applications can tolerate a certain level of loss or delay bound violations. For example, a typical voice transmission is still understandable – albeit at reduced quality – if some packets are lost or arrive too late. Therefore, *normal inelastic traffic* tolerates a certain amount of loss or delay bound violations. *Adaptive traffic* is similar to normal inelastic traffic but can adapt its required rate to the network conditions and is thus assumed to be extremely flexible. *Elastic traffic* represents file transfer traffic like WWW, FTP or peer-to-peer traffic. The utility of the elastic traffic is a concave function of its throughput as the throughput determines when the transfer is finished; the loss probability does not directly influence the utility.

Because of the complexity of the models, the analysis is focused on a single bottleneck. The next chapter deals with larger topologies, more realistic traffic, and so on using simulations.

The first set of models (Section 7.1) used is based on Breslau and Shenker (1998); Shenker (1995). As is common and good practice in sciences, we first reproduce the results of Breslau and Shenker (1998); Shenker (1995); then we give some further insights.

*Lecture Notes in Computer Science, 3552, 2005, 151-163, Best-Effort Versus Reservations Revisited, Oliver Heckmann and Jens B. Schmitt, copyright 2005. With kind permission of Springer Science and Business Media.

In these works, a single type of traffic (elastic or strict inelastic or adaptive inelastic) uses the bottleneck. The expected total utility is analysed by assuming a probability distribution for the number of arriving flows. The main issues investigated with these models are admission control and bandwidth guarantees.

The second set of models (Section 7.2) is a contribution of this book. Contrary to the other models, they analyse a given load situation and a traffic mix consisting of elastic and inelastic flows filling the link at the same time. By using the queueing theory and the TCP formula, more sophisticated utility functions and more realistic network behaviour than in the first set of models can be modelled. The main effects investigated with these models are scheduling and service differentiation.

When we compare the QoS and the BE system, it is quite obvious that for the same capacity (e.g. bandwidth) the QoS system will offer better QoS. But it also has a higher complexity that leads to higher costs. For judging which of the two systems is 'better', a way has to be found to put the QoS and the costs in a relationship. For the additional costs of the QoS system, more bandwidth could be bought for the BE system, improving its QoS. To compare the two systems, we have to make sure that either the costs of the two considered systems or the QoS are equal. The costs are hard to predict[1] while the QoS is measured in the models anyway. Therefore, we bring the QoS levels in line and use the *overprovisioning factor* as metric to compare the systems: A specific QoS system leads to a certain level of QoS; its overprovisioning factor is the factor with which the capacity (bandwidth) of the BE system has to be multiplied so that it offers the same level of QoS. A high overprovisioning factor indicates that QoS system is the preferable choice while an overprovisioning factor close to one indicates that the QoS system is not worth its additional complexity. The factor for which the QoS system becomes the preferable choice depends on the exact costs. With the knowledge of the overprovisioning factor and an estimation of costs for its network, an Internet Network Service Provider (INSP) can therefore make the correct decision.

7.1 On the Benefit of Admission Control

Breslau and Shenker (1998); Shenker (1995) analyse two fundamentally different QoS systems in their works:

1. A BE system without admission control where all flows admitted to the network receive the same share of the total bandwidth.
2. A reservation-based QoS system with admission control, where only the flows are admitted to the network that optimally (w.r.t. total utility) fills the network. Their bandwidth is guaranteed by the system. This system can be built using the Intserv/RSVP architecture and to a certain extent using a Diffserv/bandwidth broker architecture.

---

[1] The technical costs like memory usage or used CPU cycles could be predicted. However, networking has seen many technological breakthroughs in the last years, for example, for packet classification (see Section 6.3.1.1) and scheduling (see Section 6.2.1). The prediction could therefore become insignificant quickly. Furthermore, the finally relevant costs are monetary costs of the systems and they depend among many other things on business policies and marketing decisions which are – besides being almost impossible to predict – completely out of scope of this technical work.

We start with a fixed load model that assumes a given traffic load for the network; next, a variable load and finally variable load and capacity are analysed.

7.1.1 Fixed Load

The fixed load model from Shenker (1995), also published in Breslau and Shenker (1998), assumes that there are a number of identical flows requesting service from a link with capacity C. The utility function $u(b)$ of a flow is a function of the link bandwidth b assigned for that flow with:

$$\frac{du(b)}{db} \geq 0 \ \forall b > 0, \ u(0) = 0, \ u(\infty) = 1 \tag{7.1}$$

A flow rejected by the admission control is treated as receiving zero bandwidth, resulting in zero utility. The link capacity is split evenly among the flows so that the total utility U of k admitted flows is given by

$$U(k) = k \cdot u\left(\frac{C}{k}\right) \tag{7.2}$$

If there exists some $\epsilon > 0$ such that the function $u(b)$ is convex but not concave[2] in the neighbourhood $[0, \epsilon]$, then there exists some k_{max} such that

$$U(k_{max}) > U(k) \ \forall k > k_{max} \tag{7.3}$$

In this case, the network is overloaded whenever more than k_{max} flows enter the network; the system with admission control would yield the higher total utility for because it could restrict the number of flows to k_{max}.

If the utility function $u(b)$ is strictly concave, then $U(k)$ is a strictly monotonically increasing function of k. In that case, the total utility is maximised by always allowing flows to the network and not using admission control.

Elastic applications typically have a strictly concave utility function as additional bandwidth aids performance but the marginal improvement decreases with b. Therefore, if all flows are elastic, the BE system without admission control would be the optimal choice.

Looking at the other extreme of the spectrum, there are *strictly inelastic applications* like traditional telephony that require their data to arrive within a given delay bound. Their performance does not improve if data arrives earlier, they need a fixed bandwidth $\tilde{b}$ for the delay bound (see Section 6.2.2.4). Their utility function is given by

$$u(b) = \begin{cases} 0 & b < \tilde{b} \\ 1 & b \geq \tilde{b} \end{cases}, \tag{7.4}$$

which leads to a total utility of

$$U(k) = \begin{cases} 0 & k > C/\tilde{b} \\ k & k \leq C/\tilde{b} \end{cases} \tag{7.5}$$

---

[2] This rules out functions simple linear functions $u(b) = a_0 + a_1 \times b$ which would, by the way, also violate (7.1).

In this case, admission control is clearly necessary to maximise utility. If no admission control is used and the number of flows exceeds the threshold $C/\tilde{b}$, the total utility $U(k)$ drops to zero.

The two extreme cases of elastic and strictly inelastic applications show that the Internet and telephone network architectures were designed to meet the needs of their original class of applications.

Another type are the *adaptive applications*; they are designed to adapt their transmission rate to the currently available bandwidth and reduce to packet delay variations by buffering. Breslau/Shenker propose the S-shaped utility function with parameter κ

$$u(b) = 1 - e^{-\frac{b^2}{\kappa+b}} \tag{7.6}$$

to model these applications (see Figure 7.1). For small bandwidths, the utility increases quadratically $\left(u(b) \approx \frac{b^2}{\kappa}\right)$ and for larger bandwidths it slowly approaches one $\left(u(b) \approx 1 - e^{-b}\right)$. The exact shape is determined by κ.

For these flows, the total utility $U(k)$ has a peak at some finite k_{max} but the decrease in total utility for $k > k_{max}$ is much more gentle than for the strictly inelastic applications. The reservation based system thus has an advantage over the BE system, but two questions remain: The first is *whether that advantage is large enough to justify the additional complexity* of the reservation based QoS system and the second is, *how likely is the situation where $k > k_{max}$*. These questions are addressed in the next section with the variable load model.

7.1.2 Variable Load

7.1.2.1 Model

The previous section showed that in an overload situation where $k > k_{max}$, the reservation-based QoS system offers a certain advantage over the plain BE system for some utility functions. Breslau and Shenker (1998) analyse the likelihood of the overload situation

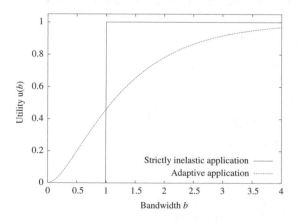

Figure 7.1 Utility Functions for $\tilde{b} = 1$, $\kappa = 0.62086$

for the strictly inelastic and adaptive applications (see Figure 7.1) by assuming a given probability distribution $P(k)$ of the number of flows k. They use two models, a model with a discrete and one with a continuous number of flows k. We base our following analysis on the discrete model[3], assuming three different load distributions (see Figure 7.2):

$$\text{Poisson:} \quad P(k) = \frac{v^k e^{-v}}{k!} \tag{7.7}$$

$$\text{Exponential:} \quad P(k) = \left(1 - e^{-\beta}\right) \cdot e^{-\beta k} \tag{7.8}$$

$$\text{Algebraic:} \quad P(k) = \frac{v}{\lambda + k^z} \tag{7.9}$$

The *Poisson load distribution* describes a scenario where the load is tightly controlled within the region around the average v. Large or small loads are extremely rare. For the *exponential load distribution*, the load is not peaked around the average but instead decays at an exponential rate over a large range. The decay is determined by β; the expected number of flows for the exponential distribution is $E(k) = 1/\left(e^{\beta} - 1\right)$. The *algebraic load distribution* is similar but decreases slower than the exponential load distribution. It has three parameters v, λ and z[4]. The algebraic distribution is normalised so that $\sum_{k=0}^{\infty} P(k) = 1$; we analyse $z \in \{2, 3, 4\}$.

Similar to Breslau and Shenker (1998), for the following analysis we choose the parameters of the probability distributions so that the expected number of flows $E(k) = \sum_{k=0}^{\infty} k \cdot P(k)$ is 100. Figure 7.2 depicts the probability density and distribution functions. For the utility functions, $\tilde{b} = 1$ in (7.4) and $\kappa = 0.62086$ in (7.6) this parameter setting yields $k_{max} = C$ for both utility functions.

The two utility functions analysed should be seen as the extremes of a spectrum. The strictly inelastic utility function does not tolerate any deviation from the requested minimum bandwidth $\tilde{b}$ at all, while the adaptive utility function embodies fairly large

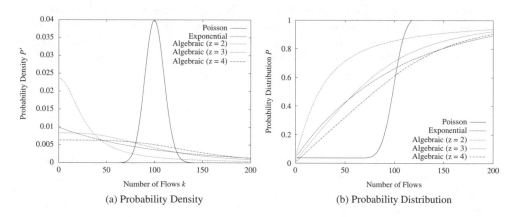

(a) Probability Density   (b) Probability Distribution

Figure 7.2 Load Distribution Functions (Continuous)

---

[3] That the number of flows increases in discrete steps seems more realistic. However, the continuous model is easier to solve in many cases and generally leads to similar results, see Breslau and Shenker (1998).

[4] λ is introduced so that the distribution can be normalised for a given asymptotic power law z.

changes in utility across a wide range of bandwidths above and below C/k_{max} (the level the reservation-based approach would assign to an adaptive flow).

The expected total utility $\overline{U}_{BE}$ of the BE system is

$$\overline{U}_{BE}(C) = \sum_{k=1}^{\infty} P(k) \cdot U(k) = \sum_{k=1}^{\infty} P(k) \cdot k \cdot u\left(\frac{C}{k}\right) \qquad (7.10)$$

The QoS system can limit the number of flows to a k_{max}. The expected utility $\overline{U}_{QoS}$ of the QoS system is

$$\overline{U}_{QoS}(C) = \sum_{k=1}^{k_{max}(C)} P(k) \cdot k \cdot u\left(\frac{C}{k}\right) + \sum_{k=k_{max}(C)+1}^{\infty} P(k) \cdot k_{max} \cdot u\left(\frac{C}{k_{max}(C)}\right) \qquad (7.11)$$

To compare the performance of the two QoS systems, Breslau and Shenker (1998) propose the bandwidth gap as a performance metric. The bandwidth gap is the additional bandwidth Δ_C necessary for the BE system so that the expected total utilities are equal:

$$\overline{U}_{QoS}(C) = \overline{U}_{BE}(C + \Delta_C) \qquad (7.12)$$

As argued in the beginning of this chapter, we propose a different metric: the unit-less *overprovisioning factor* OF. It puts the bandwidth gap in relation to the original bandwidth

$$OF = \frac{C + \Delta_C}{C} \qquad (7.13)$$

The overprovisioning factor expresses the bandwidth increase necessary for a BE based QoS system to offer the same expected total (and average) utility as the reservation based one. The higher the overprovisioning factor, the more attractive the reservation-based approach becomes; if the overprovisioning factor is close to unity, however, the additional complexity of the reservation-based approach is not justified.

7.1.2.2 Evaluation

We now determine the overprovisioning factors. The results for the strictly inelastic and the adaptive utility function and for all three load distributions over a wide range of link bandwidths C are shown in Figure 7.3. The reader is reminded of the fact that the expected number of flows $E(k)$ is 100 in all cases.

The *Poisson load distribution* (Figure 7.3 (a)) describes a situation where the load is fairly tightly controlled within a region around the average; excursions to large and small loads are extremely rare. If the link capacity is small compared to the bandwidth required by the average number of strictly inelastic flows, the overprovisioning factor is very high. It drops down to 1.2 if the link capacity equals the expected bandwidth demand and for higher bandwidths, it quickly approximates to 1.0.

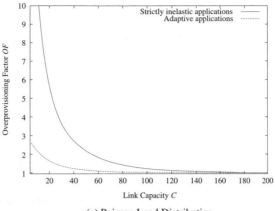

(a) Poisson Load Distribution

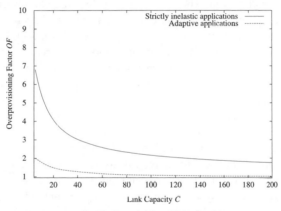

(b) Exponential Load Distribution

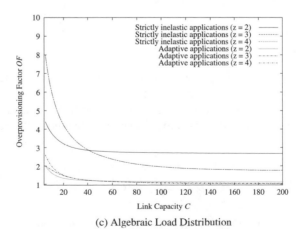

(c) Algebraic Load Distribution

Figure 7.3   Results of the Variable Load Model

In contrast to the strictly inelastic application, the overprovisioning factor is much more controlled and smaller for the adaptive application. It is lower than 3.0 even if the link bandwidth is only 5% of the expected bandwidth demand and below 1.1 as soon as the link capacity exceeds 50% of the expected bandwidth demand. This demonstrates that the adaptive utility function (7.6) allows very large changes in utility across a wide range of bandwidths.

The results for the *exponential load distribution* (Figure 7.3 (b)) represent a situation where the load is not peaked around the average and decays over the whole range at exponential rate. For the strictly inelastic application, the overprovisioning factor for low capacities is lower and for higher capacities higher than the factor of the Poisson distribution. It is 2.2 if the capacity equals demand and 1.8 if the capacity is twice the demand.

For adaptive applications, the overprovisioning factor is again close to one (roughly 1.1 if capacity equals demand).

The *algebraic load distribution* also decays over the whole range but at a lower rate than the exponential distribution. The lower the z value, the slower the decay. The overprovisioning factor is quite similar to the exponential case but decreases more slowly for higher capacities. The very slow decay for $z = 2$ results in a significantly higher overprovisioning factor (2.70 if capacity equals demand and 2.67 if capacity equals twice the demand in the strictly inelastic case). For adaptive applications, the overprovisioning factor is again close to one (between 1.05 and 1.14 if capacity equals demand).

The results show that the overprovisioning factor is close to unity for adaptive applications and significantly higher than unity for the inelastic applications. The link capacity significantly influences the performance of both QoS systems and the overprovisioning factor. The capacity of the network is determined by the network design and the engineering process of the INSP. Therefore, these results are another indication that it is important to look at the QoS problem from a system-oriented point of view.

The reservation-based QoS system can provide significant advantages over the pure BE system in a well dimensioned network for strictly inelastic applications. For adaptive applications, the advantage is rather low in a well dimensioned network.

7.1.3 Variable Capacity

7.1.3.1 Model

The results above depended strongly on the relationship of the link capacity to the average number of flows and the flow/load distribution. One can further analyse the capacity level C_{opt} that maximises social welfare for both QoS systems. The social welfare W is the total utility minus the costs of the capacity C that are assumed as linear functions here:

$$W_{QoS}(C, \ p_R) = \overline{U}_{QoS}(C) - p_{QoS} \cdot C \tag{7.14}$$

$$W_{BE}(C, \ p_{BE}) = \overline{U}_{BE}(C) - p_{BE} \cdot C \tag{7.15}$$

If the provider uses a tariffing scheme that allows him to charge the users full utility, then the capacity maximising social welfare also maximises the provider's profit.

The bandwidth price of the reservation-based QoS system can be assumed to be a factor ρ higher than that of the plain BE system because of the additional complexity involved:

$$p_{QoS} = \rho \cdot p_{BE}, \; \rho \geq 1 \tag{7.16}$$

Now, the *equalising price factor* ρ' can be analysed as a function of the best-effort bandwidth price p_{BE} for the following situation: The reservation-based system is operated at the capacity C_{QoS}^{max} that maximises social welfare W_{QoS}. It yields the same social welfare as the BE system that is operated at the (different) capacity C_{BE}^{max}, which maximises social welfare W_{BE} in the BE case:

$$W_{QoS}\left(C_{QoS}^{max}, \rho' \cdot p_B\right) = W_{BE}\left(C_{BE}^{max}, p_{BE}\right) \tag{7.17}$$

If the real price factor for reservation-based capacity is higher than ρ', then the BE system offers higher social welfare (correspondingly, profit for the provider) than the reservation-based system and vice versa.

7.1.3.2 Evaluation

The equalising price factors for strictly inelastic and adaptive applications and the three different load distributions are depicted in Figure 7.4. If a certain BE price p_{BE} is exceeded, the social welfare profit becomes negative. In that case, not investing in network capacity is the optimal choice. The x-axis of Figure 7.4 only contains values of p_{BE} that lead to a positive profit.

Similar to the overprovisioning factor, the equalising price factor is significantly higher for the strictly inelastic application than for the adaptive application. This holds true for all load distributions. For the adaptive applications, the equalising price ratio is below 1.25 for all distributions and p_{BE}. Thus, if the price for providing bandwidth with the reservation-based system is more than 25% higher than that of the BE system, it is in no case worth it.

The cheaper the bandwidth is (p_{BE}), the lower the equalising price factor for all load distributions. The conclusion is that the cheaper the bandwidth gets, the more attractive the BE system becomes.

In the Poisson load distribution case, the equalising price factor is below 1.25 over a wide range of prices for both application types. For the strictly inelastic application and the exponential load distribution, the equalising price ratio is significantly higher than unity unless the BE price approaches zero. In the latter case, the equalising price ratio converges to one. For the algebraic load distribution, the equalising price ratio does not converge to one. This is shown analytically in Breslau and Shenker (1998). In these cases, the reservation-based system is preferable even if it is significantly more expensive than the BE system.

7.1.4 Summary and Conclusions

The models presented in this section help in understanding whether a reservation based or a pure BE QoS system is better. The overprovisioning factors express the amount of additional bandwidth necessary for the BE QoS system to offer the same utility as the reservation-based system. The costs of the additional bandwidth – expressed by the

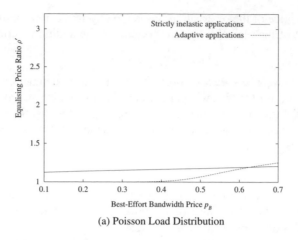

(a) Poisson Load Distribution

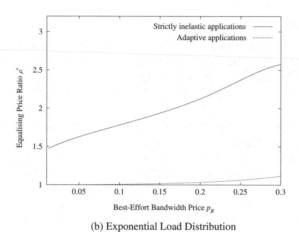

(b) Exponential Load Distribution

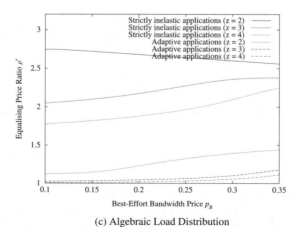

(c) Algebraic Load Distribution

Figure 7.4   Equalising Price Factors

overprovisioning factor – have to be weighted against the costs of the additional complexity of the reservation based system. For linear bandwidth costs, we have seen that the bandwidth price of the reservation-based system can be twice the price of the BE system and still the reservation-based system would be the preferable choice for the strictly inelastic applications in many cases. However, as the price for bandwidth drops, the BE system generally becomes more attractive even for these types of applications.

The results indicate that for strictly inelastic applications, the reservation-based approach is probably more efficient while this is very doubtful for the discussed adaptive applications.

The above analysis in Breslau and Shenker (1998) gives valuable insights but can also be criticised in some points:

- It assumes that only a single type of application utilises the network. If different applications with different requirements utilise a network at the same time (Multiservice network), QoS systems that know the QoS requirements of the flows and can differentiate between them – for example, by protecting loss sensitive flows or by giving delay sensitive flows a higher scheduling priority – offer a further advantage over the BE system. This advantage is not included in the overprovisioning factors obtained with the models above.
- The load distributions (Poisson, exponential, algebraic) used in the models above to derive the expected utility for a given bandwidth are not based on empirical studies.
- In addition, it is doubtful whether this expected utility really represents the satisfaction of the customers with the network performance:

  If the network performance is very good most of the time but regularly bad at certain times (e.g. when important football games are transmitted), this might be unacceptable for customers despite a good *average* utility.

  Instead of assuming a load distribution and optimising for the whole range of the distribution, a provider would probably base its decision on the performance of the network in a high-load situation.

In the next section, we use a novel approach to avoid these drawbacks and shed more light on the comparison of the two QoS systems.

7.2 On the Benefit of Service Differentiation

When analysing a mix of different traffic types competing for bandwidth, it is not trivial to determine the amount of bandwidth the individual flows will receive and the delay it experiences. In this section, we present an analytical approach that – contrary to the previous approach – uses queueing theory and the TCP formula as a foundation to calculate the overprovisioning factor for a traffic mix of elastic TCP-like traffic flows and inelastic traffic flows.

7.2.1 Traffic Types

We assume that two types of traffic – elastic and inelastic – share a bottleneck link of capacity C. For *inelastic traffic*, we use index 1 and assume that there are a number of inelastic flows sending with a total rate r_1. The strictly inelastic traffic analysed in Section 7.1 did not tolerate any loss. Most multimedia applications, however, can tolerate a certain level of loss. For example, a typical voice transmission is still understandable if some packets are lost – albeit at reduced quality. We model this behaviour here by making the utility of the inelastic traffic degrading with the packet loss[5] and with excessive delay.

For the *elastic traffic,* we use index 2; it represents file transfer traffic with the characteristic TCP 'sawtooth' behaviour: the rate is increased proportional to the round-trip time (RTT) and halved whenever a loss occurs. We use the TCP formula (4.2) to model this behaviour; the two main parameters that influence the TCP sending rate are the loss probability p_2 and the RTT delay q_2. We assume there are a number of greedy elastic flows sending as fast as the TCP congestion control is allowing them to send; their total rate is $r_2 = f(p_2, q_2)$. The utility of the elastic traffic is a function of its throughput.

7.2.2 Best-Effort Network Model

A BE network cannot differentiate between packets of the elastic and inelastic traffic flows and treats both types of packets the same way. The loss and the delay for the two traffic types is therefore equal:

$$p_{BE} = p_1 = p_2 \tag{7.18}$$

$$q_{BE} = q_1 = q_2 \tag{7.19}$$

Let μ_1 be the average service rate of the inelastic flows, μ_2 the one for elastic flows, λ_1 the arrival rate of the inelastic traffic and λ_2 the arrival rate of the elastic traffic. The total utilisation ρ is then given by

$$\rho = \rho_1 + \rho_2 = \frac{\lambda_1}{\mu_1} + \frac{\lambda_2}{\mu_2} \tag{7.20}$$

and the average service rate $\overline{\mu}$ by

$$\overline{\mu} = \frac{\rho_1 \mu_1 + \rho_2 \mu_2}{\rho_1 + \rho_2} = \frac{\lambda_1 + \lambda_2}{\rho_1 + \rho_2} \tag{7.21}$$

In the BE model, the loss probability p_{BE} is the same for both traffic types and can be estimated with the well-known *M/M/1/B* loss formula for a given maximal queue length of B packets assuming Markovian arrival and service processes:

$$p_{BE} = \frac{1 - \rho}{1 - \rho^{B+1}} \cdot \rho^B \tag{7.22}$$

For the queueing delay q_{BE} of the bottleneck link, the *M/M/1/B* delay formula is used:

$$q_{BE} = \frac{1/\overline{\mu}}{1 - \rho} \cdot \frac{1 + B\rho^{B+1} - (B+1)\rho^B}{1 - \rho^B} \tag{7.23}$$

---

[5] It can be seen as an intermediate application between the strictly inelastic and the adaptive traffic of Section 7.1.

The arrival rate λ_1 of the inelastic traffic is given by the sending rates r_1 of the inelastic flows (7.31) while the arrival rate λ_2 of the elastic traffic depends on the TCP algorithm and the network condition. As explained in Section 4.1.3, there are many works like Cardwell *et al.* (2000); Floyd (1991); Mathis *et al.* (1997); Padhye *et al.* (1998) that describe methods for predicting the average long-term TCP throughput, depending on the loss and delay properties of a flow. For our high-level analysis, we are not interested in details like the duration of the connection establishment and so on. Therefore, we use the plain square-root formula (4.2) for this analysis; it allows us to keep the complexity of the resulting model low:

$$\text{throughput} = \frac{\text{MSS}}{\text{RTT} \cdot \sqrt{2/3} \cdot \sqrt{p_2}} \tag{7.24}$$

with MSS as maximum segment size and RTT as the round-trip time. RTT is assumed to be dominated by the queueing delay q_2. The throughput of the queue can also be expressed as a function of the arrival process λ_2 and the loss probability p_2:

$$\text{throughput} = \lambda_2(1 - p_2) \tag{7.25}$$

Introducing parameter t that we call *flow size factor*, (7.24) and (7.25) can be simplified to

$$\lambda_2 = \frac{t}{q_{BE} \cdot \sqrt{p_{BE}}} \cdot \frac{1}{1 - p_{BE}} \tag{7.26}$$

t encompasses the MSS$/\sqrt{2/3}$ part of (7.24) and part of the RTT and is used to put the TCP flows in correct dimension to the inelastic flows, which are dimensioned by their fixed sending rate r_1.

The resulting best-effort network model is summarised in Model 7.1. As λ_2 is a function of p_{BE} and q_{BE} and at the same time influences p_{BE} and q_{BE}, the network model is a non-linear equation system. It can be solved with numerical methods. For individual equations, methods like the fixed point iteration method, the bisection or secant method, regula falsi, the Newton or the Newton–Raphson method can be used, see, for example, Press *et al.* (1992). For whole equation systems, the Gauss–Newton and the modified Newton–Raphson method can be used. Mathematical libraries like JMSL (Visual Numerics (2004)), MatLab (Mathworks (2004)) and Maple (Maplesoft (2004)) offer sophisticated non-linear equation solvers. We used the Maple 9 tool *fsolve* to solve the equation system.

7.2.3 QoS Network Model

To model a QoS system that differentiates between the inelastic and elastic traffic, we use priority queueing. The inelastic traffic receives strict non-preemptive priority in time and (buffer) space over the elastic traffic.

Using the *M/M/1* queueing model, the expected waiting time $E(W_1)$ for a packet of an inelastic flow depends on the expected number of packets waiting to be served $E(L_1)$ and the residual service time of the packet currently in the queue. Because non-preemptive queueing is used, the latter can be a type 1 (inelastic flow) or type 2 (elastic flow) packet;

Model 7.1 Best-effort Network Model

Parameters

r_1 Total sending rate of the inelastic flows [pkts/s]

t Flow size factor of the elastic flows [pkts]

μ_1 Service rate of the inelastic traffic [pkts/s]

μ_2 Service rate of the elastic traffic [pkts/s]

B Queue length [pkts]

Variables

p_{BE} Loss probability

q_{BE} Queueing delay [s]

λ_1 Arrival rate of the inelastic traffic at the bottleneck [pkts/s]

λ_2 Arrival rate of the elastic traffic at the bottleneck [pkts/s]

ρ Utilisation of the queue

$\overline{\mu}$ Average service rate [pkts/s]

Equations

$$\overline{\mu} = \frac{\lambda_1 + \lambda_2}{\rho} \tag{7.27}$$

$$\rho = \frac{\lambda_1}{\mu_1} + \frac{\lambda_2}{\mu_2} \tag{7.28}$$

$$p_{BE} = \frac{1 - \rho}{1 - \rho^{B+1}} \cdot \rho^B \tag{7.29}$$

$$q_{BE} = \frac{1/\overline{\mu}}{1 - \rho} \cdot \frac{1 + B\rho^{B+1} - (B+1)\rho^B}{1 - \rho^B} \tag{7.30}$$

$$\lambda_1 = r_1 \tag{7.31}$$

$$\lambda_2 = \frac{t}{q_{BE} \cdot \sqrt{p_{BE}}} \cdot \frac{1}{1 - p_{BE}} \tag{7.32}$$

because the exponential service time distribution is memoryless, the expected residual service time is $\sum_{i=1}^{2} \rho_i \frac{1}{\mu_i}$:

$$E(W_1) = E(L_1)\frac{1}{\mu_1} + \sum_{i=1}^{2} \rho_i \frac{1}{\mu_i} \qquad (7.33)$$

By applying Little's Law (see Section 3.1.3)

$$E(L_i) = \lambda_i E(W_i) \qquad (7.34)$$

we get

$$E(W_1) = \frac{\sum_{i=1}^{2} \rho_i \frac{1}{\mu_i}}{1 - \rho_1} \qquad (7.35)$$

To determine the average queueing delay q_1, we need the expected sojourn time $E(S_1) = E(W_1) + 1/\mu_1$

$$q_1 = E(S_1) = \frac{1/\mu_1 + \rho_2/\mu_2}{1 - \rho_1} \qquad (7.36)$$

For the second queue, the determination of the expected sojourn time is more complicated. The expected waiting time $E(W_2)$ and the sojourn time $E(S_2) = q_2$ for a packet of type 2 is the sum of

- the residual service time $T_0 = \sum_{i=1}^{2} \rho_i \frac{1}{\mu_i}$ of the packet currently in the queue because the queue is non-preemptive,
- the service times $T_1 = E(L_1)/\mu_1$ for all packets of priority 1
- and the service times $T_2 = E(L_2)/\mu_2$ for all packets of priority 2 that are already present waiting in the queue at the point of arrival of the new packet of type 2 and are therefore served before it
- plus the service times $T_3 = \rho_1(T_0 + T_1 + T_2)$ for all packets of priority 1 that arrive during $T_0 + T_1 + T_2$ and that are served before the packet of type 2 because they are of higher priority.

The waiting time is $E(W_2) = T_0 + T_1 + T_2 + T_3$, for the sojourn time; the queueing delay service time has to be added $q_2 = E(S_2) = E(W_2) + 1/\mu_2$. By applying (7.33) and (7.34), we get

$$q_2 = E(S_2) = \frac{(1 + \rho_1) \sum_{i=1}^{2} \rho_i \frac{1}{\mu_i}}{(1 - \rho_1 - \rho_1\rho_2)(1 - \rho_1)} + \frac{1}{\mu_2} \qquad (7.37)$$

A packet of type 1 is not dropped as long as there are packets of type 2 waiting in the queue that could be dropped instead. With respect to loss, the arrival process 1 with arrival rate λ_1 thus experiences a normal $M/M/1/B$ queue with a loss probability for a packet of type 1 of

$$p_1 = \frac{1 - \rho_1}{1 - \rho_1^{B+1}} \cdot \rho_1^B \qquad (7.38)$$

We make the simplifying assumption that λ_1 is small enough so the loss for queue 1 is negligible $p_1 \approx 0$. For the low priority queue, the loss probability is then given by

$$p_2 = \frac{(1 - \rho_1 - \rho_2)}{1 - (\rho_1 + \rho_2)^{B+1}} \cdot (\rho_1 + \rho_2)^B \cdot \frac{\lambda_1 + \lambda_2}{\lambda_2} \qquad (7.39)$$

The first part of (7.39) represents the total loss of the queueing system; the second part $\frac{\lambda_1 + \lambda_2}{\lambda_2}$ is necessary because the packets of type 2 experience the complete loss.

The priority queueing based QoS network model is summarised in Model 7.2. Like the BE network model, it is a non-linear equation system.

7.2.4 Utility Functions

Before we compare the performance of the BE and QoS network models, we have to address the question as to which performance metrics is to be used. From the Models 7.1 and 7.2, it follows that the loss probability and queueing delay for inelastic flows are strictly smaller in the QoS model while for the elastic flows they are smaller in the BE model.

We now introduce utility functions for both types of traffic that transform the technical parameters loss and delay into a utility value.

7.2.4.1 Inelastic Traffic

The inelastic traffic represents multimedia or other real-time traffic that is sensitive to loss and delay. Therefore, the utility u_1 of the inelastic flows is modelled as strictly decreasing function of the loss probability p_1 and the deviation of the delay q_1 from a reference queueing delay q_{ref}:

$$u_1 = 1 - \alpha_p p_1 - \alpha_q \frac{q_1 - q_{ref}}{q_{ref}} \qquad (7.40)$$

As a reference queueing delay q_{ref}, we use the queueing delay (7.44) of the QoS network model as that is the minimum queueing delay achievable for this traffic under the given circumstances (number of flows, link capacity, non-preemptive service discipline, etc.).

Please note that because $p_1 \approx 0$ for the QoS model, $u_1 = 1$ when the QoS model is used.

7.2.4.2 Elastic Traffic

The elastic traffic represents file transfer traffic. The utility of this traffic depends mostly on the throughput as that determines duration of the transfer. The utility u_2 is therefore modelled as a function of the throughput d_2:

$$u_2 = \beta \cdot d_2 = \beta \cdot \frac{t}{q_2 \cdot \sqrt{p_2}} \qquad (7.41)$$

We determine the parameter β so that $u_2 = 1$ for the maximum throughput that can be reached if $\lambda_1 = 0$; both network models lead to the same β if there is no inelastic traffic.

Model 7.2 QoS Network Model

Parameters

r_1	Total sending rate of the inelastic flows [pkts/s]
t	Flow size factor for the elastic flows [pkts]
μ_1	Service rate for the inelastic traffic [pkts/s]
μ_2	Service rate for the elastic traffic [pkts/s]
B	Queue length [pkts]

Variables

p_1	Loss probability of the inelastic flows
q_1	Queueing delay of the inelastic flows [s]
λ_1	Arrival rate of the aggregate of inelastic flows [pkts/s]
p_2	Loss probability of the elastic flows
q_2	Queueing delay of the elastic flows [s]
λ_2	Arrival rate of the aggregate of elastic flows [pkts/s]
ρ_1	Utilisation of the queue with inelastic flows
ρ_2	Utilisation of the queue with elastic flows

Equations

$$\rho_1 = \lambda_1/\mu_1 \tag{7.42}$$

$$\rho_2 = \lambda_2/\mu_2 \tag{7.43}$$

$$q_1 = \frac{1/\mu_1 + \rho_2/\mu_2}{1 - \rho_1} \tag{7.44}$$

$$q_2 = \frac{(1 + \rho_1) \sum_{i=1}^{2} \rho_i \frac{1}{\mu_i}}{(1 - \rho_1 - \rho_1\rho_2)(1 - \rho_1)} + \frac{1}{\mu_2} \tag{7.45}$$

$$p_1 = \frac{(1 - \rho_1)}{1 - \rho_1^{B+1}} \cdot \rho_1^B \approx 0 \tag{7.46}$$

$$p_2 = \frac{(1 - \rho_1 - \rho_2)}{1 - (\rho_1 + \rho_2)^{B+1}} \cdot (\rho_1 + \rho_2)^B \cdot \frac{\lambda_1 + \lambda_2}{\lambda_2} \tag{7.47}$$

$$\lambda_1 = r_1 \tag{7.48}$$

$$\lambda_2 = \frac{t}{q_2 \cdot \sqrt{p_2}} \cdot \frac{1}{1 - p_2} \tag{7.49}$$

7.2.5 Evaluation

The default parameter values we use for the following evaluation are depicted in Table 7.1. The effect of parameter variation is analysed later. The motivation behind the utility parameter α_p is that the utility of the inelastic flows should be zero for 10% losses (if there is no additional delay); for the parameter α_q the motivation is that the utility should be zero if the delay doubles compared to the minimal delay of the QoS system. β is chosen so that the utility of the elastic flow is 1 for the maximum throughput as explained in Section 7.2.4.2.

During the evaluation, we vary w_1, r_1 and t. For the choice of w_1, we assume that for the total utility evaluation, the inelastic flows are more important than the elastic flows because they are given priority over the elastic flows and it seems reasonable to expect users to also have a higher utility evaluation for one real-time multimedia flow (e.g. a phone call) than for a file transfer. An indication for that is the fact that the price per minute for a phone call nowadays is typically much higher than the price per minute for a dial-up Internet connection used for a file transfer.

To derive an anchor point for t, we arbitrarily determine a t_0 that leads to $\rho_1 = 20\%$ and to $\rho_2 = 60\%$ using the QoS network model. This represents a working point with $\lambda_1 = 0.2 \cdot \mu_1$ with a total utilisation of 80%. Every fourth packet is a multimedia packet, creating a typical situation where a QoS system would be considered. If t is increased to $t = 5t_0$ and λ_1 kept constant, then the proportion of multimedia packet to file transfer packet drops to 1:3.4 and for $t = 10t_0$ it drops to 1:3.8. At the same time, the aggressiveness of TCP against the inelastic flows increases in the BE network model as can be seen in the evaluation results below (e.g. Figure 7.5).

As evaluation metric we again use the *overprovisioning factor*; it is determined as follows:

- For a given r_1 and t, we determine the solution vector (p_1, q_1, p_2, q_2) of the QoS network Model 7.2.

Table 7.1 Default Parameter Values for the Evaluation

Parameter	Value
μ_1	1Mbps/(1500 bytes/pkt) = 83.3 pkts/s
μ_2	Same as μ_1
α_q	1
α_p	10
β	See Section 7.2.4.2
B	10 pkts
t	t_0, $5t_0$, $10t_0$
r_1	[0, ..., 40] pkts/s
w_1	[1, 2, 5]
w_2	1

- The utility values $u_1 = f(p_1, q_1)$ and $u_2 = f(p_2, q_2)$ and the weighted average utility U_{ref} are derived from the solution vector with $w_1, w_2 > 0$

$$U_{ref} = \frac{w_1 u_1(p_1, q_1) + w_2 u_2(p_2, q_2)}{w_1 + w_2} \tag{7.50}$$

- For the best-effort Model 7.1, we can now also derive the solution vector (p_1, q_1, p_2, q_2) and calculate the weighted average utility U_{BE}. Unless the parameters $\alpha_p, \alpha_q, w_1, w_2$ are set to extreme values[6], the utility of the BE system is smaller than that of the QoS system ceteris paribus: $U_{BE} < U_{ref}$.
  - The BE system based on Model 7.1 is overprovisioned by a factor OF. The bandwidth respectively service rates μ_1 and μ_2 are increased by that factor OF. Additionally, the buffer space B is increased by the same factor:

$$\mu_i = OF \cdot \mu_i^{original} \tag{7.51}$$

$$B = OF \cdot B^{original} \tag{7.52}$$

  - U_{ref} is used as a reference value and OF is increased by a linear search algorithm until $U_{BE}(OF^*) = U_{ref}$.
  - OF^* is the overprovisioning factor and represents the resource increase in bandwidth and buffer space necessary for the BE system to perform as well as the QoS system w.r.t. the total utility U.

7.2.5.1 Basic Results

The overprovisioning factors OF for different flow size factors t and for different weight ratios $w_1 : w_2$ are depicted on the y-axis in the graphs of Figure 7.5. The total sending rate r_1 of the inelastic flows is shown on the x-axis.

As can be seen from all three graphs, the higher the ratio $w_1 : w_2$ is – that is, the more important the inelastic flows are for the overall utility evaluation – the higher the overprovisioning factor becomes. This can be expected, because for small overprovisioning factors the utility u_1 of the inelastic flows is smaller in the BE system than the QoS system where they are protected from the elastic flows because they experience more loss and delay. Thus, the higher u_1 is weighted in the total utility function U, the more bandwidth is needed in the BE system to compensate this effect.

Comparing the three graphs, it can be seen that as the flow size factor is increased more overprovisioning is needed. Increasing the flow size factor represents increasing the number of elastic (TCP) senders and the aggressiveness of the elastic flows. In the BE system where the inelastic flows are not protected, a higher flow size factor increases the sending rate of the elastic flows on cost of additional loss and delay for the

---

[6] Assuming $\lambda_1 = 10$, $U_{BE} < U_{ref}$ no longer holds true for example, if $w_2 > 4.58 \cdot w_1$ using the default α_i values or for $w_1 : w_2 = 2 : 1$ if the α_i are $\alpha_p < 0.05 \wedge \alpha_q < 0.005$. These values, however, are unrealistic and therefore not considered in our approach.

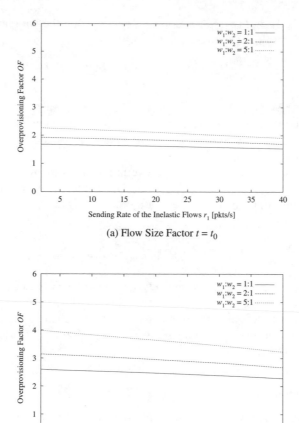

(a) Flow Size Factor $t = t_0$

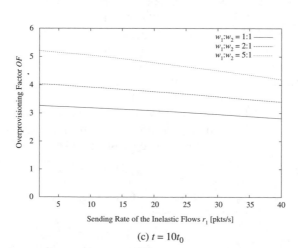

(b) $t = 5t_0$

(c) $t = 10t_0$

Figure 7.5   Overprovisioning Factors for the Configuration of Table 7.1

inelastic flows that in return has to be compensated by more capacity leading to a higher overprovisioning factor.

Keeping the flow size factor constant, with an increase of the sending rate r_1 the overprovisioning factor decreases; the decrease is stronger when the flow size factor is higher. For a weight ratio of $w_1 : w_2 = 2 : 1$, for example, the overprovisioning factor drops from $r_1 = 2$ to 40 by 12.0% for $t = t_0$, 14.9% for $t = 5t_0$ and 15.6% for $t = 10t_0$. This phenomenon can be explained in the following way: When comparing the resulting utility values u_1 and u_2 of the QoS system with the BE system ($OF = 1$), the utility value of the inelastic flows u_1 drops because they are no longer protected. At the same time, the utility value of the elastic flows u_2 increases because they no longer suffer the full loss.

The increase of u_2 is stronger than the decrease of u_1 the higher r_1 is, therefore for higher r_1 less overprovisioning is needed.

7.2.5.2 Modification of the Utility Functions

The following graphs – unless stated otherwise – are based on a weight ratio $w_1 : w_2 = 2 : 1$ and a flow size factor of $t = 5t_0$.

If we increase or decrease the utility function parameters α_p and α_q of the inelastic traffic, the overprovisioning factor changes as shown in Figure 7.6.

A decrease of α_p and α_q represents more loss in delay tolerance of the inelastic flows as their utility is decreasing more slowly if the loss in delay increases. The lower the utility decrease is, the less additional bandwidth is needed for the BE system as compensation; therefore, the overprovisioning factor is lower.

Arguing vice versa, a higher α_i leads to a higher overprovisioning factor.

7.2.5.3 Different Bottleneck Resources

Figure 7.7 shows the overprovisioning factors if the reference buffer space B of the systems is increased from $B = 10$ to $B = 20$ while the bandwidth is kept constant ($w_1 : w_2 = 2 : 1$, $t = 5t_0$, and $\alpha_p = 10$ respectively $\alpha_q = 1$).

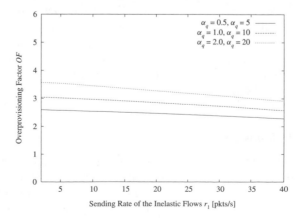

Figure 7.6 Overprovisioning Factors for Different Utility Parameters

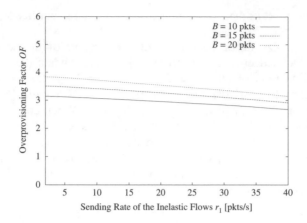

Figure 7.7 Overprovisioning Factors for Different Buffer Spaces

Increasing the buffer space B has two adverse effects; it decreases the loss rate and increases the potential queueing delay. As can be seen from the figure, an increase of B results in an increase of the overprovisioning factor OF. This is an indication that for the utility calculation, the queueing delay has a stronger effect than the loss rate. This is not surprising because for the $M/M/1/B$ formulas, the loss becomes quickly negligible for larger B.

To confirm this, we reduced the queueing delay effects by setting $\alpha_q = 0.05$ and repeated the experiment. Now, with an increase of B from 10 over 15 to 20 the adverse effect can be observed: the overprovisioning factor drops from 1.76 over 1.68 to 1.66 for $r_1 = 10$.

To conclude, the effect of the buffer size depends on the ratio of α_p to α_q in the utility function.

Next, the reference buffer space B and at the same time the bandwidth (the service rates μ_1 and μ_2) are doubled; r_1 was increased accordingly. Figure 7.8 shows the results.

Compared to Figure 7.7, the overprovisioning factors only increased insignificantly for $t = 5t_0$. In the BE system – as can be seen from (7.30) – for large B, the queueing delay q_{BE} becomes inverse proportional to the service rate $\overline{\mu}$ and therefore the bandwidth. For large B, the loss p_{BE} exponentially approaches zero as can be seen from (7.29). Via (7.32), this leads to a massive increase in the elastic rate λ_2 and overall utilisation ρ. This explains why the buffer space has a larger influence than the service rate. Similar arguments hold true for the QoS system.

7.2.5.4 Different Packet Sizes

Real-time multimedia traffic like voice or video traffic usually has significantly smaller packet sizes than file transfer traffic that are mostly Maximum Transmission Unit MTU sized. The effect of the smaller packet size can be represented in the models by increasing the average service rate μ_1 of the inelastic flows. Figure 7.9 shows the results for an

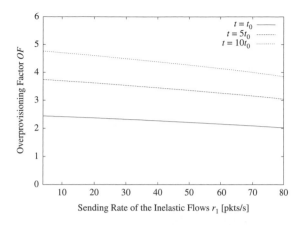

Figure 7.8   Overprovisioning Factors for an Increase in Bandwidth and Buffer Space

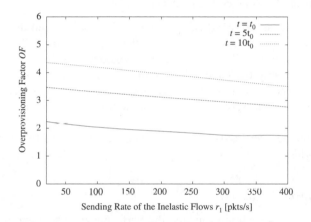

Figure 7.9   Overprovisioning Factors for Different Packet Sizes

decrease of a factor of 10 in the packet size for the inelastic flows compared to the default experiment of Figure 7.5. In this experiment, the sending rate r_1 was also increased by a factor of 10 to keep the average traffic volume constant.

As one can see, the difference in service rate increases the overprovisioning factors. This effect can be explained by the fact that the queueing theory based approach chosen in our models cannot handle different space requirements of the packets. The buffer space is limited to B packets irrespective of their type or size in our models. As the number of inelastic packets now significantly increases, the loss increases, too, and is compensated only by a further increase in bandwidth and buffer space that leads to higher overprovisioning factors. In the basic experiment of Section 7.5, the loss rate p_2 for $\lambda_1 = 10$ was 2.79%. In this experiment, for a comparable value of $\lambda_1 = 100$ the loss rate p_2 is 5.25% which confirms our explanation.

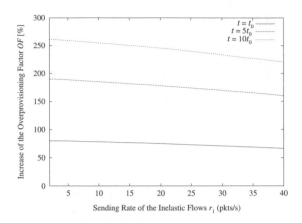

Figure 7.10   Isolation of the Service Rate Effect

7.2.5.5 Isolation of the Service Rate Effect

In the experiments so far, the bandwidth of the bottleneck link and the buffer space were overprovisioned equally. We now try to answer the question, what effect overprovisioning bandwidth alone has. Figure 7.10 depicts relative increase of the overprovisioning factor if for the BE system only the bandwidth – represented by the service rates μ_1 and μ_2 – but not the buffer space B is multiplied with the overprovisioning factor OF.

As we can see from the results, 60 to 200% additional bandwidth is needed to compensate the now missing buffer space. As a result, when overprovisioning a network the buffer space should be overprovisioned, too, unless it is significantly more expensive than additional bandwidth.

7.2.6 Summary and Conclusions

The experiments of this section evaluated the performance advantage of a priority based QoS system over plain BE system. The systems have two resources: buffer and bandwidth. We used two types of traffic – elastic and inelastic traffic – that share a bottleneck link. The evaluation is based on the aggregated utility function. Our results are overprovisioning factors. They show how much the resources of the BE system that cannot differentiate between the traffic classes have to be increased to offer the same total utility that the QoS system provides.

Compared to the approach in the previous Section 7.1, the overprovisioning factors of the models in this section are generally higher. This is explained by the fact that the models of Section 7.1 do not consider different traffic types sharing the bottleneck resources. Therefore, they miss one very important aspect of QoS systems: the service differentiation between traffic classes.

In today's Internet, the overwhelming part of the traffic is TCP based file transfer traffic, especially peer-to-peer and web traffic, see Chapter 5. In the beginning, when real-time multimedia applications spread, their initial share of traffic will be low. In our models this can be represented by rather low sending rates r_1 (few inelastic flows), and a high flow size factor t (many elastic flows). Unfortunately, our results show that especially for this combination, the overprovisioning factors are the highest. Therefore, to support the *emerging* real-time traffic applications, QoS architectures have their greatest advantages.

The two approaches in this chapter have their limitations because they are based on analytical models that by nature only allow a certain degree of complexity to be still solvable. Our analysis is based on a single bottleneck link; the influence of the network topology has been neglected so far. We turn to simulations in the next chapter to shed more light on the question, how different QoS approaches perform. The simulations allow us to analyse more complex topologies and to employ more sophisticated traffic models.

8

Experimental Comparison of Quality of Service Systems

In the previous chapter we investigated with the help of analytical methods the potential benefit of a Quality of Service (QoS) system over a plain Best-effort (BE) system. In the analytical approaches, a single bottleneck was assumed. Also, the QoS systems are modelled in an abstract way (e.g. with strict priority queueing in Section 7.2). To work out the differences between real QoS systems (e.g. Intserv and different Diffserv systems) that use more sophisticated admission control and scheduling algorithms, actual implementations of the systems should be used. We do so in this chapter, using packet-level, event-based simulations. The following *QoS systems* based on the main Internet Engineering task Force (IETF) architectures were implemented and used for these simulations:

- **Integrated Services (Intserv)**
  The Intserv QoS architecture was presented and discussed in Section 6.2.2. Intserv guaranteed service (GS) allows deterministic loss and delay guarantees. In that sense, it is the 'strongest' service we are investigating.
  The Stateless Core (SCORE) architecture with Dynamic Packet State (DPS) (see Section 6.2.3) can be used to offer a scalable GS; it therefore leads to results very similar to those of Intserv and can be evaluated on the basis of the Intserv results in this chapter.
- **'Standard' Differentiated Services (Diffserv)**
  The Diffserv QoS architecture was discussed in Section 6.2.4. We name the Diffserv systems that use the expedited and Assured Forwarding (AF) behaviour from RFC 2597 (see Heinanen *et al.* (1999)) and RFC 2598 (see Jacobson *et al.* (1999)) 'standard' Diffserv.
  For resource management and admission control in the Diffserv systems, we consider three different types of bandwidth brokers (BBs):
  o Centralised Bandwidth Broker
    The centralised BB has full knowledge of the routing by keeping track of the paths that the different flows take through the network. We designed and implemented a very sophisticated centralised BB that can also guarantee the delay bounds for

The Competitive Internet Service Provider: Network Architecture, Interconnection, Traffic Engineering and Network Design
Oliver Heckmann   © 2006 John Wiley & Sons, Ltd

admitted flows, thus mimicking the Intserv GS behaviour while still maintaining the low Diffserv per-class scheduling complexity. To increase the efficiency of the Diffserv system, we allow relaxing the service guarantees to stochastic guarantees and investigate overbooking of the service classes.

○ Decentralised Bandwidth Broker

The decentralised BB is a simplified version of the central one. It uses a decentralised admission control algorithm that is based on information *locally* available at the ingress node. Thus, it is easier to implement and maintain than the centralised broker, but it is less efficient. In addition, it cannot give delay bound guarantees along a path.

○ No BB/No Admission Control

A Diffserv network can also be operated without admission control if it is well dimensioned and relying on mid-term and long-term traffic and network engineering. These methods are discussed in Part IV of this book. In our experiments, a system without BB and other admission control mechanism is therefore included as reference.

• Olympic Differentiated Services

Contrary to the 'standard' Diffserv systems, Olympic Diffserv systems are based on a very low number of Per Hop Behaviours (PHBs) (in our case three) that are differentiated by strict priority queueing. The three services built on these PHBs are called *gold, silver* and *bronze*, hence the name 'Olympic'[1].

We use the same BB types that we use for standard Diffserv with adaptation to the Olympic service scheme.

• Overprovisioned Best-Effort

As the QoS of a system can be expected to be satisfying if it is dimensioned well enough, we use plain BE networks that are overprovisioned with different overprovisioning factors (similar to the previous chapter) as reference. This allows us to determine overprovisioning factors and compare the results with the analytical results of the previous chapter.

As defined in Section 6.2, a *QoS system* consists of the *QoS architecture* that describes the general technical foundation of the QoS system and the *QoS strategy* that determines how an Internet Network Service Provider (INSP) exploits the technical features offered by the chosen architecture. The strategy includes the configuration of the architecture.

In the experiments of this chapter, we show how different QoS systems perform when facing a certain traffic mix and a certain network topology. The *performance* is evaluated by technical criteria like the dropping probability or the throughput and by application-specific utility functions. Utility functions are important because different applications of the traffic mix have different QoS requirements. For TCP-based file transfer applications, the utility largely depends on the overall throughput as they can recover from losses and delay variations (jitter) to a certain extent. For multimedia applications that are – at the timescale of the experiment – not rate adaptive, the loss and the delay will typically be more important. Utility functions are therefore necessary to evaluate the benefit a *user* has if a certain QoS system is used.

We developed and implemented an *experimentation environment* on top of the packet-level network simulator NS2 (see NS2 (2004)). NS2 is commonly used for QoS

---

[1] Please note that the term 'Olympic' in the context of Diffserv services is in other works sometimes used for a cascade of AF services, see Heinanen *et al.* (1999).

experiments. For an experiment, a certain traffic mix plus a network topology is used as input. The experiment is conducted in several steps, in each step a different QoS system is used and a complete packet-level simulation is performed. All steps use exactly the same traffic, allowing us to directly compare their results.

We consider different *traffic* mixes that consist of different types of traffic, for example, Constant bit-rate (CBR) and Variable bit-rate (VBR) traffic. For our experiments, we considered using traffic sessions or direct individual flows as traffic input. A *session* consists of a number of closely related and interdependent flows. For example, a World Wide Web (WWW) session could represent a series of webpages[2] a user is reading with short variable reading times after each page is downloaded. It can be represented as a series of Transmission Control Protocol (TCP)/Hypertext Transfer Protocol (HTTP) flows, each transferring a potentially different amount of data. For this example, in an experiment that uses traffic session semantic as traffic input, a flow would not start until the previous flow of the same session is finished plus possibly a certain variable 'reading' time. Because the starting times of flows depend on the network condition, it is not possible to generate the traffic flows off-line. If traffic is modelled on the session layer, the application behaviour can be modelled more realistically. The traffic emulator[3] GenSyn (see Heegaard (2000)) is an example for a session-based traffic emulator. It models user behaviour with different state machines for different application types (WWW, File Transfer Protocol (FTP), video streaming, voice over Internet Protocol (VoIP), etc).

Alternatively, the individual *flows* could be specified directly and used as traffic input. They can be generated off-line from session models. However, as the network conditions (loss rates, delays, etc.) are not known in advance, certain aspects of the application/user behaviour will then not be modelled as nicely as when using sessions as input with online flow generation.

For the purpose of our experiments, however, using flows instead of sessions has one crucial advantage in that it allows a direct comparison: If flows are specified and used as input, the amount of load 'offered' to the network remains constant in each step of an experiment   that means for each evaluated QoS system. If sessions would be used where a second flow is only started once the first is finished, a QoS system offering poor throughput performance for the first flow would in fact be 'rewarded' with less traffic as the second flow would start delayed or not at all. This would not only seem unfair, it also makes the direct comparison of technical parameters like loss and throughput impossible because large variations in the network load would occur. The overall evaluation would then only be possible based on 'session' utility functions that evaluate the overall utility of a session. We want to avoid this for the following reasons: Utility functions that evaluate the performance of a single flow can be based directly on the technical parameters like loss and delay of the flow. Few assumptions have to be made for these 'flow' utility functions (see Section 8.2.3.2). For the higher-level 'session' utility functions, however, more assumptions are necessary and therefore more subjectivity would be introduced.

Because of these reasons, we chose to use the session concept for off-line flow generation and use flows as input for the simulations and as a basis for the evaluation; the

---

[2] Each consisting of a Hypertext Markup Language (HTML) file plus possibly some graphics.
[3] We use the term traffic emulator for software/hardware that generates artificial traffic for a physical network and traffic simulator for software that generates traffic for simulations.

evaluation can thus be based on flow utility functions backed by the technical parameters as 'hard' facts.

We start with describing the technical details of the admission control mechanisms and implementation details for the QoS systems. Then, Section 8.2 sheds light on the experiment set-up. A fairly sophisticated experimentation environment is used to run the experiment with the same traffic flows using different QoS systems; this approach allows us to directly compare the results obtained from the simulations. Section 8.2 also describes the experimentation and evaluation parameters, for example, the chosen topologies, traffic mixes and utility functions. Finally, the different *experiments* and their results are presented as follows.

- In the first set of experiments (Section 8.3), the QoS systems that can give loss and delay-bound guarantees are compared: the Intserv system using per-flow scheduling and the Diffserv systems with the centralised BB and per-class scheduling.
  The experiments shed light on the trade-off between additional data-path complexity and more efficient resource allocations. In addition, it sheds light on the overbooking potential of the Expedited Forwarding (EF) service class when using the central BB for stochastic service guarantees.
- For Diffserv systems, a decentralised admission control decision promises less computational complexity and communication overhead. However, as it has no control of the interior of the network, the risk of service disruptions (packet drops, delay-bound violations) increases. This effect is investigated in Section 8.4.
- In the direct comparison experiments of Section 8.5, the QoS systems that performed best in the previous experiments are pitted against each other directly. Different traffic mixes and topologies are evaluated. These experiments display and quantify the individual strengths and weaknesses of the QoS systems. In addition, we determine the range of overprovisioning factors for the QoS systems and compare them with the analytical results of the previous chapter.

This chapter concludes with a summary and conclusion.

8.1 QoS Systems

First, we describe the implementations of the admission control mechanisms for the QoS systems. While the design space of admission control mechanisms for Intserv is limited by the according Request for Comments (RFCs), there are almost no restrictions for admission control in Diffserv. The central BB we specify below for Diffserv is able to give very strong guarantees on one side and allows for overbooking and efficient network usage on the other side. The admission control algorithms introduced in this section were implemented for the experiments of this chapter and those in Chapter 13 (Section 13.1).

8.1.1 Intserv/RSVP QoS Systems

For our experiments, we use the traditional Intserv/Resource Reservation Protocol (RSVP) QoS architecture as discussed in Section 6.2.2. We use Intserv/RSVP as reference for the 'strongest' service, the GS, as it is a deterministic service with per-flow guarantees; therefore, we focus on GS within the Intserv/RSVP architecture. The controlled load service is not evaluated, as it does not promise significant advantages over the various Diffserv services.

As it is also possible to provide the same GS service guarantees with a core stateless architecture, the performance of a core stateless architecture like DPS (see Section 6.2.3) can be evaluated on the basis of our results for Intserv/RSVP.

8.1.1.1 Admission Control

The Intserv/RSVP admission control is to a large extent specified in RFCs (e.g. RFC 2212 for GS). In terms of the classification of Section 6.6, it is a hop-by-hop network-based admission control system with deterministic guarantees based on worst-case descriptions of the flow and networking behaviour. The traffic description uses a TSpec; the allocated network resources are buffer and bandwidth. The basic granularity is fine (microflows) although approaches exist for aggregation of flows. Intserv/RSVP has explicit support for multicast. Our implementation is non-preemptive, does not support reservation in advance and no end-time is specified by a flow during the reservation, as these points are also not mentioned in RFC 2212.

The Intserv per-flow admission control is used for GS flows based on the token bucket descriptor (r_f, b_f) of the arrival curve[4]. The Intserv/RSVP reservation process allows the explicit declaration of a queueing delay bound; it influences the amount of resources that have to be allocated for a flow. Our admission control manages two resources for an outgoing link l at a router: the available bandwidth bw_l and the buffer space bf_l. For all our analysed QoS systems, these resources were set to equal values to allow a fair comparison.

For a GS flow f with a queueing delay bound d_f^q, the admission control has to allocate the rate R_f and the buffer space B_f for each link l along the path P (see Section 6.2.2.4):

$$R_f = \max \left\{ \frac{b_f + \sum_{l \in P} C_{fl}}{d_f^q - \sum_{l \in P} D_l}, r_f \right\} \tag{8.1}$$

$$B_f = b_f + \sum_{l \in P} C_{fl} + \sum_{l \in P} D_l \cdot R_f \tag{8.2}$$

We do not need to make use of the slack term S of RFC 2212 (see Shenker *et al.* (1997)). C_{fl} and D_l are the scheduling error terms of flow f on link l. Set ϑ_l contains all other currently accepted and active GS flows passing through link l. A flow f is only admitted if R_f and B_f can be allocated for each link l of the path P and do not exceed a given maximal share α_{GS} of that link's bandwidth bw_l and buffer resources bf_l:

$$R_f + \sum_{g \in \vartheta_l} R_g \leq \alpha_{GS} \cdot bw_l \qquad \forall l \in P \tag{8.3}$$

---

[4] We simplified the TSpec to a token bucket. A small additional efficiency gain can be achieved by using the TSpec as basis for the admission control algorithm, see Section 6.2.2.4.

$$B_f + \sum_{g \in \vartheta_l} B_g \leq \alpha_{GS} \cdot bf_l \qquad \forall l \in P \tag{8.4}$$

As mentioned above, we do not use the Intserv *Controlled Load* service class (see Wroclawski (1997)). BE flows are not admission controlled at all in the Intserv system.

8.1.1.2 Scheduling

Weighted Fair Queueing (WFQ), (see Demers *et al.* (1989)) is used for the scheduling of the flows. WFQ has the following scheduling error terms

$$C_{fl} = \text{maximum packet size of flow } f \tag{8.5}$$

$$D_l = \frac{MTU}{bw_l} \tag{8.6}$$

In Intserv, per-flow scheduling is used for all GS flows (contrary to the Diffserv per service class scheduling); the WFQ weight w_{fl} assigned to a GS flow f on link l is

$$w_{fl} = R_f / bw_l \tag{8.7}$$

All BE flows share a single queue that is assigned the remaining weight

$$w_{BEl} = 1 - \sum_{g \in \vartheta_l} R_g / bw_l \tag{8.8}$$

8.1.2 Standard Diffserv QoS Systems

We name the Diffserv approach with EF/AF PHB *'standard' Diffserv*. As described in Section 6.2.4, Diffserv is more of a QoS system framework than an exact specification of a certain QoS system, so there cannot be a real 'standard' Diffserv. However, the EF/AF PHBs are up to now the only PHBs in the standardisation process of the IETF and the ones most commonly found in Diffserv related works, which justifies our choice of name.

In this set-up, we proceed according to the RFCs; they prescribe two PHBs:

- Expedited Forwarding (EF) and
- Assured Forwarding (AF).

The EF PHB is intended for traffic with low delay requirements. We refrain from using all three drop precedences from Heinanen *et al.* (1999) to keep the complexity of the experiments low. Further, preliminary experiments showed that their influence on the results of the entire system is negligible for the purpose of our evaluation.

A key issue is whether and what type of admission control is conducted. We evaluate three different types of 'standard' Diffserv QoS systems that differ in their *admission control bandwidth broker*. A BB is *an entity that manages and configures the network devices of a Diffserv domain and keeps state in terms of how loaded the network is and whether a new flow is admissible.* The three different types are as follows.

- **Centralised (global) Bandwidth Broker**
  Please note that the goal of the BB is to show the 'best-you-can-do' approach; this is why it checks and guarantees the delay bounds for individual flows throughout their complete network path.

The global BB checks the entire path throughout the network before admitting a flow. Consequently, it has to keep state about the routes of the network as well as the load throughout the network. It has to find out which routes the new flow will take through the network and check resource availability along each hop.

Additionally, the global BB keeps track of the resource allocations of the individual flows that make up one forwarding class. This allows the BB to check whether the delay bounds of the flows can be guaranteed as we demonstrate below. As our experiments will show, it is possible to reduce the amount of state of this bandwidth broker on the control path without disrupting the service.

In terms of the classification of Section 6.6, the central BB has a centralised network-based admission control system based on worst-case descriptions of the flow and network behaviour that gives deterministic stochastic guarantees.

- **Decentralised (local) Bandwidth Broker**
  We define a local BB as one that operates on each edge node and checks only whether this edge node has the capacity to admit the flows. This is a low complex operation, not much state has to be kept.

  In terms of the classification of Section 6.6, the decentral BB is also network-based but located at the edge; more specifically at the ingress node. It uses a contingent-based algorithm based on worst-case behaviour. It cannot give better than stochastic guarantees.

- **No Bandwidth Broker and no Admission Control**
  The easiest solution is, of course, to refrain from using a bandwidth broker and admission control and rely on a well-dimensioned network.

8.1.2.1 Centralised Bandwidth Broker

Admission Control   We assume that the centralised BB has perfect knowledge of the network state at each point in time: It knows all routes through the Diffserv domain and keeps track of the aggregate bandwidth and buffer allocations of each link. It knows which route a newly arriving flow will take through the Diffserv domain. Such a central bandwidth broker is complex to develop and maintain for a large network but represents the 'best-you-can-do' approach in a Diffserv network.

The knowledge of the Diffserv central BB allows it to also check whether it is possible to guarantee delay bounds for EF flows and in this aspect mimic the service guarantees of Intserv guaranteed service.

Because the individual flows that are merged into a single Diffserv class are not protected against each other inside that class, the resource management in the Diffserv network is less efficient than for Intserv. However, this leads to less complexity on the data path, which usually is more important.

Before a new flow f can be admitted, the BB has to check the availability of bandwidth and buffer space along the path P_f of the flow through the network. In addition, the bandwidth broker has to check whether the delay bound of that flow can be guaranteed or not.

Because the flows inside a class are not protected against each other, admitting a new flow to a Diffserv service class C can degrade the quality of the other flows in that class. Therefore, before admitting a new flow f, it has to be checked whether the delay bound of all flows already admitted to the Diffserv service class C that share at least one hop with the new flow f can still be guaranteed after admitting the new flow. The advantage of our BB approach is that it keeps track of the path a flow takes through the network and that it can thus determine easily which other flows the admittance of the new flow would affect. Without the path information, a worst-case assumption would have to be made about how the flows affect each other, leading to a lower admittance quota.

In order to fulfil these tasks, the admission control of the central BB works in three steps when a new flow with token bucket arrival curve (r_f, b_f) and path P through the Diffserv domain requests admittance to service class C:

1. For service class C, a proportion α_C of the link bandwidth bw_l of each link l is assigned off-line. Service class C is overbooked with an overbooking factor ob_C (see below). Let ϑ_l be the set of all currently active flows passing through link l. The new flow f is only admitted to the network if the *bandwidth* limit on each link along its path is not exceeded:

$$r_f + \sum_{g \in \vartheta_l \wedge g \in C} r_g \leq \alpha_C \cdot ob_C \cdot bw_l \qquad \forall l \in P \qquad (8.9)$$

2. Similarly, the availability of *buffer space* bf_l has to be checked. The new flow f is only admitted to the network if the buffer limit on each link along its path is not exceeded:

$$\beta \cdot (b_f + \sum_{g \in \vartheta_l \wedge g \in C} b_g) \leq \alpha_C \cdot ob_C \cdot bf_l \qquad \forall l \in P \qquad (8.10)$$

The problem with the buffer space management is that flows entering the network can become more bursty as they share transmission capacities with other flows. The same holds true for protected Intserv flows (RFC 2212, see Shenker *et al.* (1997)) and is expressed by the error terms in (8.2).

For the Diffserv central admission control, we have to take into account that – contrary to Intserv – the burstiness of the flows sharing a class mutually influences each other. We introduce the error factor β that captures the increase in burstiness of the flows. For *feed-forward networks*, the burstiness can be calculated exactly (see Le Boudec and Thiran (2001)) but not for arbitrary network topologies. Feed-forward networks are networks in which routes do not create cycles of interdependent packet flows. A typical example for feed-forward networks are access networks; for these networks the central bandwidth broker can thus directly give the same *deterministic* service guarantees that Intserv/RSVP or SCORE architectures with DPS can give.

For arbitrary non-feed-forward networks, the Charny bound (see Section 6.2.4.2) could be used as a delay bound. However, it does not use the full information that is available to our central BB (e.g. the paths of the microflows through the network) and is therefore not efficient in our context. It leads to very low link utilisations for networks of medium to large diameters, as shown in Figure 6.7.

Exploiting the knowledge about the routing of microflows for non-feed-forward topologies is generally very complex, see, for example, Charny and Le Boudec (2000);

Starobinski *et al.* (2002) and the works cited therein. One possible approach is to use the turn-prohibition algorithm from Starobinski *et al.* (2002) to change the routing in an arbitrary topology to avoid cycles so that the feed-forward properties hold true for that network and the traffic flows in the network. In that case, deterministic guarantees can be given, see above. A similar approach is used in Fidler (2003). The drawback of that approach is that it influences the routing by extending the length of some paths (causing additional delay), depends on an explicit routing mechanism and limits as well as complicates traffic engineering and load balancing. For the purpose of these experiments, it also would introduce a bias towards this special Diffserv system because the routing would be either optimised specifically for this system in all experiments or different in the Diffserv experiments.

The goal of the turn-prohibition routing is to make the network calculus apply to general topologies. This makes it possible to give relatively efficient deterministic guarantees for general topologies with aggregate scheduling. However, these guarantees are only deterministic within the mathematical models themselves and do not take possible failure reasons outside these models like link failures, router misconfigurations, or packet losses and delays caused by routing changes into account. Because of that, a provider would normally be allowed a limited amount of guarantee violations in a Service Level Agreement (SLA) anyway; even the offered service is a 'deterministic' one.

Additionally, our experiments in Section 8.3 show that the EF service can be overbooked quite massively, especially for realistic topologies. Therefore, it can be assumed that most providers overbook the EF class to a certain extent to efficiently use their network. Then, there is no need to determine the worst-case burstiness exactly, especially if it complicates routing and traffic engineering. In these cases, we make the simplifying feed-forward assumption to determine a base value for the error term β. If delay bound violations or packet drops are observed if the class is not overbooked, we increase β until they disappear. Throughout all experiments in Chapter 8, this was necessary only in very extreme experiment set-ups. β is never set to a value below the feed-forward value. Concluding, we adjust the error introduced by applying the feed-forward formulas to non-feed-forward networks with the error term β. This leads to the admission control being based on a statistically relaxed deterministic model controlled by measurements.

3. The *delay bounds* are only checked for the premium service class based on the EF PHB. A flow is only admitted if its delay bounds can be guaranteed.
The delay bounds of the new flow f and all already admitted flows of service class C that share at least one hop with flow f have to be checked as follows.
The maximum queueing delay d_{fl}^q for flow f on link l is

$$d_{fl}^q = \frac{\sum_{g \in \vartheta_l \wedge g \in C} \beta \cdot b_g}{R_l} + \frac{C_{fl}}{R_l} + D_l \qquad (8.11)$$

where R_l is the link bandwidth $R_l = bw_l$ and C_{fl} and D_l are the scheduling error terms of link l; the rate-dependent term C_{fl} typically also depends on the maximum packet size of flow f.

For each flow, the maximum queueing delays d_{fl}^q along the path have to be added, the propagation delay d_l^p has to be taken into account, and the result has to be compared with the absolute delay bound D_f of that flow:

$$\sum_{l \in P} d_{fl}^q + \sum_{l \in P} d_l^p \leq D_f \qquad (8.12)$$

Please note that the 'pay-burst only once' property holds true in networks where flows are protected against each other (e.g. Intserv) but it does not hold true in Diffserv networks. Therefore, the delay bound check in a Diffserv network is much more conservative than in a comparable Intserv network. Our experiments in Section 8.3 demonstrate that.

Implementation Issues   From the complexity with respect to control/admission control *information*, the Diffserv central BB is roughly as complex as an Intserv/RSVP implementation. This, however, is not surprising as the Diffserv central BB represents the 'best-you-can-do' approach. The advantage of the Diffserv central BB implementation over an Intserv implementation is that the complexity is located at one point and not distributed amongst the routers. It can thus be handled by a dedicated machine that – contrary to a router – does not have to perform other time-critical tasks as well. Moreover, a provider offering premium services will typically have to use a centralised system for authentication and accounting anyway, which also has to be involved in the admission control process.

The third step of the admission control decision above is the most problematic operation the central BB has to make: For one arriving flow, a possibly large number of other flows have to be analysed with respect to their delay bound. The actual implementation of this mechanism, however, offers a great deal of optimisation potential. For each flow, for example, it could be noted by how much slack Δb_g the flow has until its delay bound is violated. For each newly admitted flow it is sharing a link with, that slack would be reduced accordingly. New flows that would make the slack negative have to be rejected.

In addition, our experiments show that the Diffserv QoS systems can be overbooked significantly before anything goes wrong. Because of that potential, it is not necessary to perform the admission control decision for each flow in real-time and fully exact. It could, for example, be replaced in many cases with simpler heuristics because the risk of wrong decisions is very small.

Scheduling   Pseudo-priority queueing is used as first scheduling discipline with Weighted Round Robin (WRR) as the implementing scheduler: The EF packets obtain a higher non-preemptive priority than the AF packets by assigning the EF queue in each hop a very large weight w_{EF}. Because the error terms of WRR depend on the number of service classes, WRR is generally not the most preferable scheduler. In our case, however, this drawback does not weigh very much because the number of service classes is very small for the Diffserv QoS systems (four classes). Moreover, NS2 contains a working and tested Diffserv WRR implementation. Another reason is that WRR is also used in related experiments, for example, those in the original EF PHB RFC (RFC 2598, see Jacobson *et al.* (1999)). The WRR weights are shown in Table 8.1.

Table 8.1   Default Scheduling and Configuration

PHB	EF	AF-1	AF-2	AF-3
WRR Weight w_C	1500	1	1	1
Admission control parameter α_C	0.5	0.25	0.25	n.a.
Overbooking factor ob_C	varies	1.0	2.0	n.a.

We used three AF service classes (AF-1 to AF-3); AF-3 being used as a BE service class upon which no admission control is exerted. The bandwidth, remaining after the pseudo-prioritised EF traffic is served, is by default split up 1:1:1 among the AF service classes.

The parameter α_{EF} for EF traffic of 0.5 was based on prior calibration experiments; because EF traffic is treated with priority, we limited its basic resources to 50% of the available resources to avoid starvation of the other classes. α_C is kept constant throughout the experiments but the resources allocatable to EF traffic are varied through the experiments using the EF overbooking factor ob_{EF}.

The scheduling error term C_{fl} for our WRR implementation is the maximum packet size of the flow f and, with n_l denoting the number of queues of link l ($n_l = 4$), the error term $D_l = (n_l - 1)\frac{MTU}{bw_l}$.

Overbooking   To differentiate the QoS in the different AF classes, the overbooking factor ob_C is introduced for each class C. The AF-1 class is not overbooked while the AF-2 class is overbooked by 100%. Therefore, more traffic is admitted to the AF-2 class than it can theoretically handle. The quality of AF-2 therefore should be lower than that of AF-1.

The EF class is overbooked with varying overbooking factors.

Random Early Detection   Active queue management algorithms like Random Early Detection (RED, see Floyd and Jacobson (1993)) are often used in conjunction with the three different levels of drop precedences of the AF services. In our experiments, we did not activate RED or a similar algorithm (see Section 6.2) for the Diffserv queues, as we do not use active queue management and different levels of drop precedences for the other QoS systems. We do not want to give Diffserv an unfair advantage and we do not want to mix the effect of active queue management with our comparison of QoS systems.

Policing   For the EF traffic, we police strictly at the ingress nodes dropping out-of-profile packets. As there are no misbehaving flows in our experiments and as the token buckets in our traffic specification are dimensioned large enough (see Table 8.2), there were no out-of-profile EF packets.

For the AF-1 and AF-2 traffic, out-of-profile packets are put into the same physical queue as the in-profile packets to avoid packet reordering. However, out-of-profile packets are dropped with a higher probability than in-profile packets. Out-of-profile packets are always dropped if the queue is filled by 80% or more while in-profile packets are only dropped when the queue is completely full.

AF-3 packets are not policed.

8.1.2.2 Decentral Bandwidth Broker

The difference to the central BB approach is that the decentral BB does not need complete knowledge of the network. The admission control decision is based only on local knowledge at the ingress node and not on knowledge about traffic in the whole domain. Each ingress link l is assigned a certain contingent of bandwidth Γ_{lC}^{bw} and buffer Γ_{lC}^{bf}. This assignment is done prior to the experiment. The BB admits EF, AF-1 and AF-2 flow only up to this limit. The decentral algorithm does not keep track of the state in the network, therefore the delay bound constraint (8.12) cannot be checked for the flows.

For comparison, we set the contingent proportional to the maximum admissible amount of bandwidth and buffer for that link l in the central BB approach multiplied with a scaling factor $\gamma \leq 1$:

$$\Gamma_{lC}^{bw} = \gamma \cdot \alpha_C \cdot ob_C \cdot bw_l \tag{8.13}$$

$$\Gamma_{lC}^{bf} = \gamma \cdot \alpha_C \cdot ob_C \cdot bf_l \tag{8.14}$$

8.1.2.3 No Admission Control

The performance of a Diffserv network that does not use per-flow admission control is also evaluated. It relies on other methods to (roughly) control the traffic to bandwidth ratio, for example, on long-term service-level agreements or on network-engineering methods. For the purpose of these experiments, all flows are accepted and assigned their initially requested Differentiated Services Codepoint (DSCP). Policing is not used.

8.1.3 Olympic Diffserv

The Olympic service Diffserv approach uses strict priority queueing with three priority classes implemented by a simple non-preemptive priority scheduler. The same BB/admission control approaches (central, decentral, none) as in Section 8.1.2 are used. Admission control is imposed on the gold service in the same way as for the premium service (EF PHB) of the standard Diffserv approach described in Section 8.1.2. Policing for gold service is also the same as in the standard Diffserv approach.

All flows requesting silver or bronze service are admitted to the network without admission control and policing.

8.1.4 Overprovisioned Best-Effort

Overprovisioned BE networks with different overprovisioning factors are used. The overprovisioning factor OF describes how the bandwidth bw_l of each link l is increased

$$bw_l^{BE} = OF \cdot bw_l \qquad \forall l \tag{8.15}$$

All links of a network are increased by the same overprovisioning factor OF in the experiments.

8.2 Experiment Setup

For the experiments presented in this chapter, we used NS2. Among other things, it contains a load generation module that allows repeating an experiment in different contexts – in our case, with different QoS systems. For one experiment, traffic flows are generated off-line, and the experiment is then repeated with the different QoS systems and evaluated. The results like loss rate, acceptance rate and utility can therefore be compared directly. The NS2 simulator (see NS2 (2004)) is used to conduct the packet-level simulations.

8.2.1 Traffic

8.2.1.1 Traffic Types

For the experiments, we use different traffic mixes consisting of two types of elastic and two types of inelastic sessions. Elastic sessions produce elastic TCP-based traffic flows that react to congestion in the network (indicated by packet loss[5]) by reducing their rate. We use the following two elastic traffic types.

- Short-lived TCP flows ('s_TCP') resemble small file transfers like most WWW traffic. Short-lived TCP flows rarely spend much time in the TCP congestion-avoidance phase.
- Long-lived TCP flows ('l_TCP') resemble larger file transfers, for example, peer-to-peer traffic. Long-lived TCP flows spend much time in the TCP congestion-avoidance phase. Because of the relatively small duration of our individual experiments (a couple of minutes simulation time, at most), the size of the long-lived TCP flows can be kept relatively small, too.

Inelastic sessions consist of inelastic flows that do not adjust their rate to the network condition. We use constant and VBR flows as inelastic flows and assume that these types of flows represent real-time multimedia traffic as follows.

- Constant bit-rate flows ('CBR') resemble Voice-over IP (VoIP), Game and similar real-time traffic. Our CBR traffic is not perfect CBR as that is unlikely to occur in reality. The sending times of the individual packets are randomised. The arrival curve of the randomised CBR traffic can be described by a token bucket with bucket depth $b = 600$ bytes and a rate $r = 78688$ kbps.
- Variable bit-rate flows ('VBR') represent video conferences and similar applications. We generated three different tracefiles (called L, M, H) in a loss-less testbed environment without interfering with background traffic using Microsoft NetMeeting Version 3.01[6]. After initialising the connection, 180 seconds of video were recorded. Three different set-ups (L, M, H) were considered:
  - M is a normal video-conference, that is, a talking head with an average amount of voice traffic.

---

[5] We do not use Explicit Congestion Notification (ECN) to signal congestion by packet marking as described in Durham *et al.* (2000).

[6] The traces for video-conferences were recorded using Ethereal Version 0.9.12. The bandwidth settings were set to 'Local Area Network'. The image size was set to medium and the quality controller that allows a step-less adjustment for 'Faster video' vs. 'Better Quality' was set fully to 'Better Quality'.

Table 8.2 Trace File Parameters

Parameter	Trace L	Trace M	Trace H
Average rate	83.46 kbps	171.99 kbps	456.54 kbps
Average packet size	929.8 bytes	871.9 bytes	782.2
Average number of packets per second	11.2	24.6	73.0
Token bucket parameter r	128 kbps	320 kbps	640 kbps
Token bucket parameter b	3980 bytes	12520 bytes	14610 bytes

○ L is a still picture with hardly any voice.
○ H consists of constant talking and an always moving camera.

The trace file statistics are listed in Table 8.2, the parameters of the traffic types are summarised in Table 8.3.

8.2.1.2 Traffic Mix

We use three different traffic mixes A, B, and C. A and B contain a relatively large amount of inelastic flows with B containing twice the amount of inelastic flows than A. Mix C contains a low amount of inelastic flow and a lower amount of short elastic flows than A and B. Table 8.4 lists the number of flows that are started on average in each edge node of a topology within a time window of one minute of simulation time. These numbers are scaled with the available bandwidth in our experiments to adjust the point of operation to the type of experiment.

The amount of sent or received packets and the transfer volume for these two traffic mixes depend strongly on the bandwidth and the used QoS system. QoS systems protecting, for example, the inelastic flows obviously increase the transfer volume of these flows at the cost of the TCP throughput. Table 8.5 lists the transfer volume obtained with the two traffic flows in a best-effort network[7]; this bandwidth led to an average dropping probability of 2–4%. For each traffic type, its average percentage of the total amount of received bytes over five simulation runs is depicted. The 95% confidence interval for every value is below +/− 2%.

8.2.1.3 Token Bucket Parameters for Admission Control

For the admission control in Intserv / Diffserv, a token bucket traffic specification is used. The token bucket parameters for the *VBR traces* are listed in Table 8.2. For the VBR traces, the number of possible token bucket parameters is infinite, as there is a trade-off between the rate r and the bucket depth b. We chose the smallest possible parameter combination (r, b) so that all packets of the three minute trace conform to the token bucket. The rate r was chosen as a whole-numbered multiple of 64 kbps and set to the lowest possible value that keeps the buffer size below 15 KByte.

---

[7] DFN topology with a bandwidth of 30 Mbps.

Table 8.3 Session and Flow Parameters

Parameter	Short TCP	Long TCP	CBR	VBR
Flow distribution within Session	Exponentially distributed inter-arrival time between the start of two flows with $1/\mu = 3.0$.	Exponentially distributed inter-arrival time between the start of two flows with $1/\mu = 30.0$.	Flows have exponentially distributed duration with an expected duration of $1/\mu = 30$. Two successive flows have opposite directions.	One flow per session, played back endlessly from a 180-second trace file with a random starting point.
Flow size	Pareto distributed with $\alpha = 0.63$. Expected size 10 packets each 15 kB .	Pareto distributed with $\alpha = 0.63$. Expected size 100 packets each 150 kB .	The transmission rate is set fixed to 78,688 bps (equals 64,000 bps without UDP and IP header).	The average transmission rate depends on the trace file, see Table 8.2.
Average flow duration	Strongly depends on network conditions.	Strongly depends on network conditions.	30 s.	Duration of the experiment.
Packet size	Constant 1500 bytes	Constant 1500 bytes	Constant 150 bytes	Variable, average 875 bytes
Packet inter-arrival Time (Time between Two Generated Packets)	According to the TCP greedy algorithm, (there is always data available from the application layer if the transport layer is allowed to send a packet).	According to the TCP greedy algorithm, (there is always data available from the application layer if the transport layer is allowed to send a packet).	Constant	Variable, determined by trace (see Table 8.2).

Table 8.4 Traffic Mix–Number of Started Flows in a Time Window of One Minute

Traffic Mix	Short TCP	Long TCP	CBR	VBR
A	100	50	40	5
B	100	50	80	10
C	50	50	10	2

Table 8.5 Traffic Mix–Percentage of Transfer Volume

Traffic Mix	Short TCP	Long TCP	CBR	VBR
A	37.46	40.27	10.02	12.24
B	28.18	31.96	17.89	21.95
C	25.87	63.10	3.61	7.41

The token bucket parameters r for the CBR flows were set to the average rate of the flows that leads to a buffer depth b of four packets with 600 Bytes as the smallest possible value.

For the TCP flows, we set the token bucket parameter r to 100 kbps, which is slightly below the maximal throughput assumed for TCP in the utility function (see below) and the bucket depth b to the default receiver window size of the NS2 TCP implementation of 20 packets of 30 Kbyte each.

8.2.2 Topologies

8.2.2.1 Used Network Topologies

For our experiments, we used the following topologies; they are depicted in Appendix A. Their basic graph properties are presented there, too.

Star The star topology has a single node in the centre where all cross-traffic will occur, see Figure A.3. The analysis of cross-traffic is important for investigating overbooking for Diffserv, see Section 8.3.

Cross The cross topology is depicted in Figure A.3. It is a variation of the star topology where cross-traffic will occur not only in a single node but also in all three central nodes. For the star and the cross topology, the edge nodes that are marked grey in Figure A.3. are the only nodes sending and receiving traffic flows.

DFN For most of the experiments, we use a real-world topology as the basic topology. We chose the DFN GWiN backbone topology as it is a medium sized real-world topology. The DFN GWiN backbone is the backbone network of the German research network that is connecting most universities and research labs in Germany. The topology is depicted in Figure A.1. For the DFN topology, we assumed that every node is a source of traffic flows and can act as traffic sink for any traffic flow.

Table 8.6 Bandwidth Settings and Average Path Length

	Low Bandwidth	High Bandwidth	Av. Path Length
Cross	20 Mbps	–	4
Star	20 Mbps	–	5.07
DFN	10 Mbps	15 Mbps	2.6
Artificial-3	10 Mbps	15 Mbps	4.2

Artificial-3 In addition, an artificially created topology is used, see Figure A.2. The topology generator Tiers (see Tiers (2004)) was used with the parameters listed in Table A.2 to generate that artificial topology. It is roughly 50% bigger than the DFN topology and has different graph properties (as shown in Table A.1).

8.2.2.2 Bandwidth and Buffer Dimensioning

For the experiments, it is important to adapt the link bandwidth to the configuration and purpose of the experiment. For the ease of implementation, we assumed equal link bandwidths in the network. For the same reasons, the buffer space resources are assigned statically to the outgoing links of a router.

The bandwidth setting is shown in Table 8.6. The low bandwidth setting creates a scenario with high load in the network and is used, for example, in the experiments of Section 8.3 where it is very important that the amount of inelastic flows is very high so that the system can be massively overbooked without the acceptance rate reaching 100%.

The high bandwidth setting was set to 1.5 times the low bandwidth value, which creates a congested but not extremely overloaded network (see e.g. Section B.13). Please note that for the BE reference architecture, we further increased the bandwidth by upto a factor of 8 on top of that.

The buffer space bf_l of one link is set proportional to the link bandwidth bw_l, so the given maximum queueing delay $d_l^{q,\,max}$ for that link l is fix. A maximum queueing delay of 50 ms is used as default value.

Please note that the number of flows (Section 8.2.1.2), the bandwidth and the buffer space is rather low compared to, for example, the actual DFN GWiN network. This is necessary because we are using packet-level simulations. They allow us on one side to obtain realistic flow, delay and dropping behaviour but on the other side are not scalable enough to simulate much larger networks with more flows in reasonable time. This is why packet-level simulations with a halfway realistic amount of traffic flows are practically never found in the literature. The experiments already took much longer than two weeks on 2.2 GHZ Pentium 4 machines with 1 GB RAM.

Individual experiments were repeated with a higher bandwidth setting and a corresponding increase in the number flows; they did not lead to fundamentally different results.

8.2.2.3 Mapping of Sessions to Network Nodes

We distinguish between edge nodes and core nodes in a topology. Only edge nodes are sources and sinks for the traffic sessions respectively flows. For each edge node, a

Table 8.7 Traffic Weight Distribution

Weight	Probability
0.5	25%
1.0	50%
2.0	25%

weight is selected as shown in Table 8.7. The number of sessions instantiated in a node is multiplied by the weight of that node. The probability for a node being selected as the communication partner of a session starting in another node is also proportional to the node's weight.

8.2.3 Utility

8.2.3.1 Delay Bounds

Real-time multimedia applications are typically delay sensitive. We model this by giving each inelastic flow a delay bound. Packets that are exceeding the delay bound are treated the same as dropped packets for the purpose of calculating the utility function (see below).

For the experiments, we varied the delay bound for these flows. In an experiment, each inelastic flow is assigned the same end-to-end queueing delay bound. Because the size of the topology influences the average path length, the influence of the topology has to be accounted for when setting the delay bound. Therefore, the end-to-end queueing delay bound of all flows is set to $\lambda \cdot d^e$ where λ is the average path length of the topology and d^e is the average per hop delay bound specified by the experimenter for a certain experiment. The default value of d^e is 20 ms.

8.2.3.2 Utility Functions

Figure 8.1 shows the utility functions for the different traffic types. With respect to utility, each flow is evaluated individually. Later on in our experiments, we evaluate the average utility of each traffic type; each flow is weighted the same.

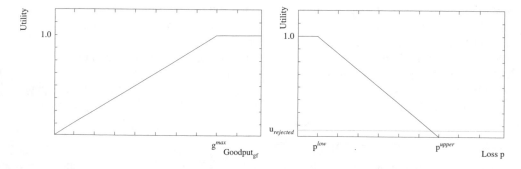

Figure 8.1 Utility Functions

Elastic Flows The elastic flows represent data transfer applications. The utility of data transfer applications mostly depends on the transmission time: the time it takes from the start to the end of the data transfer. The start of the data transfer is the time when the sender sends the first packet to open the TCP connection using the three-phase handshake. As the end of the data transfer, we count the time the receiver has received all bytes of the data transfer. The time until the last packets are acknowledged and the sender disconnects is not counted, because the receiver can use the data before that time.

In the experiments, not all elastic flows finish during the experiment time. Therefore, the transmission time is not known for all flows and approximated through the goodput instead. As the amount of data that a flow is transmitting is given, the transmission time is inversely proportional to the goodput g_f of the flow f. The goodput is defined as:

$$g_f = \frac{\text{number of correctly received packets}}{\text{elapsed time}} \tag{8.16}$$

The goodput can be determined if a data transfer is not complete. For the elastic flows, we use the utility function (a) of Figure 8.1. The utility is a linear function of the goodput g_f upto a certain maximal goodput g^{max}. We assume that once the maximal goodput is reached, the application or the user no longer benefits from a shorter transmission time. We chose a default maximal goodput of 10 pkts/s with 120 kbps, which lies in the same order of magnitude as the transmission rate of the inelastic applications. The utility function is normalised to 1.0.

Inelastic Flows For the inelastic multimedia applications, the loss probability influences the perceived utility. The utility function (b) of Figure 8.1 is used for the inelastic flows. We count packets that are *dropped* (because of congestion) and packets that are not dropped but arrive later than their delay bound (*delayed*) both as *lost* packets. A certain amount of loss can be tolerated (lower threshold p^{low}), then the utility decreases and reaches zero for the upper loss threshold p^{upper}.

As default values, we chose $p^{low} = 1\%$ and $p^{upper} = 10\%$. The utility function is normalised to 1.0.

If an admission control rejects a flow, the information that there are not enough network resources available to transport the flow can be deemed worth a certain amount of utility, especially when compared to a flow that is accepted at first but receives such a high loss probability that its utility is reduced to zero. Therefore, flows that were rejected by the admission control and did not transmit data at utility value $u_{rejected} \geq 0$; the default value for $u_{rejected}$ is 0.05.

8.2.3.3 Assignment of Flows to Services

We assign the delay-sensitive inelastic flows to the 'best' service a QoS system can offer. The elastic applications are assigned to the other services. If there are several alternatives, the short-lived flows are assigned to the higher-quality service. The motivation behind that is that the short-lived flows represent interactive traffic (e.g. web traffic) while the longer flows stand for file sharing (e.g. peer-to-peer traffic) that is supposed to be less time-critical.

Intserv   For the Intserv QoS system, all inelastic flows use GS. If they are rejected by the admission control, they are not transmitted. The rate R of the Intserv FlowSpec is set at the assigned rate necessary for guaranteeing the delay bound of the inelastic flow, the slack term S is set to zero.

All elastic flows use the BE service without admission control.

Standard Diffserv   All inelastic flows use the premium service using the EF PHB. EF flows rejected by the admission control are not transmitted.

Short-lived TCP flows are assigned the AF-1 class if there are resources available, otherwise they are downgraded to AF-2 and AF-3. As there is no admission control for AF-3, they will always be transmitted.

One-third of the long-lived TCP flows are assigned to AF-2 and downgraded to AF-3 if the resources are not available. The rest is assigned to AF-3 from the beginning.

Olympic Diffserv   The inelastic flows use the gold service or – if rejected – do not transmit at all. Short-lived TCP flows are assigned to the standard BE class while the long-lived flows are assigned to the low-priority bulk transfer class. Admission control is imposed only on the gold service.

Overprovisioned Best-Effort   In the overprovisioned BE QoS system, all flows use the BE service. There is no admission control that would stop any sources from sending.

8.2.4 Evaluation Metrics

In order to evaluate the QoS systems, the utility and other performance criteria can be measured. Throughout the experiments, the following evaluation metrics are used.

- **Average Utility**
  This is the average utility for each flow of a traffic type.
  All flows of the same type have equal weights irrespective of their actual size.
- **Average Utility of the Accepted Flows Only**
  If admission control is used, only flows that have not been rejected by the admission control are counted for determining this second utility average.
  If no admission control is imposed on a traffic type, this criterion yields the same result as the pure 'Average Utility' criterion above.
  Note: Flows that are downgraded to a lower service class, nevertheless, count as accepted.
- **Acceptance Rate**
  This is the percentage of the flows that were accepted by the admission control.
  If no admission control is imposed on a traffic type, the acceptance rate is automatically 100%.
- **Dropping Probability**
  This is the probability that a packet of a certain traffic type gets dropped because of a full queue before it reaches its destination.
- **Delay Bound Violation Probability**
  This is the number of packets arriving later than their delay bound permits relative to the total number of received packets.

Table 8.8 Abbreviations for the Different Quality of Service Systems

QoS System	Abbrev.	Parameters
Intserv	$IS - \alpha_{GS}$	α_{GS} = Maximum proportion of the link resources available for the guaranteed service class
Standard Diffserv	$sDS - bb - p$	bb = Bandwidth broker type (c = central, d = decentral, n = none) p = Bandwidth broker parameters for the central BB: p = overbooking factor ob for the decentral BB: p = overbooking factor times scaling factor $(ob \cdot \gamma)$
Olympic Diffserv	$oDS - bb - p$	bb = Bandwidth broker type (c = central, d = decentral, n = none) p = Bandwidth broker parameters, same as above
Best-effort	$BE - OF$	OF = Overprovisioning factor

- **Throughput**
  This is the average per-flow throughput of a traffic type in kbps.
- **Traffic Volume**
  This is the amount of volume of correctly received traffic of one traffic type divided by the total received traffic volume of all traffic types.

The graphs depicting the results also always contain the 95% confidence intervals for the different metrics; because of their size, the results are presented in Appendix B. Each experiment was repeated a number of times with new flows but the same bandwidth and topology. The number of repetitions was dynamically increased until the confidence intervals were satisfactory low. The typical number of repetitions is between 5 and 15.

For ease of presentation, abbreviations were assigned to the different QoS systems, see Table 8.8.

8.3 Per-Flow versus Per-Class Scheduling

In the first experiment, we compare the strongest QoS systems of our complete evaluation: The systems using Intserv/RSVP to offer guaranteed service and the Diffserv systems with EF PHB and a central BB. Both systems use per-flow admission control and allocate resources along the path – in the Diffserv case, resources are only allocated within the Diffserv domain. To make sure there are no bottlenecks outside the Diffserv domain, the bandwidth of the links outside the Diffserv domain is set to 10 times the bandwidth of the links inside the domain.

There are two central differences between Intserv and the Diffserv approaches which are as follows.

1. The first difference between the two approaches is that Intserv uses a per-flow scheduler while Diffserv schedules per-class. This also influences the admission control decision as discussed in Section 8.1.1.
2. The second difference between the two approaches is that the Diffserv QoS system also differentiates the non-EF flows into three service classes while the Intserv system treats them all as BE traffic. This difference will only influence the performance of the elastic flows.

The bandwidth for the experiment was set to the lower values of Table 8.6 to allow for massive overbooking of the Diffserv premium and gold service class. The results are depicted in Figures B.1 to B.3 for the DFN topology.

We first focus on the performance of *Intserv*. Configurations with a different parameter α_{gs} are shown. α_{gs} is the maximum proportion of the total link resources available for the GS flows (the inelastic flows). The following things can be noticed.

- The utility of all the accepted *inelastic* guaranteed service flows (CBR and VBR) is 1.0 – the maximum possible value (see Figure B.1) as can be expected from the fact that Intserv offers strict loss and delay guarantees.
- The utility of the *elastic* flows increases slightly (Figure B.1) if the parameter α_{gs} is decreased; in that case more flows requesting GS are rejected, as can be seen in Figure B.2.

We now look at the performance of the *Diffserv* systems (sDS, oDS). The figures show the performance metrics for overbooking factors *ob* from 1 to 8:

- The admission control decision for Diffserv has to be more conservative than that of Intserv because the flows within one service class are not protected against each other. The conservativeness of the decision is visible when comparing the acceptance rate (Figure B.2) of Diffserv with an overbooking factor *ob* of up to 3 (sDS-c-3) with those of Intserv IS-0.9:
  Despite the fact that the bandwidth and buffer *assumed* available for the admission control decision is significantly higher for sDS-c-3 than for IS-0.9, IS-0.9 can still admit slightly more flows to the network than Diffserv because of the flow protection. On the basis of the worst-case assumption in the admission control decision, the Intserv approach has to allocate fewer resources than the Diffserv system to guarantee the same delay bound. To quantify this, Intserv has to allocate only about 44% of the resources that the standard Diffserv system needs in this experiment.
- The same conclusion holds true for the Olympic Diffserv (oDS). The only difference is that the Olympic Diffserv admission control decision can admit slightly more flows than sDS if everything else is the same because of the smaller error terms of the scheduling algorithm (priority versus WRR); see Figure B.2. Intserv has to allocate about 47% of the resources that the Olympic Diffserv system needs in this experiment.

Because of the conservativeness of the sDS and oDS admission control decision, an important question to ask is *how much can the EF-based (premium respectively gold) service class be overbooked?*

- As can be seen from Figure B.3 (also reflected in the utility values of Figure B.1) for the DFN topology, the sDS/oDS systems can be massively overbooked. The first packet drops and delay-bound violations occur at an overbooking factor $ob = 4$ but only on a very small scale (significantly less than 1 per 10^6 packets and therefore hardly noticeable in the figures).
  However, even for an overbooking factor $ob = 8$, the dropping and delay-bound violation probabilities are still very small and for most applications acceptable. For the oDS

Table 8.9 Per-Flow vs. Per-Class Scheduling, Cross and Star Topology, Dropped or Delayed Packets [%], Summary

QoS System	Over-booking	Cross Topology CBR	Cross Topology VBR	Star Topology CBR	Star Topology VBR
sDS-c	1	0	0	0	0
	2	0	0	0	0
	3	0	0	0	0
	4	0.012	0.010	0.194	0.115
	6	4.34	3.58	8.26	7.24
	8	17.16	16.41	20.86	20.85
oDS-c	1	0	0	0	0
	2	0	0	0	0
	3	0	0.0004	0.0003	0
	4	0.020	0.006	0.097	0.052
	6	3.646	2.97	6.80	5.67
	8	16.16	15.23	19.55	19.41

systems, they are lower than for the sDS systems, which is explained by the stricter priority scheduling discipline.

On the basis of the results discussed so far, an overbooking factor $ob = 4$ to 8 can be recommended for the DFN topology when using a central BB for statistical guarantees, or $ob = 3$ if the EF traffic is extremely sensitive to loss respectively delay.

- The Charny bound (see Section 6.7) for this experiment set-up and traffic predicts a maximal utilisation of 7.98%. The centralised BB can raise the utilisation in the experiment to 13.13% without overbooking and with an overbooking factor of 4 to more than 27%.

- The overbooking potential is not as high as demonstrated above for all types of topologies. We repeated the same experiment for the artificial *Cross and Star topologies*. The resulting dropping and delay probabilities are summarised in Table 8.9.

For the *star topology*, an overbooking factor $ob = 3$ already leads to some dropped respectively delayed packets for Olympic Diffserv. For an overbooking factor $ob = 6$, the loss is already higher than 5.67% for both Diffserv flavours, surely unacceptable for a premium service.

Comparing the standard Diffserv with the Olympic Diffserv, the latter has a generally lower loss ratio despite the fact that it is accepting more CBR and VBR flows because of the smaller error terms. This can be explained with the strict priority scheduler that empties the EF queues quicker than the WRR-based pseudo-priority scheduler, leading to less loss for small overbooking factors. The additional amount of admitted flows in the oDS systems, however, is noticeable by the fact that for oDS-c-3 the losses are higher than for sDS-c-3, where in fact no loss was observed.

For the *cross topology,* similar arguments hold true. However, the dropping and delay-bound violation probabilities are generally lower than for the star topology. At the central node of the star topology, all flows cross paths; this creates a lot of cross-traffic within the EF service class on the outgoing links of that node. For the cross topology, the cross-traffic is distributed among the three central nodes where the paths of the flows cross. For the DFN topology, there are no clear bottlenecks. The cross-traffic creates additional delay and in an overbooked system increases the dropping probability.

Looking back generally at the dropping probabilities shown in Figure B.3, one can notice two phenomena as follows.

- The dropping probabilities for the elastic flows are extremely high (>10%) for all QoS systems. They are higher than one would observe in a physical network with real users. There are several reasons contributing to this extremely high loss rate here, as listed below.
  - First of all, in this experiment the system is extremely loaded and thus bandwidth and buffer are very scarce. This is necessary to analyse overbooking because not all inelastic flows will be accepted even with an overbooking factor $ob = 8$.
    As the number of elastic flows is proportional to the number of inelastic flows, their number is extremely high contributing to the high losses.
  - Second, the elastic flows are treated with lower priority by the Diffserv systems and receive only a small share of the bandwidth than in the fully loaded Intserv systems.
  - The third reason is the experiment set-up itself. In reality, some users would back off in a network loaded as highly as assumed in this experiment. Throughout our experiments, we do not model this type of user behaviour because we want to maximise the comparability of the results for different QoS systems. If a system providing poor QoS would be 'rewarded' with less traffic after some time (by users backing off), this would seem 'unfair' and the comparability would not be warranted.
  These three reasons explain the high dropping probability that would probably not be observed in reality. As ECN is not used in the experiments, the elastic TCP flows rely on packet drops as congestion indication, therefore even for very high bandwidths the dropping probability is significantly above zero; this can be seen, for example, for the BE-8 results in Figures B.15, see also Section 8.5.
- Another phenomenon that can be observed throughout the experiments is that the dropping probability is usually orders of magnitude higher than the delay-bound violation probability.
  This is explained by the relationship between the delay bound and the available buffer space. The star topology has an average path length of 4 hops. In the experiments, we set the delay bound of the flows proportional to the average number of hops that lead to an end-to-end queueing delay bound of 80 ms. The cross topology has an average path length of 5.07 hops, leading to a queueing delay bound of slightly more than 100 ms. For the DFN topology, every node acts as source and sink, which leads to a rather short average path length of 2.6 hops and a queueing delay bound of 52 ms for the flows.
  The available buffer space of an sDS EF queue allows a maximum queueing delay of 25 ms for a conforming packet in a single EF queue before the packet is dropped.
  Comparing these numbers, for a delay-bound violation an EF packet has to traverse several congested queues in a row. When it does so, it automatically has a high dropping probability. For the cross and start topology, the number of congested queues is mostly limited to the central links, so that delay-bound violations are unlikely.

Next, we analyse the effect of changing the delay bound from 20 ms to 10 ms and 40 ms per average number of hops. The effect on the acceptance rate is depicted in Figures B.4 and B.5 as follows.

- Decreasing the delay bound increases the amount of bandwidth and buffer resources allocated to a flow by the Intserv admission control, see (8.1) and (8.2). The resource allocations are largely influenced by the token bucket depth b; this explains why the effect is stronger for the VBR flows.
- For Diffserv, the acceptance rate for sDS and oDS drops by around 20% for the CBR flows and 40% for the VBR flows because the delay-bound check (8.12) more often fails for the lower delay bound. As the delay bound (8.11) depends on the burstiness of the flow, the effect is again stronger for the more bursty VBR flows.
- As the acceptance rate for 10 ms is really low, the dropping probability drops to zero for sDS and oDS even for an overbooking factor $ob = 8$. Delay-bound violations can be observed at $ob = 6$ and 8 but are less than 4 per 10^6 packets.
- If the delay bound is increased, the opposite effects can be observed (Figure B.5).

To conclude, our analysis of per-flow and per-class scheduling showed the following things: Our central Diffserv BB can give Intserv-like deterministic loss and delay guarantees for individual flows despite the fact that these flows are aggregated into classes when routed through the Diffserv domain. However, Intserv-like per-flow scheduling is more efficient than the per-class scheduling.

Because of this and the worst-case decision made by the central Diffserv BB, the Diffserv system can be overbooked. The overbooking factor depends on the topology, especially on the amount of cross traffic. A well-connected topology like the DFN topology can be safely overbooked by a factor of three or four.

8.4 Central versus Decentral Admission Control

The central BB in the Diffserv systems can become a bottleneck itself. If resource allocations are made on a small timescale (e.g. per flow), centrally managing a larger network can quickly become an impossible task. We have already argued that there is some optimisation potential for the BB that could be used. Reservation thresholds could be introduced as described in Schmitt *et al.* (2002).

Another solution to this problem is to decentralise the admission control completely and base the admission control decision purely on local information at the edge. In this section, we evaluate this approach. A contingent-based admission control algorithm is used; each edge node is assigned a contingent of resources (bandwidth and buffer) for each ingress link. As link bandwidths and buffer spaces are equal within the Diffserv domain in the experiments, we made the contingent proportional to the link bandwidth and buffer of the ingress link. This also allows for a better comparison with the central BB approach, where the admission control decision is also based on the link bandwidth and buffer (in the case of all Diffserv domain links on the path).

The efficiency of the decentral algorithm will depend strongly on the correct setting of the contingents. If the contingent assigned to an edge node is too low, then too many flows will be rejected or degraded. If it is too high, then too many flows are admitted and the QoS suffers. On a medium timescale[8], the INSP can react and reassign the contingents. This, however, does not guarantee a good performance for the future as the traffic patterns can change.

We evaluate the following two situations.

- *Situation A* represents a situation where the contingents match the traffic patterns very well.

  For situation *A,* the traffic sources and sinks are distributed evenly among all edge nodes by assigning each node the node weight 1.0 (see Section 8.2.2.3). Equal contingents are also assigned to all Diffserv ingress links.

- *Situation B* emulates the case in which the traffic prognosis and contingent assignment are not matched with the distribution of the traffic sources and sinks.

  For situation B, we distribute the traffic sources and sinks non-uniformly with the method described in Section 8.2.2.3, while we assign equal contingents to the ingress links. Thus, a mismatch is created.

The results are shown in Figures B.6 to B.11 in the Appendix. We start by analysing *situation A*. Several effects can be noticed as follows.

- The acceptance rate increases massively from the central bandwidth broker/admission control (sDS-c and oDS-c) to the decentral one (sDS-d, oDS-d). The following two reasons can be given for that.
  - First, the delay-bound check is not performed by the decentral but by the central BB. The central BB thus performs a stricter admission control per se.
  - The second reason is that the decentral BB only checks a single ingress link. The central BB checks the complete path through the Diffserv domain that – for the DFN topology – consists of 2.4 links on average. As in situation A, the flows arrive in random order, the used link resources differ from link to link. Thus, when a check for available resources fails, the more likely that more links are part of the check.
- The acceptance rate of the CBR and especially of the VBR flows increases, the higher the parameter $ob \cdot \gamma$ of the decentral BB becomes – that is, the more the decentral BB overbooks.

  This is obvious because overbooking increases the *assumed* amount of available resources for the inelastic flows.

- Evaluating the dropping and delay-bound violation probability for the inelastic flows, one notices that even for a small overbooking factor ($ob \cdot \gamma = 1.5$), loss occurs and the utility drops below 1 because the inelastic flows experience packet drops and delay-bound violations. Please note that for $ob \cdot \gamma = 3$ the losses would be even higher had the acceptance rate not already reached 100%.

  These losses are interesting because our previous experiments of Section 8.3 show that the central BB can be overbooked for the DFN topology by more than a factor of $ob = 3$ until this occurs.

---

[8] Outside the scope of a single simulation run.

For the decentral BB approach, this obviously no longer holds true. The missing delay-bound check and the fact that only the resources of the ingress links and not the core links are managed leads to a more generous and less controlled admission of flows. The chance of failures increases, thus the network cannot be overbooked so much.

- The loss of the elastic flows drops with an increase of inelastic acceptance rate and so does traffic volume as can be expected. Also, the utility of the short TCP flows is generally higher than that of the long TCP flows because they are preferably assigned to better service classes in sDS.
- Comparing the loss probabilities of sDS and oDS, those of oDS are generally slightly smaller for the higher overbooking factors. This is the same behaviour as observed in Section 8.3 and can be explained by the strict priority scheduler.

Next, we compare the effect of a mismatch of the assigned link contingents of the decentral BB algorithm (*situation B*):

- The acceptance rate effects visible for situation A are also visible for situation B. The general acceptance rate in situation B is similar to situation A, only for high $ob \cdot \gamma$ the acceptance rate is slightly lower. This is explained by the random influence of the scenario generation method.
- The dropping and delay violation probabilities are larger by roughly a factor of 5 in situation B than in situation A. The mismatch of situation B increases the risk of dropped and delayed packets.

  This behaviour shows the additional risk of the decentral bandwidth broker when the contingents are not well matched with the arriving flows at the edge of the network.

To conclude, using the decentral BB, the admission control decision becomes inexact. The risk of losing EF packets increases. The system should no longer be overbooked. If the system is not overbooked (for $ob \cdot \gamma = 1$), we did not observe any packet drops or delay-bound violations.

8.5 Direct Comparison

We next compare the different QoS systems directly. With the previous two experiments, we have already narrowed down the choice of sensible Diffserv configurations. On the basis of the results of these experiments, we evaluate all mentioned QoS systems by comparing their performance for different traffic mixes and different topologies in this experiment. Also, the overprovisioning factors are determined now. First, the DFN topology and later an artificial topology are analysed.

DFN Topology

The results for traffic mix A are depicted in Figures B.12 to B.17; for traffic mix B and C the main results are summarised in Tables B.2 and B.3 in Appendix B. As many of the effects visible in these graphs have already been discussed in the previous sections, we focus on the general performance evaluation here.

The traffic situation analysed in this section is a high-load situation in which the network is significantly congested. But, still, in the analysed situations, the available bandwidth is not too small. The network is more or less well dimensioned (when a QoS system is used). This is reflected itself in the results for sDS-n and oDS-n, where all inelastic flows are admitted to the network (as there is no admission control) and still experience practically no drops or delay-bound violations.

We start by looking at the average *throughput* of the different types of traffic flows, as shown in Figure B.17.

- The throughput of the inelastic flows equals their sending rate very closely for the Intserv and Diffserv systems. This results from their low dropping probability (Figure B.15).
- The average throughput of the elastic flows exceeds that of the inelastic flows, as there are a number of paths through the network that are only lightly loaded, at least for a period of the simulation time. The elastic flows adapt their transmission window to make use of the available bandwidth.
  A throughput of 120 kbps yields a utility of 1.0 for both elastic flow types. As can be seen from the utility results (Figure B.13), a significant number of elastic flows do not reach this throughput because they are on a congested path through the network.

We now analyse the *inelastic flows* admitted to the network. Figure B.12 shows that they achieve maximum utility in all Intserv (IS) and Diffserv (sDS, oDS) configurations. The acceptance rates between these systems, however, differ greatly (Figure B.14) and thus also the overall utility of the admitted *and* rejected inelastic flows (Figure B.13).

- For the given load situation and with respect to the overall utility of the inelastic flows, the sDS and oDS systems without admission control (sDS-n, oDS-n) perform best. The reason is that they admit all inelastic flows to the network and the network just has enough resources to serve them. Also, these systems are the QoS systems that have the lowest implementation complexity, as they require no signalling and admission control and only rely on long-term network engineering.
  It has to be mentioned that using these systems, however, leads to a certain risk. If the number of inelastic flows increases, the inelastic flows experience a service degradation that all the other QoS systems can avoid because they are using admission control (see the results for traffic mix B below and the results of Section 13.1).
- Deterministic service guarantees (with delay-bound guarantees) can only be given by the IS and the non-overbooked sDS-c and oDS-c systems. Looking at Figure B.13, IS-0.9 performs best. It offers the highest utility of the mentioned systems for the inelastic flows and only a slightly lower utility for the elastic flows than IS-0.6.

Next, we evaluate the performance of the *elastic flows*.

- The oDS systems assign higher priority to the short-lived elastic flows than the long-lived ones. This is clearly visible from the dropping probability and utility.
- The sDS systems also give preferential treatment to the short-lived flows. The short-lived flows are assigned preferably to service classes that are not (or not so much)

overbooked. Therefore, for the sDS system, the short-lived flows receive better performance than the long-lived ones. The difference between both flow types is smaller than for the oDS systems.

It must be pointed out that the performance difference for the elastic flows is much more controlled for the sDS system than for the oDS system. A provider has much more adaptation possibilities for sDS by assigning flows to a number of service classes and assigning the weights and overbooking factors to these classes than he has for oDS where strict priority scheduling is used. This advantage of sDS over oDS, however, also depends on whether the provider can explain and sell its customers the more complicated sDS differentiated services.

In addition, it should be stressed that when not using a bandwidth broker and admission control (sDS-n, oDS-n), the service degradation of the elastic flows is not controlled because the higher-priority flows are uncontrolled. This effect is partly visible for sDS-n and oDS-n that offer the lowest utility for the elastic flows.

- The IS systems do not support differentiation of the elastic flows. The long-lived flows receive a slightly higher throughput and fewer packet drops within the same service class as the short-lived ones. The same holds true for the BE systems where also both types of TCP flows are treated equally.

  o The explanation for the higher throughput is that because the flows are long-lived, they are dominated by the congestion-avoidance phase. If a short-lived and a long-lived flow have the same path through the network and that path is only lowly congested for a certain time, both flows will increase their rate. The short-lived flow finishes after transmitting a low number of packets and stops. The long-lived flow continues increasing its rate so that the average rate of the long-lived flows can be expected to be higher.

  o The short-lived flows are dominated by the slow start phase during which they double their congestion window while the long-lived flows are more likely dominated by the congestion-avoidance phase during which they linearly increase the congestion window. Additionally, long-lived flows are typically more aware of the congestion situation along their path than short-lived ones because the latter rarely transmit long enough – as their name implies – to experience loss, go through slow start again and switch to congestion avoidance mode to slowly approach to congestion point of the network path. This explains not only why the dropping probability of the short-lived flows is higher but also why it does only drop insignificantly for the extremely overprovisioned networks BE-4, BE-8; see Figure B.15.

Next, we evaluate how much a BE network has to be overprovisioned to offer the same performance as a network using a QoS system.

- The results for BE-1 show that without a QoS architecture and without overprovisioning, it is mainly the inelastic flows that suffer (see Figure B.13). The reason is not so much the dropping probability as the delay-bound violations (see Figures B.16 and B.15). The performance increases when the BE system is overprovisioned. The dropping probability is always significantly higher than zero because TCP is using packet drops as congestion indication and increases its rate until it experiences packet loss.

- Evaluating the performance of the inelastic flows, massive overprovisioning by a factor of 4 to 8 is necessary to compete with the best QoS systems. Only for an overprovisioning factor of 6 in our experiments, exactly the same utility is reached. For that overprovisioning factor, however, the elastic flows show a much better performance in the overprovisioned BE system than in any of the other (non-overprovisioned) QoS systems. The performance is probably acceptable for most inelastic applications with an overprovisioning factor of 4. This result is consistent with the analytical results of the previous chapter.
- The performance of the elastic flows is generally better for the BE systems than for the QoS systems because the latter systems degrade the service of the elastic flows for protecting the inelastic ones. The utility of the elastic flows in the BE-1 system is worse than in most QoS systems; for an overprovisioning factor of 1.5, however, the utility is already better than for most QoS systems (this depends on how the long- and short-lived flows are weighted).

  The reader should not get the idea from these observations that TCP performance is generally bad in the QoS systems. It is just that first, in our experiments, we assigned purely non-TCP flows to the premium service classes (GS, EF, Gold) and second, the load in the premium classes is very high.

The impact of the traffic mix on the performance can be seen by comparing the results for *traffic mix B and C* (summarised in Table B.2 and B.3) with the results for traffic mix A as discussed above.

- In traffic mix B, the number of inelastic flows is doubled compared to traffic mix A while the amount of elastic flows remains equal. The utility of all types of accepted flows remains roughly the same for the systems with admission control (IS, sDS-c, oDS-c) while it drops significantly for the Diffserv system without BB (sDS-n, oDS-n) and the BE systems. This effect is caused by the increased amount of traffic against which the systems without admission offer no protection.
- The acceptance rate of the admission-controlled systems decreases as can be expected by the increased number of flows upon which admission control is exerted. However, it does not halve, as one might expect from the fact that the number of inelastic flows doubles. This effect is explained by the fact that the different flows differ in their starting times, duration, their target nodes and (for VBR) in their size. With an increasing number of offered flows, it is more likely that the admission control can fit in a flow on a path where a certain amount of resources is left. Therefore, more flows fit into the same network when more flows are offered and the acceptance rate does not drop fully by 50%.
- For traffic mix C, the number of inelastic flows is drastically reduced (see Table 8.4). The number of short TCP transfers is halved. As can be expected, the acceptance rate of the inelastic flows increases because less flows are competing for the resources. The performance of the elastic flows generally improves, which can be attributed to the fact that there are less short-lived elastic flows. Short-lived flows are less reactive to congestion than the long-lived flows because of their short lifetime; their reduced number therefore leads to significantly less congestion and better performance of the elastic flows.

- The recommended overprovisioning factor for the BE architectures remains relatively independent of the traffic mix, which is again consistent with the analytical results of the previous chapter.

The resulting overprovisioning factor of four for the different traffic mixes already indicates that the Intserv and Diffserv QoS systems can offer a significant advantage over BE systems. That advantage, however, comes at the cost of increased complexity. After the BE systems, the least complex QoS system is the Olympic Diffserv system without BB (oDS-n). It uses simple priority queueing that is implemented in practically all modern router operating systems; no BB or admission control is required. The only requirement is that packets are marked at the ingress nodes. If the marking is based purely on packet header information[9], not even SLAs have to be negotiated and managed. This QoS system can – despite its simplicity – offer a *tremendous advantage* over a BE system with the same bandwidth and therefore similar costs in times of a high utilisation; this can be seen very clearly in Figure B.14 and Table B.2. The utility[10] of the inelastic (multimedia) flows and the short-lived TCP flows (representing e.g. web traffic) is much higher for oDS-n than for BE-1. This comes at the cost of a reduced performance of the long-lived TCP flows in the experiments that were assumed to be representing, for example, P2P traffic. Considering the fact that the willingness-to-pay of a customer for the high-quality transmission of a multimedia flow is probably significantly higher than that of a P2P flow, the simple oDS-n QoS system can be expected to improve the profitability of a network massively. It has to be kept in mind, however, that it still shares one disadvantage with the BE systems – the lack of admission control – and thus depends on traffic and network engineering measures. This is further elaborated in Part IV of this book, especially Section 13.1.

Artificial Topology

We next analyse the influence of the topology. The DFN topology has a relatively low diameter and a low node-degree (see Table A.1 in Appendix A). The average path length through the topology is 2.6. For comparison, we chose a topology that is quite different, the Artificial-3 topology of Appendix A. First of all, it is created artificially with the topology generator Tiers (see Appendix A for details). The average path length is 4.2; a flow needs significantly more hops through the topology as for the DFN (or a similar) topology. The average node-degree is much higher and much more unevenly distributed, which is reflected itself in the much higher standard deviation of the node-degree (see Table A.1). By visual comparison, the DFN and Artificial-3 topology are also quite different (see Figure A.1 and A.2).The artificial topology has more distinctive star-shaped connections than the DFN topology. As our experiments for the Star and Cross topology in Section 8.3 showed, the cross-traffic occurring at these nodes is more challenging for the admission control and reduces the overbooking potential. This is also reflected in the results for the Artificial-3 topology. They are depicted in Figures B.19 to B.16.

---

[9] The application a packet belongs to can be guessed by looking at the protocol number and port numbers of the TCP/IP respectively UDP/IP headers.

[10] As there is no admission control, the overall utility and the utility of the accepted flows only is equivalent.

- First of all, one notices the increased dropping probability (Figure B.21 compared to B.15) for the elastic flows. There is more congestion in this network. One explanation for this follows: The absolute number of flows in our experiments is proportional to the number of edge nodes, which are almost 50% higher for the artificial topology than for the DFN topology. At the same time, the average path length increases by more than 60%. The traffic is distributed among an increased number of links. The number of links increases by 85%. So, the average number of flows passing through a single link increases by roughly 30%. This leads to increased congestion, visible for the elastic flows in all architectures.
- The QoS systems with a non-overbooked exact admission control (IS, sDS-c, oDS-c) lead to zero dropped or delayed packets for the inelastic flows because of their admission control and resource management. The overbooked central BB systems (sDS-c-3, oDS-c-3) lead to very few packet drops (1 per 10^5 packets). Compared to the DFN topology, the dropping probability is increased, which can be explained by the cross-traffic effects that were also observed for the Star and Cross topologies (see Section 8.3).
- The decentral BB (sDS-d, oDS-d) and the Diffserv systems without BB (sDS-n, oDS-n) fail and lead to extreme losses for elastic and inelastic flows. The similar performance of these systems is explained with Figure B.20; the acceptance rate of all these systems is very similar and close to 100%, which explains the small differences.

  Obviously, the decentral BB is not well suited for this topology. From the structure of the artificial topology, it can be expected that the load of the individual nodes and links varies a lot. Links connecting two star-shaped subnetworks can be expected to be loaded much higher than the average link. As the decentral BB bases its decision purely on the situation of the ingress links, it admits too many flows to the network. This becomes visible when comparing the acceptance rates of sDS-c-3 and sDS-d; sDS-c-3 checks every link along the path. For the DFN topology, the sDS-d acceptance rates are only 5–10% higher than the sDS-c-3 ones (Figure B.14), whereas for the artificial topology they are around 35% higher (Figure B.20), clearly an indication that most bottlenecks are not at the ingress link.

  The performance of the decentral BB improves significantly if the threshold of the decentral BB is reduced to 0.5 instead of 1 or 3. Still, this result stresses the advantages of the central BB approach.
- The delay-bound violation probability for the BE systems shows a different behaviour for the artificial topology compared to the DFN topology. While it continuously drops for the DFN topology (Figure B.16), it first increases with an increased overprovisioning factor and only later decreases for the artificial topology (Figure B.22). We already argued above why the artificial topology is more congested. Therefore, it shows the same behaviour as the less-congested DFN topology once the bandwidth was increased enough to compensate for the additional congestion. The seemingly illogical increase of the delay-bound violations when the bandwidth is increased by the overprovisioning factor is explained as follows: The overprovisioning factor is applied to the bandwidth *and* the buffer space of the routers. As the buffers increase, the possible queueing delay and thus the chance of a delay-bound violation increases, too. Only if the additional increase in bandwidth is enough to empty the queues and reduce the congestion, the queueing delay and with it the delay-bound violation probability are reduced.

- For the artificial topology, an overprovisioning factor of two is necessary for the elastic flows to show the same average utility as for the sDS-c/oDS-c. For the inelastic delay-sensitive flows, the situation is dramatically different. Even an overprovisioning factor of 8 is not enough to offer the inelastic flows the same utility in the BE systems as in the best QoS systems.

8.6 Summary and Conclusions

In this chapter, QoS systems based on the Intserv, Diffserv and a plain BE architecture were evaluated in a series of experiments. Also, a central BB for Diffserv was developed. It can give Intserv guaranteed service-like guarantees for individual flows without needing per-flow complexity in the core network. Our experiments demonstrate that while it can offer the same QoS as Intserv, it is not efficient without overbooking. Generally, our experiments show that the different Diffserv systems with central BBs can be overbooked significantly. The exact amount of overbooking depends on the topology.

If a decentral admission control is used instead of a central one, significant control over the network is lost, resulting, for example, in the loss of overbooking potential.

The overprovisioning factors determined in our experiments are similar to those found in our analytical study of Section 7.2. The experiments also demonstrate that the BE systems perform generally by for flows with the delay requirements – even with high overprovisioning factors – because of their inability to differentiate between the different service classes.

The performance of the lightweight Diffserv QoS systems without any admission control was generally very good, especially as their implementation and administration costs can be expected to be relatively low. Because of the absence of admission control, however, they – and the BE systems – rely on in-time capacity expansion and other traffic and network engineering measures. This is being further investigated in Part IV of this book, especially in Chapter 13.

Part III
Interconnections

9

Interconnections Overview

> *The community believes that the goal is connectivity*
> RFC 1958 (see Carpenter (1996))

The Internet consists of a variety of interconnected heterogenous networks (autonomous systems, AS[1]), managed by multiple independent INSPs. Despite the competitive INSP market, each INSP must interoperate with its neighbouring Internet networks to provide efficient connectivity and end-to-end service. No INSP can operate in complete isolation from others; therefore, every INSP must not only coexist with other INSPs but also cooperate with them. Contrary to the situation in most telecommunication markets, there is no central authority in the Internet that enforces cooperation.

Both the number of networks and ASes as well as the average number of ASes a given AS is connected to are increasing at a fairly high rate. The number of ASes rose from 909 in 9/95 to 4427 in 12/98, 7563 in 10/00 to over 30,000 in 2004; see CAIDA – Cooperative Association for Internet Data Analysis (2004); Fang and Peterson (1999); IANA (Internet Assigned Numbers Authority). Similarly, the average interconnection degree, that is, the number of providers a certain provider has interconnection agreements with, rose from 2.99 in 9/95 to 4.12 in 12/98. It is also notable that a single provider may interconnect with up to 1000 other providers; see Fang and Peterson (1999).

Considering this, and the fact that the highest cost factors of Internet Service Providers (ISPs) are, typically, the interconnection costs and line costs, it is obviously important to study the effect of the interconnections on the network efficiency and Quality of Service (QoS) of an INSP.

In this book, we define a *network edge* as a connection between two different networks; there are two types of network edges:

- **Homing** describes the connection of an end-user and End-User Network Operator (ENO) to an Access Internet Service Provider (AISP) network.
- An **interconnection** is the connection between the networks of two different AISPs/Backbone Service Providers (BSPs).

---

[1] An autonomous system (AS) is a group of IP networks operated by one or more Internet Network Service Providers (INSP(s)), which has a single and clearly defined exterior routing policy.

The Competitive Internet Service Provider: Network Architecture, Interconnection, Traffic Engineering and Network Design
Oliver Heckmann © 2006 John Wiley & Sons, Ltd

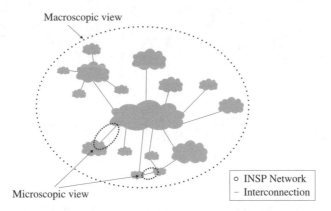

Figure 9.1   Macroscopic and Microscopic View on Interconnections

In this and the next chapter, we focus on the latter type of network edge: interconnections. There are two different ways of looking at interconnections as shown in Figure 9.1. The macroscopic view focuses on the large-scale connection structure of many networks as a whole, while the microscopic view analyses a single interconnection.

We start with a macroscopic view on interconnections in Section 9.1. In Section 9.2, we look at individual interconnections (microscopic view). Peering and transit interconnections are elaborated in that context; they form the basis of the analysis in the next chapter.

One important aspect of an interconnection is the interconnection method. Different methods are discussed in Section 9.3.

Real INSPs almost always use a mix of different (peering and transit) interconnections as discussed in Section 9.4 and further analysed in the following chapter.

9.1 A Macroscopic View on Interconnections

There are many different ways to connect a given set of networks with each other. The two extreme structures are the strictly hierarchical and the fully meshed structures as shown in Figure 9.2. The Internet is a heterogenous network of networks and follows neither of these two structures. However, as aspects of both the structures can be found in real connection structures (see e.g. Huston (1999a)) and as they are often referenced in literature, we investigate them first. Towards the end of this section, we look at empirical results about the real interconnection structure.

9.1.1 Strictly Hierarchical Structure

A strictly hierarchical structure, also called *tier structure*, consists of a small number of global INSPs at the 'top' that are referenced as *tier 1 INSPs*. Kende (2000) specifies five tier 1 providers in 2000, also called the *Big Five*: Cable&Wireless, WorldCom, Sprint, AT&T, and Genuity[2]. These few large backbones interconnect solely by peering and do

---

[2] Genuity is now a member of Level 3. WorldCom filed the largest bankruptcy in the US history in 2002.

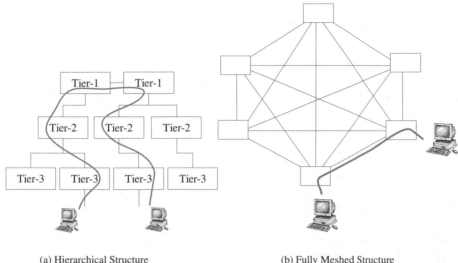

(a) Hierarchical Structure   (b) Fully Meshed Structure

Figure 9.2   Archetypical Structures

not need to purchase transit from any other backbone; they incorporate a pure BSP role. The *tier 2* consists of national INSPs; they have a smaller presence than tier 1 INSPs and may lease part of the network structure of tier 1 INSPs. The INSPs that are considered as tier 2 can be big AISPs like America Online (AOL) as well as national BSPs like Deutsches Forschungsnetz (DFN). Local AISPs are considered as *tier 3* INSPs. At each tier, the INSPs are clients of the tier above, see Figure 9.2 (a).

The microscopic interconnection relation between two INSPs of different tiers in that structure is typically the *classical transit relation* that we describe and analyse in Section 9.2.

If the hierarchical model is strictly enforced, the traffic between two local INSPs may need to transit all the way through a tier 1 provider. Such extended paths are inefficient, because they generate extended transfer delays and increased costs. In a competitive market like the INSP market, there is strong pressure to reduce costs, which explains why the reality does not match the strictly hierarchical structure. A typical modification is a local interconnection between two neighbouring tier 3 or tier 2 INSPs. However, a benefit of the hierarchical structure is the relatively small number of interconnections needed for each INSP to establish end-to-end connectivity.

9.1.2 Fully Meshed Structure

The other extreme is the fully meshed structure of Figure 9.2 (b) which shortens the path length compared to that of the hierarchical structure. The transmission in such an environment is fast, because the AS-level distance is always one hop[3].

However, the fully meshed structure does have obvious scaling issues if the number N of providers interconnected in that way becomes large, because the total number of interconnections is $N \cdot (N - 1)/2$.

---

[3] Of course, one AS system level hop can consist of many IP level hops.

The microscopic type of interconnection in the fully meshed structure is typically a *classical peering relation* as we define and analyse it in Section 9.2.

9.1.3 Realistic Structures

The two structures above are archetypical structures that do not represent the true structure of the Internet as is shown by a number of works discussing the properties of the Internet structure, for example, Aiello *et al.* (2001); Bu and Towsley (2002); CAIDA – Cooperative Association for Internet Data Analysis (2004); Chen *et al.* (2002); Faloutsos *et al.* (1999); Medina *et al.* (2000); Palmer and Steffan (2000); Spring *et al.* (2002); Tangmunarunkit *et al.* (2001); Zegura *et al.* (1997).

In Faloutsos *et al.* (1999), power law relationships are found in three inter-domain (AS-level) topologies of the Internet, which were constructed from Border Gateway Protocol (BGP) data. This paper started a discussion on power law AS-level topologies: Medina *et al.* (2000) investigate, on the basis of the work of Barabasi and Albert (1999), possible origins of these power laws using topology generators to create artificial topologies. Aiello *et al.* (2001); Bu and Towsley (2002); Palmer and Steffan (2000) are based on the power law relationship.

However, Chen *et al.* (2002) show that during the process of constructing the topologies of Faloutsos *et al.* (1999) from BGP data, 20% to 50% of the physical links are missed and that more exact topology graphs do not follow the power law relationship found in Faloutsos *et al.* (1999). The authors also show that works based on Barabasi and Albert (1999), for example, Medina *et al.* (2000), are not supported by the more exact topologies.

A nice macroscopic visualisation of the Internet, based on measurements by CAIDA, is shown in CAIDA – Cooperative Association for Internet Data Analysis (2003).

Besides that, there are a number of different topology generators, for example, BRITE (2004), Tiers (2004), Georgia Tech Internetwork Topology Models (2004), INET (2004) that can be used to generate artificial topologies that are deemed realistic by their authors. An evaluation of topology generators with respect to power law AS-level graphs is presented in Tangmunarunkit *et al.* (2001). A node in an AS level graph represents one AS and a link, an interconnection. The AS level graph is thus a graph representing what we call the macroscopic view in interconnections, see Figure 9.1. A similar study but for the topologies of INSP networks (one node representing a POP) is presented in Heckmann *et al.* (2003). Li *et al.* (2004) present an innovative new approach to understanding the structure of INSP networks. Contrary to the previous works that focus mainly on graph theoretic properties of topologies (e.g. the node-degree distribution), Li *et al.* (2004) take into account in their study the basic technological and economical trade-offs[4] that network designers face. The authors show that topologies that have the same graph theoretic properties (e.g. node-degree distribution) can have very different throughput performance. They further show that high-performance topologies are not likely obtained by any random graph generation method.

Spring *et al.* (2002) present the tool 'Rocketfuel' for measuring router-level topologies based on traceroutes, and BGP and DNS data: Using publicly available traceroute servers,

---

[4] A technological constraint is for example the bandwidth over degree function of actual switches/routers as it is determined by the cross-connection fabric. Economical considerations show that the costs of wiring can dominate the infrastructure costs, which gives a practical incentive to wiring networks such that they can support traffic using the least number of links.

the topology of a network can be revealed. Rocketfuel uses BGP data to calculate those traceroutes that most likely traverse the target network, at the same time redundant traceroutes are discarded. DNS information is finally used to cluster the IP addresses of the router interfaces to routers.

9.2 A Microscopic View on Interconnections

For our interest of investigating interconnections from the point of view of a single INSP, the microscopic structure of the interconnections is very important because it directly influences the QoS, cost structure, and transmission capacity of an INSP. Also, the microscopic structure of these interconnections – the mix of different interconnection types – is finally the decision of the INSP.

The literature typically distinguishes only two types of interconnections; see for example, Kende (2000); McGarty (2002); Songhurst (2001); Weiss and Shin (2002). This is not enough. The Internet service market is the outcome of business and technology interaction, rather than a planned outcome of some regulatory process. This leads to the appearance of a wide and diverse variety of interconnection types in reality. Therefore, we start by deriving a more detailed definition and classification of interconnections than typically found in related works.

9.2.1 Taxonomy and Classification of Interconnections

Huston (1999a,b) divides interconnections into physical and financial interactions. He describes the different possible connections. The routing entries that are exchanged at interconnections are called 'the currency of interconnection'. Also, different financial settlement options of the telephony industry (bilateral settlement, sender keeps all and transit fees) and possible options for the Internet industry (e.g. per packet or session accounting) are discussed. Section 9.3 of this chapter is based on Huston's physical interaction.

A more economical focused point of view can be found in Bailey (1997). It analyses the different economic incentives associated with different types of Internet interconnection arrangements; it does not consider regulatory issues. Friedmann and Mills-Scofield (1997), from a purely economic perspective, examine the optimal settlement pricing strategies for INSPs.

Kende (2000) gives insights into the market development of interconnection and the two most dominant interconnection forms: classical peering and transit. Kende (2000) also examines interconnection policies and regulatory issues as well as international interconnection.

Looking at the interconnections observed in the real world and learning from the works cited above, we find that the following three aspects comprehensively classify the variety of the existing interconnections (see Figure 9.3):

- **Route Advertisement**
  The route advertisement can be symmetrical or asymmetrical. *Symmetrical* in this context means that both clients exchange their own and their direct customers' routes. *Asymmetrical* in this context means that one INSP (the upstream provider) offers the other INSP (the downstream provider) access to all destinations in its routing table

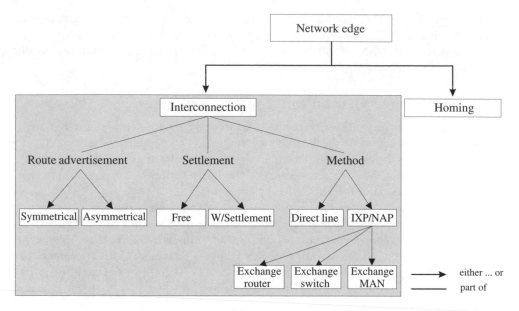

Figure 9.3   Network Edges and Interconnection Types

and advertises the downstream provider's network entries in its routing table, while the
downstream provider only advertises its own and its direct customer's networks.

- **Settlement**
  The settlement aspect of an interconnection is concerned with whether one INSP com-
  pensates the other for the exchanged traffic. In *settlement-free* interconnections, the
  providers might share the costs for the connection method but do not pay each other
  for the traffic itself, while in interconnections *with settlement*, one provider pays the
  other for the exchanged traffic. The price usually depends on the volume of the exchanged traffic and typically decreases with the volume (concave cost function), see
  Norton (2002).

- **Interconnection Method**
  The interconnection method describes how the physical interconnection between the
  two providers is realised: Either through one or more direct connections between the
  two providers' networks (*direct-line* method), or through an Internet Exchange Point
  (*IXP*). An Internet exchange point is typically used by a larger number of INSPs that
  are connected to
  - a central router (exchange router structure),
  - a central switch (exchange switch structure) – also called *exchange Local Area Net-
    work, LAN* – or
  - a Metropolitan Area Network, MAN (exchange MAN structure, distributed ex-
    change).

  The exchange switch and exchange MAN (distributed exchange) methods are typically
  found in large IXPs (like LINX, DE-CIX, Parix). The exchange router method is not
  very common because the routing between the INSPs is performed by the central router
  and managed by a central instance, leaving the INSPs with too little influence on the
  routing and exchange policies.

The settlement, route advertisement, and interconnection methods can be freely combined to $2^3 = 8$ combinations[5]. Two out of these eight combinations make up most of the interconnections currently found in the Internet. We call them *classical peering* and *classical transit* interconnections.

9.2.2 Peering

At the beginning of the commercial Internet, interconnection agreements evolved from the informal interactions that characterised the Internet at the time the National Science Foundation (NSF) was running the backbone. The commercial backbones developed a system of interconnections known as *peering*. Although the term 'peering' is used frequently, it rarely has a uniform meaning. There is no set definition for it in the Internet Engineering Task Force (IETF). RFC 1983 (see Malkin (1996)), which provides definitions to important Internet-related terms, has no entry for 'peering' or 'interconnection'.

We use the term *classical peering* for the most common form of an interconnection relationship that treats both INSPs more or less equally:

> In *classical peering,* two INSPs use settlement-free symmetrical route advertisement and interconnect at an IXP.

Therefore, classical peering has the following distinctive characteristics:

1. Peering INSPs exchange traffic on a settlement-free basis; they do not charge each other for the transfer volume between them as in a transit relationship.
2. Peering INSPs use symmetrical route advertisement. They exchange traffic that originates with the customer of one INSP and terminates with the customer of the other peered INSP. To enable this, they exchange their own and their direct customer's routes. As part of the classical peering arrangement, an INSP would not, however, act as an intermediary and accept the traffic of a peering INSP and transit that to another connected INSP.
3. The classical peering interconnection exchanges traffic via an IXP. The peering INSPs have to own or lease lines to the access point of the IXP. The connection to the IXP gives the connecting INSP also access to a wide number of other possible peering and transit partners.

Initially, most peering traffic took place at IXPs as it was efficient for each INSP to interconnect with as many INSPs as possible at the same location. The rapid growth in the Internet traffic caused the IXPs to eventually become congested; see for example, Robertson (1997). This lead to the situation that some INSPs avoided IXPs and peered directly with each other. This kind of peering differs from classical peering by using a direct-line interconnection method instead of an IXP and is known as *private peering*.

Badasyan and Chakrabarti (2003) describe a game-theoretic model in which INSPs decide on private versus classical peering agreements as a multistage game; their result is that a mixed approach of connection via private peering and classical peering has the most advantages.

---

[5] If we ignore for the moment that there are different realisations of IXPs. They are discussed in Section 9.3 and Appendix C.

Nowadays, the congestion problem at IXPs seems to be solved. There are recent indications that a large proportion of the Internet traffic is again exchanged via classical peering interconnections. For example, Boardwatch (2003, viii) state that 90% of UK Internet traffic is routed through the IXP LINX in London and that LINX provides access through its memberships to around 50% of the world's Internet networks.

Another atypical peering interconnection type is *peering with settlement*. It has all the properties of the classical peering arrangement as described above with the sole exception that one INSP receives a financial compensation from the other INSP, usually because the peered traffic is unbalanced in favour of the second INSP.

9.2.3 Transit

Because each peering arrangement only allows INSPs to exchange traffic destined for each other's customers, INSPs would need to peer to a significant number of other INSPs to gain access to the full Internet. One alternative to classical peering is the classical transit interconnection:

> In a *classical transit* interconnection, the two INSPs can be clearly distinguished as a *customer INSP* and a *transit INSP*. They are sometimes also called *downstream and upstream INSPs*. Asymmetrical route advertisement is used, the customer INSP pays the transit INSP for the exchanged traffic (settlement) and a direct-line connection is used.

The main differences between classical transit and classical peering are thus:

- In a transit interconnection, one INSP pays another INSP for the exchanged traffic; the amount of settlement typically depends on the exchanged traffic volume.
- The transit INSP advertises the customer INSP's routing table entries and routes its traffic to all its peering and transit partners, thus connecting the customer INSP to 'the rest of the world'.
- The customer INSP, on the other hand, only advertises its own routes and thus only receives traffic from the transit INSP that ends in its own network.
- Transit agreements often include Service Level Agreements (SLAs); see below.

There are several *non-classical transit-like* interconnections imaginable. Sometimes, for example, an IXP could be used instead of a direct-line. However, not all IXPs allow these type of agreements over their infrastructure.

9.2.4 Service Level Agreements

Service Level Agreements (SLAs) are bilateral contracts at a network edge between an INSP and a customer that can be either another INSP or an end-user. RFC 3198 defines a SLA as *the documented result of a negotiation between a customer/consumer and a provider of a service, that specifies the levels of availability, serviceability, performance, operation, or other attributes of the service*; see Westerinen *et al.* (2001). SLAs are typically used in transit-like interconnections agreements. They contain a *Service*

Level Specification (SLS). A SLS is a set of parameters and their values, which together define the service offered to the customer. Besides the SLS, a SLA can contain pricing, contractual and other information.

An example for a SLA is the one MCI/UUNET is offering for Internet services, see MCI (2004):

- A 100% *network availability* is promised. For each cumulative hour of network un-availability or fraction thereof, the customer is credited one day of charges.
- Within the United States, *latency* guarantees of 55 ms between MCI's inter-regional transit backbone routers (hubs) are guaranteed; transatlantic latency guarantees of 95 ms are guaranteed. Also specified in the SLA is how the latency is measured: by averaging sample measurements taken during a calendar month.
- Packet *delivery* guarantees of at least 99.5% are given between hub routers. Again, this is a monthly average. The credit is one day of the MCI monthly fee.
- MCI notifies customers by email or pager within 15 minutes after it is determined that their *service* is *unavailable*. Unavailability is assumed if the edge router does not respond after two consecutive 5-minute ping cycles.
- A scheduled *maintenance* notification is specified to reach the customer 48 hours in advance; maintenance is performed during a standard maintenance window as also specified by the SLA.
- A response time of maximal 15 minutes is guaranteed for *denial of service* attacks.
- If acts of God, embargoes, terrorism, fires, sabotage, and so on, are responsible for the breach of the SLA no credit has to be given.

Several commercial SLA management solutions exist and are deployed, e.g. CiscoWorks (2004) and Lucent (2004).

SLAs are also an integral part of the Differentiated Services QoS architecture (see Black *et al.* (1998)). In that context, they are discussed in Section 6.2.4.2.

9.3 Interconnection Method

9.3.1 Internet Exchange Points

Since the introduction of the four original Network Access Points (NAPs), aka IXPs, in the NSF-proposed post-NSFNET architecture in 1995, the IXP market has developed significantly. More IXPs have emerged all over the world, enabling local as well as global interconnection.

Europe's leading IXPs have set up the European Internet Exchange (EURO-IX) Association with LINX in London, United Kingdom and AMS-IX in Amsterdam, Netherlands as the biggest members (see EURO-IX (2004)). As mentioned above, more than 90% of UK's Internet traffic is routed through the LINX exchange, Europe's largest IXP, which provides access through its memberships to around 50% of the world's Internet networks, see Boardwatch (2003, viii).

INSPs (especially AISPs) may connect to more than one IXP to ensure better connectivity and gain access to more peering partners. 47% of the INSPs connected to an IXP in the EURO-IX Association are also connected to at least one other IXP within

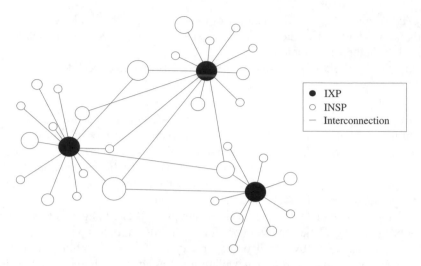

Figure 9.4 IXP Network Structure

the EURO-IX Association. Besides connecting via IXPs, AISPs typically also have direct interconnections with other INSPs. The resulting network structure evolving around IXPs is depicted in Figure 9.4.

A problem with respect to IXPs is that small regional providers that join an IXP will have the same reach as large INSPs that have invested in the national backbones. However, small regional networks will have smaller sunk costs that might lead to prices lower than that of the national providers. Cost recovery may become a significant issue for the larger providers. However, if the national backbones cannot recover cost and go out of business, the small regional INSPs lose the connectivity their customers want.

Because of this problem, membership to IXPs is sometimes restricted to national backbones, or small regional INSPs are required to have settlements with the larger INSPs. Both approaches can be found in the IXP market. All IXPs in the EURO-IX Association have the prerequisite that candidates for membership shall have an ASN (Autonomous System Number), see e.g. AMS-IX Membership (2004); German Internet Exchange DE-CIX (2004); LINX (2003). The DE-CIX IXP in Frankfurt, Germany, requires its members to peer with at least 10 other DE-CIX members after 6 months of membership, see German Internet Exchange DE-CIX (2004).

9.3.2 Evaluation

If we analyse the connection structures that are dominated by peering-like interconnections, it is quite intuitive that using IXPs can save costs compared to a connection structure that purely relies on a larger number of direct connections as shown in Figure 9.2 (b). A simple analytical model in Appendix C sheds some light on this intuition and shows that already for a very few number of providers within a city, a central exchange point is more cost efficient than direct connections.

There are different types of IXP structures. Our classification of interconnections shown in Figure 9.3 and Huston (1998) distinguish between the exchange router, exchange LAN, and exchange MAN (distributed exchange) structures. The first IXPs were built using the exchange router structure. Nowadays, this structure is not used anymore owing to its technical limitations. A simple analytical comparison of the exchange MAN and LAN structures is also given in Appendix C and suggests that the LAN structure is more cost efficient with a relatively small number of connected INSPs. The MAN structure gains cost advantages with a rising number of INSPs. However, these are just theoretical structures and in reality the distinction between MAN and LAN structures is fluid. The most dominant structure for bigger IXPs like LINX and DE-CIX is the combination of both structures; they use multiple colocation facilities containing a LAN (exchange switch) that are connected via a MAN.

9.4 Interconnection Mix

Almost all providers use a combination of different interconnection types. Most INSPs make use of an interconnection mix as they peer with other local INSPs and transit with at least one BSP to ensure global connectivity.

9.4.1 Negotiation Process

An interconnection arrangement is based on a negotiation process that results in the general framework for the interconnection. Usually all parties want to act as transit INSP, as it is preferable to be paid for an interconnection. The decision on which party takes on the role of a transit INSP and which one the role of a customer INSP is not easy to make. Generally, the closer the INSPs are, considering their size, their customer base, and their infrastructure, the more difficult is the negotiation process and the more likely is a peering arrangement. The negotiation over being the transit/customer INSP is often based on the greater geographical coverage criterion. However, this factor is not the only possible criterion, as one INSP may host valuable content and argue that the access to this content adds value to the other INSPs network. Also, an INSP with a very large client population with a limited geographical coverage may argue that this large client base offsets other possible criteria.

Huston (1999b) describes this negotiation as *two animals meeting in the jungle. Each animal sees only the eyes of the other, and from this limited input they must determine which animal should attempt to eat the other!*. After deciding upon which one will be the upstream INSP, the remaining parts of the contract have to be defined. The fees have to be decided upon as well as the location and the number of exchange points.

If the INSPs cannot solve the problem, the INSP may settle on a peering arrangement. Peering has some appeal to the INSPs as they do not need to track the exchanged traffic volume constantly, like in a transit relationship where the payment is typically based on the exchanged traffic volume. The tracking generates cost and the parties have to consider if transit is worth the effort. In conclusion, it can be said that peering is sustainable under the assumption of costly, unnecessary traffic measuring and mutual benefits.

There are several approaches in literature to model and solve the interconnection decision. For example, Giovannetti *et al.* (2003) identified some criteria to peer in an empirical study of the INSPs connected to the Milan IXP:

1. **Size**. The established peering points, either an IXP or private peering, entail fixed and variable technological costs. The result is that a sufficiently intense traffic flow between the customers of the two INSPs is needed for peering to be economically viable. The larger the two networks are, the more intense the traffic flow will be. WorldCom[6], for instance, published their criteria for peering in 2001. One criterion is that the traffic volume at the peering points is at least 150 Mbps.
2. **Symmetry**. Since the cost for the peering points are usually shared equally by the two peering INSPs, unbalanced traffic implies unbalanced gain from peering against a balanced distribution of costs. Such unbalanced situations have led to discontinuation of peering agreements and to its replacement with a transit interconnection. One of the criteria published by WorldCom is that the peering network has the geographical scope of at least 50% of its own. Another criterion is that the exchanged traffic volume at peering points does not exceed 1:1.5.
3. **Quality of Service**. The quality of a connection between two end-users depends crucially on the most congested network on the connection path. To ensure a certain degree of quality and to curb a potential free-riding[7] on infrastructure investments, the last criteria WorldCom sets forth is that most of the peering network has a capacity of 622 Mbps.

The study also shows that the peering decision is influenced, for example, by the proximity of the INSP's headquarters and their distance to the IXP. Other types of works for determining the interconnection mix are discussed next.

9.4.2 Determining the Interconnection Mix

There are two basic types of work that model the decision of an INSP on which interconnection type or mix to choose: game-theoretic and decision theoretic works. In decision theoretic works, the optimal decision of one INSP is analysed under a *ceteris paribus* constraint, which effectively means that possible reactions of the other parties involved are not anticipated. Game-theoretic works focus on the anticipation of possible reactions of competing INSPs and typically model the optimisation problem itself in much less detail.

Game-theoretic works are, for example Baake and Wichmann (1998); Badasyan and Chakrabarti (2003); Dewan *et al.* (1999, 2000); Giovannetti (2002); Norton (2004).

---

[6] Since 2003 WorldCom is known as MCI (www.mci.com).

[7] There is a potential free-riding on infrastructure investments as the quality of a connection between two end-users depends crucially on the most congested network in the path. When two networks peer with each other and one of them is congested, the quality of the connection does not improve when the non-congested network upgrades its infrastructure. If the congested network chooses not to upgrade its infrastructure, it would have the full cost savings, and would share the reduced performance with all the networks it peers with; see Giovannetti *et al.* (2003).

The rationales behind peering decisions for commercial INSPs and for academic research networks are analysed in Baake and Wichmann (1998); the focus lies on analysing competition and business stealing effects.

Dewan *et al.* (2000) and Dewan *et al.* (1999) concentrate on the economics of direct line interconnections, assuming that IXPs are congested and there are thus incentives to move away from them. INSPs differ on the basis of connected content providers. Dewan *et al.* (2000) discuss direct-line interconnection agreements between INSPs that compete for customers in the same area, while Dewan *et al.* (1999) discuss the same approach for INSPs that do not compete for customers in the same area. However, as the congestion problem at most IXPs seems to be solved, the results are of less interest today.

Giovannetti (2002) presents a game-theoretic analysis of the effect that offering transit for other providers (including direct competitors) has on a provider, who monopolistically controls a bottleneck, and on its competitors.

The work of Badasyan and Chakrabarti (2003) in which INSPs decide on private peering is also relevant. INSPs compete by setting capacities for their networks, capacities on the private peering links, if they choose to peer privately, and access prices. The model is formulated as a multistage game and examined from two alternative modelling perspectives – a purely non-cooperative game, where the subgame perfect Nash equilibrium is solved through backward induction, and a network theoretic perspective, where pair-wise stable and efficient networks are examined. The INSPs in this model compare the benefits of private peering relative to being connected through an IXP. The result of Badasyan and Chakrabarti (2003) was that a mixed approach of connection via private peering and IXP has the most advantages.

An interesting work related to the game-theoretic works is the Peering Simulation Game described in Norton (2004), where the participants play providers and negotiate interconnections.

Decision theoretic works are, for example Awduche *et al.* (1998); Heckmann *et al.* (2001); Hwang and Weiss (2000); Liu *et al.* (1998); Weiss and Shin (2002).

In Heckmann *et al.* (2001) a part of MPRASE (Multi-Period Resource Allocation at System Edges, see Heckmann *et al.* (2002)) is presented. It is a mathematical framework that describes and solves all kinds of resource allocation problems at the edge between two networks. Heckmann *et al.* (2001) discuss (among other things) the selection of the cheapest provider or the cheapest combination of providers from the point of view of a customer of an INSPs (which could be another INSP). A dynamic problem with multiple periods is investigated; Heckmann *et al.* (2001) make a decision in the first period about the combination of providers used for the rest of the planning horizon. The models of Heckmann *et al.* (2001) contain far less complex cost functions and do not include reliability and QoS issues.

Hwang and Weiss (2000) present an interconnection problem for a future QoS-supporting Internet, where Diffserv is used as the QoS architecture. The authors investigate how the cost of quality for different QoS networks characterises the optimal resource allocation strategies of the Diffserv bandwidth broker.

Awduche *et al.* (1998) present a mixed integer programming model for finding the cost-minimal placement of a given number of interconnection points within the topology of an INSP, once the decision to interconnect is made. Liu *et al.* (1998) take a similar approach, but additionally consider the switch/router placement (network design problem).

Weiss and Shin (2002) model the decision of two AISPs on whether to use a classical peering or a transit interconnection. They model one BSP as the upstream (transit) provider and two AISPs as the downstream providers with different market shares. The traffic is a function of the market shares of the two AISPs. The potential settlement for the transit interconnection is calculated as a function of the maximum inbound or outbound traffic volume. End-users pay a certain price for their traffic. Weiss and Shin determine the break-even price depending on the market share of the AISPs.

However, their study is fundamentally flawed in two points. First, they assume that as long as the AISPs in the model can make any profit with the transit agreement, they prefer it over peering – ignoring that peering might be more cost efficient. For an economical study this is not convincing; more convincing would be to assume that the AISPs prefer the interconnection type that offers them most profit. Second, in Weiss and Shin (2002) service requests and traffic volume are assumed to flow in the same direction, which they do not in reality for the dominant applications (WWW, P2P file transfer).

9.5 Summary and Conclusions

INSPs must connect their networks with its neighbouring networks to provide end-to-end connectivity. The connections between two networks are called *interconnections* and were discussed in this chapter. We looked at the big picture, the macroscopic topology of the Internet, and then at the individual interconnections. The most common interconnection types are classical peering and transit. How the interconnections are technically realised is also important; we call this the interconnection method. There are two basic interconnection methods, using an Internet exchange point or a direct connection. Most providers have a mix of different interconnection types with different other providers. At the end of this chapter, works related to determining the interconnection mix were discussed.

In the next chapter, the influence of the interconnection mix on efficiency and QoS is studied. Several strategies are derived for optimising the interconnection mix with respect to different goals. The next chapter of this book can be classified as a decision theoretic work.

10

Optimising the Interconnection Mix*

In this chapter, decision theory and mathematical programming methods are used to model the problem of finding the optimal set of peering and transit providers for an Internet Network Service Provider (INSP). We consider costs, reliability issues, Quality of Service (QoS) and the fact that traffic and tariffs are changing over time or that new providers enter the market. Heuristics and exact algorithms are presented and their performance is evaluated in extensive simulations. Related works were discussed in the previous chapter.

For the ease of presentation, in this chapter we assume that only the classical peering and transit interconnections are used (see Section 9.2.2 and 9.2.3). However, the models can easily be extended for nonclassical interconnection types.

We start with the presentation of an optimisation model for minimising the interconnection-related costs in Section 10.1. An exact method for solving the problem is presented; it can be used to find the optimal set of peering and transit partners for one INSP. It is compared with some heuristics in a performance evaluation. The evaluation shows that the interconnection mix significantly influences the cost structure and overall efficiency of the network of an INSP.

Minimising the interconnection-related costs, however, is not the only important goal of an INSP with respect to interconnections. The interconnection reliability and the influence of the interconnections on the QoS are important aspects, too. Therefore, we show and analyse how the original model of Section 10.1 can be extended to also consider reliability aspects in Section 10.2 and QoS requirements in Section 10.3. We evaluate the different reliability and QoS policies by simulations.

In Section 10.4, we show how previous strategies can be extended for the dynamic problem situation, which is evaluating whether a given set of peering and transit partners is still optimal considering changes in the traffic mix or the cost structure of the other providers. The administrative costs of changing peering and transit partners or Internet exchange points (IXPs) are also considered. Again, the models are evaluated using simulation.

*Reprinted from Elsevier Computer Networks Journal, 46(1), Oliver Heckmann, Jens Schmitt, Ralf Steinmetz, Optimizing Interconnection Policies, 19-39, Copyright 2004.

10.1 Costs

First, a mathematical programming model for finding the optimal set of peering and transit partners for one INSP is presented. It is important to model cost functions for peering and transit partners and for IXPs realistically. We interviewed different INSPs and IXPs and model the cost functions based on these interviews and based on Norton (2002).

We show how the model can be solved exactly and present some heuristics that closely resemble what providers do today. The heuristics are evaluated against the optimal solution that can be obtained by our model.

10.1.1 Description

Finding the optimal transit and peering partners as well as the necessary IXPs for one INSP is modelled by the following optimisation model. We assume that there are R different routes, and that the provider can predict the traffic for each route[1]. There are J transit providers offering transit service for all routes. The transit providers can be connected via a direct line. There are I peering providers offering peering only for some specific routes. Each peering provider is connected to at least one IXP and can be reached only via an IXP it is connected to.

The optimisation model tries to minimise the total costs that consist of the costs for connecting to IXPs, the additional peering costs for peering with a provider at an IXP, and the transit costs, including the direct line costs for connecting to the transit provider's closest POP (point of presence).

The costs for connecting to an IXP are largely fixed costs (leased line and backup line to the IXP, rent for rack space, costs of the exchange router, fixed IXP fees, etc.) that increase if the volume transferred via the IXP exceeds certain thresholds (representing upgrades to the leased lines and the exchange router, additional IXP fees, etc.); see Figure 10.1 (a).

A peering interconnection with provider i can only be made if there is a connection to an IXP where the peering provider i is present. The costs for peering are largely volume-independent fixed costs (transactional costs for the peering agreement, engineering costs). Some low volume dependent costs for peering[2] can also occur as some IXPs, for example LINX, also charge per volume of peering traffic, see Figure 10.1 (b). Giovannetti *et al.* (2003) presents interesting results from an empirical study of peering via the IXP in Milan. The peering decision is influenced, for example, by the proximity of the INSP's headquarters; a factor that can be modelled via the provider-specific fixed peering costs.

The costs for transit consist of fixed costs plus volume-dependent variable costs that decrease when certain volume thresholds are reached (see Figure 10.1 (c)). We assume that each transit and peering provider and each IXP can only accept traffic up to a certain maximum capacity.

---

[1] Please note that a route in the context of this book can be individual Border Gateway Protocol (BGP) routes. For performance improvement, nonoverlapping BGP routes *can* be aggregated to a single route in the optimisation problem.

[2] The variable costs for peering could also be accounted for in the IXPs cost function but to add them to the peering costs is more efficient from a modelling point of view and more flexible.

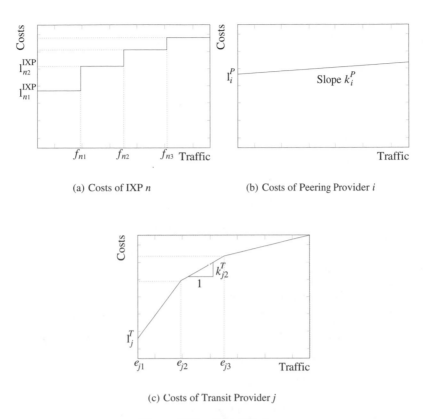

(a) Costs of IXP n (b) Costs of Peering Provider i

(c) Costs of Transit Provider j

Figure 10.1   Cost Functions

The problem is formally described by the mixed integer programming (MIP) model given with Model 10.1. The target function (10.1) minimises the total costs (IXP, peering and transit). We use the variables $\tilde{x}^T_{jm}$ to keep track of how much of the traffic of provider j is in segment m of its cost function. Constraint (10.2) connects the variables $\tilde{x}^T_{jm}$ to the routing variables x^T_{jr} of the same transit provider j: the total amount of traffic that is divided among all routes has to be equal to the traffic in all segments of the cost function. Constraint (10.3) ensures that the entire traffic demand for one route is satisfied by the combination of peering and transit interconnections.

Constraints (10.4) and (10.5) ensure that the transit cost function segments are 'filled' correctly: (10.4) limits the amount of traffic in one segment to the segment size. For the highest segment, (10.4) forms the capacity constraint of the transit provider. For concave cost functions, the higher segments would be filled first because of their lower volume costs. Therefore (10.5) is necessary; a higher segment of a cost function can only be used once the lower segment is completely full.

Constraint (10.6) is the capacity constraint for the peering providers. The other constraints are the nonnegativity and binary constraints of the variables. Constraints (10.4), (10.5) and (10.6) also connect the binary y variables to the corresponding x variables and make sure that traffic can only be routed (indicated by the x variables) through providers with which an interconnection exists (indicated by the y variables).

Model 10.1   Cost Minimising Interconnection Model

Indices

$i = 1, \ldots, I$	Peering provider i
$j = 1, \ldots, J$	Transit provider j
$r = 1, \ldots, R$	Route r
$m = 1, \ldots, M_j$	Part m of the cost function of transit provider j
$n = 1, \ldots, N$	IXP n
$s = 1, \ldots, S_n$	Step s of the cost function of IXP n

Parameters

$\widehat{x}_r$	Traffic prognosis for route r
l_i^P	Fixed costs for an interconnection with peering provider i
k_i^P	Price per unit of volume for an interconnection with peering provider i
c_i^P	Capacity of peering provider i
$\mathfrak{R}_i$	Set of routes offered by peering provider i
l_j^T	Fixed costs for an interconnection with transit provider j
M_j	Number of steps in the cost function of transit provider j
$c_j^T = e_{j\,(M_j+1)}$	Capacity of transit provider j
e_{jm}	Lower volume limit of step m of the cost function of transit provider j, see Figure 10.1 (c)
k_{jm}^T	Price per volume in step m of the cost function of transit provider j
S_n	Number of steps in the cost function of IXP n
Ψ_n	Set of peering providers that are connected to IXP n
f_{ns}	Upper volume limit of step s of the cost function of IXP n, see Figure 10.1 (a)
f_{nS_n}	Capacity of IXP n
l_{ns}^{IXP}	Costs for IXP n if it is used and the traffic volume via IXP n is in step s of the cost function of IXP n
Inf	Large number (resembling infinity)

Variables

$x_{ir}^P \forall r \in \Re_i$ Amount of traffic for route r passed through peering provider i

y_i^P Binary variable, set to 1 if an interconnection with peering prov. i is made

x_{jr}^T Amount of traffic for route r passed through transit provider j

$\tilde{x}_{jm}^T$ Traffic volume in segment m of the cost function of transit provider j

y_{jm}^T Binary variable, set to 1 if cost function segment m of transit provider j is used

y_{ns}^{IXP} Binary variable, set to 1 if IXP n is used and the traffic volume via IXP n is in step s of the cost function

$$\text{Minimise} \quad \sum_j \sum_{m \in M_j} k_{jm}^T \tilde{x}_{jm}^T + \sum_j l_j^T y_{j1}^T + \sum_i \sum_{r \in \Re_i} k_i^P x_{ir}^P + \sum_i l_i^P y_i^P + \sum_n \sum_s l_{ns}^{IXP} y_{ns}^{IXP} \quad (10.1)$$

$$\text{subject to}$$

$$\sum_m \tilde{x}_{jm}^T = \sum_r x_{jr}^T \qquad \forall j \tag{10.2}$$

$$\sum_{i \,|\, r \in \Re_i} x_{ir}^P + \sum_j x_{jr}^T = \widehat{x}_r \qquad \forall r \tag{10.3}$$

$$\tilde{x}_{jm}^T \le (e_{j\,m+1} - e_{j\,m}) \cdot y_{j\,m}^T \qquad \forall j \, \forall m \tag{10.4}$$

$$\tilde{x}_{jm}^T \ge (e_{j\,m+1} - e_{j\,m}) \cdot y_{j\,m+1}^T \qquad \forall j \, \forall m = 1, \ldots, M_j - 1 \tag{10.5}$$

$$\sum_{r \in \Re_i} x_{ir}^P \le c_i^P \cdot y_i^P \qquad \forall i \tag{10.6}$$

$$\sum_s y_{ns}^{IXP} \le 1 \qquad \forall n \tag{10.7}$$

$$\sum_{i \in \Psi_n} x_i^P \le f_{ns} \cdot y_{ns}^{IXP} + \text{Inf} \cdot \sum_{t=1,\, t \ne s}^{S_n} y_{nt}^{IXP} \qquad \forall n \, \forall s \tag{10.8}$$

$$x_{ir}^P \ge 0 \qquad \forall i \, \forall r \in \Re_i \tag{10.9}$$

$$x_{jr}^T \ge 0 \qquad \forall j \, \forall r \tag{10.10}$$

$$\tilde{x}_{jm}^T \ge 0 \qquad \forall j \, \forall m \tag{10.11}$$

$$y_i^P \in \{0, 1\} \qquad \forall i \tag{10.12}$$

$$y_{jm}^T \in \{0, 1\} \qquad \forall j \, \forall m \in M_j \tag{10.13}$$

$$y_{ns}^{IXP} \in \{0, 1\} \qquad \forall n \, \forall s \tag{10.14}$$

Constraints (10.7) and (10.8) are needed for the IXP cost function: if all y_{ns}^{IXP} for one specific n are zero, then IXP n is not used. Otherwise, exactly one variable y_{ns}^{IXP} will be 1 indicating in which step of the cost function the total traffic volume via this IXP n lies. This is ensured by constraints (10.7) and (10.8). The exact solution for this problem can be found using standard MIP solving techniques as discussed in Section 3.3.

10.1.2 Evaluation

Now, the exact solution of our model and the solutions obtained with some heuristics are evaluated in a series of simulations.

For validating the models and for meaningful results, it is important to use realistic data as input. However, providers are very reluctant to reveal information about their cost structure and explicitly do not allow publication of this information. The input data in the following experiments is based on actual but randomised data of a real (medium-sized) INSP. To cover a wider space, specific parameters are varied systematically.

10.1.2.1 Simulation Setup

We evaluate different *scenarios*. A scenario is specified by a given number of peering providers, transit providers, IXPs, routes, and an interval from which traffic and costs for these providers or routes are drawn. A *scenario instance* is created by randomly creating cost functions and traffic demand vectors from the scenario-specific parameter intervals. Per scenario, $n=100$ instances are created and solved. The averages of all instances as well as the 95% confidence interval are computed and form the basis of the following evaluation.

The parameter intervals for the basic set of scenarios are given in Tables 10.1 and 10.2. For the simulations, we assume that each peering provider offers one route and always has enough capacity for that route. The traffic demand for one route is drawn uniformly distributed from the *traffic demand for a peering provider's route* interval. The BGP routes not covered by the peering providers routes are modelled with one additional larger route. The traffic for that route is determined by the *traffic demand for the rest of the world* parameter.

The fixed costs for the peering providers are calculated as specified in the *fixed peering costs* interval in Table 10.1. We added a small amount per volume to give the fixed costs of larger peering providers the tendency to be higher than those of smaller providers. The variable peering costs are set uniformly low. The transit costs are calculated as specified in the tables; the transit capacity is drawn from the *capacity of a transit provider* interval and divided evenly across the different segments of the cost function.

The different steps of the cost function for an IXP are calculated as specified in Table 10.1. We assume that all steps are of the same size and that the costs for the later steps are significantly less than for the lower ones.

Table 10.2 lists the four scenario-dependent parameter ranges; all 16 possible combinations are evaluated. Each scenario has a number from 0 to 15. In scenario s the first

Table 10.1 Constant Parameter Intervals

Parameter Description	Parameter Value/Interval
Traffic demand for a peering provider's route	[50, 1000]
Fixed transit costs	[10,000, 50,000]
Capacity of a transit provider	[50%, 150%] of total traffic
Fixed peering costs (not including the costs to connect to the IXP)	[3000, 6000] + [0, 4] times traffic demand of peering provider's route
Variable peering costs	4
Number of steps of the transit cost function	5
Number of IXPs	4
Number of steps in the cost function of an IXP	4
Basic costs for connection to an IXP	[25,000, 60,000]
Costs for additional volume transferred to IXP	[9000, 18,000] and reduced by 10% each step
Volume that can be transferred to an IXP in one step of its cost function	Total traffic demand divided by 16

Table 10.2 Scenario-dependent Parameter Intervals

Bit	Parameter Description	Value/Range if Bit = 0	Value/Range if Bit = 1
#1	Number of peering providers	100	200
#2	Number of transit providers	10	20
#3	Traffic demand for rest of the world	20 × av. traffic demand of peering provider's route	60 × av. traffic demand of peering provider's route
#4	Variable transit costs	[20, 80], decreasing by [5%, 20%] each step	75% of the costs for bit #4 = 0

parameter from Table 10.2 is used if the corresponding bit in s is not set, otherwise the second parameter is used. For scenario $s = 7$, the second parameter intervals will be used for the number of peering and transit providers and the traffic demand for the rest of the world (bits #1, #2, #3); the first parameter interval will be used for the variable transit costs (bit #4).

The commercial MIP solver CPLEX (see ILOG CPLEX (2004)) is used to calculate the exact solution for the optimal interconnection Model 10.1 (**OPT**). We compare the solution we obtain with several heuristics.

10.1.2.2 Description of the Heuristics

Our comparison includes heuristics that can mimic the behaviour of some real-world providers:

- **Transit Heuristic (H TR)**
  The transit heuristic uses the cheapest transit provider or, if the capacity of the cheapest is not sufficient, the cheapest set of transit providers. It does not use peering.
- **Peer-With-All Heuristic (H PA)**
  The peer-with-all heuristic connects to all available IXPs and peers with every peering provider available. For the remaining traffic, it chooses the cheapest transit provider (or set of transit providers).
- **Peer-at-Selected-IXPs Heuristic (H PS)**
  The peer-at-selected-IXPs heuristic is similar but more careful in the selection of IXPs. If connected to an IXP, it peers with all available peering providers[3] at that IXP. In order to decide which IXPs to choose, it starts with the cheapest transit provider (or set of transit providers).
  It then evaluates all IXPs, starting with the one that has the highest ratio of traffic volume of the peering providers at that IXP to the costs for connecting to that IXP. Connecting to a new IXP will reduce transit costs; if the saved transit costs are greater than the additional costs required to connect to the IXP and peer with the peering providers, it connects to that IXP, otherwise it does not.
- **Evolution Heuristic (H EV)**
  The evolution heuristic describes an evolutionary approach that could describe how a real INSP has found its interconnection partners over its lifetime:
  Go with the cheapest transit provider (or set of transit providers) first and connect to the best IXP – the one with the highest volume to cost ratio (see H PS). Then, successively evaluate the peering providers available at that IXP. Peer with a new peering provider if the saved transit costs are higher than the additional peering costs.
  In the second part of the heuristic, the other IXPs are evaluated with the following method: the INSP assumes it is connected to the new IXP and chooses its peering partners at that IXP in the same way as in the first part of the algorithm. Then, the INSP compares the total costs after connecting to this IXP with the total costs when not connecting and connects only if that reduces the costs.

10.1.2.3 Performance Evaluation

We first compare the solution obtained by our model (called *OPT*) with the solution obtained by the heuristics. The results are summarised in Figures 10.2 and 10.3 for $n = 100$ instances per scenario. Each of the algorithms solved the same 100 instances per scenario. The averages over the instances and the corresponding 95% confidence intervals are shown in the figures.

The costs of the heuristics (Figure 10.2 (a)) are measured relative to the total interconnection costs of the OPT algorithm – that is, the optimal costs. The H TR and H PA

---

[3] This heuristic mimics a behaviour that is used by some INSPs. These INSPs are sometimes called 'peering-sluts' because they have peering relationships with a lot of other providers at selected IXPs.

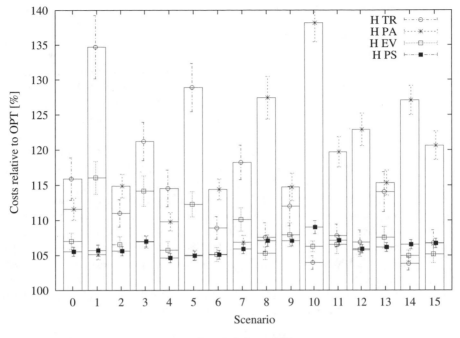

(a) Costs Relative to OPT

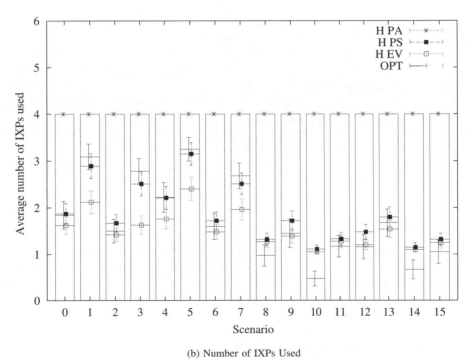

(b) Number of IXPs Used

Figure 10.2   Performance Evaluation

heuristics clearly perform worst and lead to up to 35% higher costs than the OPT algorithm. Choosing none or too many peering providers is obviously not a good idea: H PA performs systematically better than H TR for the higher transit costs (scenario 0–7) and vice versa. The best heuristics are H EV and H PS, with H PS performing better than H EV most of the time, but they still lead to 5 to 9% higher costs than the OPT algorithm. The explanation is shown in Figure 10.2 (b)[4] of Figure 10.2. In many cases, H PS chooses nearly the same number of IXPs as OPT. H EV systematically chooses too few for the higher transit costs (scenario 0–7) and thus misses some chances of saving costs by peering. It chooses too many for the lower transit costs (scenario 8–15) and thus misses out cheap transit opportunities in that case. When looking at the ratio of peering to transit providers (Figure 10.3 (a)) and peering to transit traffic (Figure 10.3 (b)), H EV systematically has too few and H PS has too many peering relationships.

The results show that the OPT algorithm presented in this chapter can save significant amounts of interconnection costs for all the different scenarios when compared with heuristics actually used in the real world. Remember that the interconnection costs are typically the largest cost factor for an INSP. The best real-world heuristic is the so-called peer-at-selected-IXPs heuristic (H PS); it peers with every possible partner at a number of carefully selected IXPs.

The next question we investigate is whether the computational complexity of the OPT algorithm might be an obstacle for using it rather than the heuristics.

10.1.2.4 Evaluation of Computational Complexity

If we define $M = \frac{1}{J} \cdot \sum_j M_j$, $R = \frac{1}{I} \cdot \sum_i \text{Size}(\Re_i)$, and $S = \frac{1}{N} \cdot \sum_n S_n$, then Model 10.1 needs $I(R+1) + J(2M+R) + NS$ variables and $I + J + 2JM + R + N(1+S)$ constraints.

The time it took to solve one problem instance of scenario 0 on a machine with a 700 MHz Pentium 3 and 256 MB RAM is depicted in Figure 10.4. The numbers of peering providers I and transit providers J were increased (x-axis) to increase the complexity of the problem. As Figure 10.4 shows, OPT can be solved in less than 10 minutes for large problems with 1100 providers. Given the fact that in the real world the problem has to be solved only rarely, the computational complexity is not an obstacle for using OPT.

A further advantage of OPT is that it is based upon a MIP model that can be further extended in different ways, as shown in the next sections. Some of these changes would be anything but straightforward to incorporate into the heuristics.

10.2 Reliability

Reliability is an important issue for INSPs. Model 10.1 can be extended in several ways to include reliability. By reliability, in this context we mean protection against the failure of one or more interconnections. For example, in all the 100 solved problem instances for the OPT algorithm, if the biggest provider selected from that strategy fails, there is not enough free capacity available from the other interconnected transit providers to compensate the

---

[4] The number of IXPs and the peering to transit ratio for H TR are zero; therefore H TR is not included in Figure 10.2 (b) and in Figure 10.3.

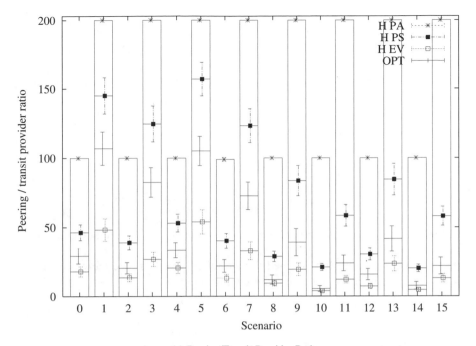

(a) Peering/Transit Provider Ratio

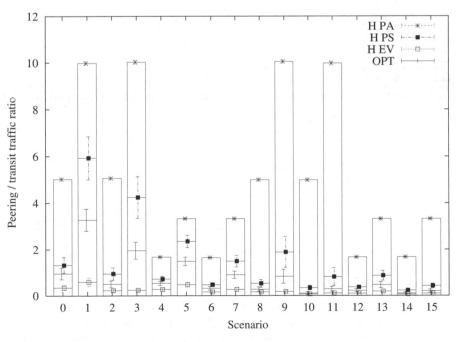

(b) Peering/Transit Traffic Ratio

Figure 10.3   Performance Evaluation

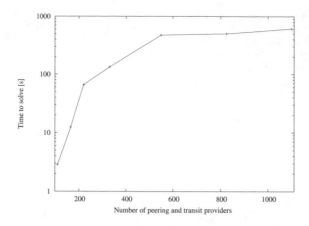

Figure 10.4   Evaluation of the Computational Complexity, Time to Solve for OPT

failure by rerouting the traffic destined for the failed provider. We therefore suggest and discuss several policies for extending the OPT strategy.

10.2.1 Policies

10.2.1.1 Minimum Number of Transit Providers Policy (MT)

One easy reliability policy is to interconnect with a minimum number of transit providers $\underline{Y}$ to reduce the dependency on each of them. This policy can be easily incorporated into the basic model (see Model 10.2). The advantage of this policy is its ease of use; the disadvantage is that it does not provide any guarantees or fine-grained control.

Model 10.2 Minimum Number of Transit Providers Policy (MT)

The following parameter and constraint are added to the otherwise unchanged Model 10.1:

$$\text{Parameter}$$

$$\underline{Y}\ \text{Minimum number of providers}$$

$$\text{Constraint}$$

$$\sum_j y_{j1}^T \geq \underline{Y} \tag{10.15}$$

10.2.1.2 Minimum Free Capacity Policy (MC)

Another reliability policy is to make sure that there is a minimum amount of free transit capacity available, for example, a percentage of the total traffic. The free transit capacity is the sum of all capacities of the transit providers minus the capacities of the providers that are already being used; see Model 10.3.

Model 10.3 Minimum Free Capacity Policy (MC)

The following new parameter, variables and constraints are added to Model 10.1. Also, we now explicitly have to assume positive fixed costs for transit providers: $l_j^T > 0$.

Parameter

Γ Required amount of free capacity

as fraction of the total traffic

Variables

f_j^T Free capacity of transit provider j

Constraints

$$f_j^T \leq c_j^T - \sum_{m \in M_j} \tilde{x}_{jm}^T \quad \forall j \tag{10.16}$$

$$f_j^T \leq c_j^T \cdot y_{j1}^T \quad \forall j \tag{10.17}$$

$$\sum_j f_j^T \geq \Gamma \cdot \sum_r \hat{x}_r \tag{10.18}$$

$$f_j^T \geq 0 \quad \forall j \tag{10.19}$$

Constraint (10.16) limits variable f_j^T to the free capacity of transit provider j, (10.17) forces f_j^T to zero if there is no interconnection with transit provider j, (10.18) enforces the minimum amount of free capacity, and (10.19) is the nonnegativity constraint for the new variables.

This policy gives the decision maker fine-grained control over the free capacity. Its drawback is that if one interconnected provider who carries more than the fraction Γ of the traffic fails, there is not enough spare capacity. This is avoided by the next policy.

10.2.1.3 Anticipating Failure Policy (AF)

Another approach would be to make sure that there is enough spare transit capacity if a single transit or peering provider fails completely. This policy is described in Model 10.4.

Constraint (10.20) anticipates the failure of each transit provider j, (10.21) does the same for each peering provider i.

10.2.1.4 Combined MC and AF Policy (MCAF)

The MC and the AF policy can be combined as they use the same variables f_j^T; we name this approach MCAF.

Model 10.4 Anticipating Failure Policy (AF)

The following new variables and constraints are added to Model 10.1:

Variables

$$f_j^T \qquad \text{Free capacity of transit provider } j$$

Constraints (10.16), (10.17), (10.19) and

$$\sum_{k \,|\, k \neq j} f_k^T \geq \sum_m \tilde{x}_{jm}^T \qquad \forall j \tag{10.20}$$

$$\sum_j f_j^T \geq \sum_{r \in \Re_i} x_{ir}^P \qquad \forall i \tag{10.21}$$

10.2.2 Evaluation

In order to evaluate the reliability policies given above we use simulations again. The results presented here are based on scenario zero of Section 10.1 but there was no fundamental difference observed for the other scenarios.

In order to evaluate the reliability performance, we calculate the free transit capacity of the solutions obtained by the different policies as a percentage of the total traffic. The higher the free capacity, the more safety buffer remains if – for example – one provider fails. For each solution, we also determine whether there would be enough free capacity to carry the traffic of the biggest (peering or transit) provider, if it fails; we call this measure *robustness*. The average results and the 95% confidence intervals are depicted in Figure 10.5, as are the average costs of the solutions obtained by the different policies. The graphs also contain the reference reliability and cost measures of the solutions obtained for the same problems by the unmodified OPT algorithm above (0% robustness, 32% free capacity).

Again, we generated $n=100$ instances that were solved by the Minimum Number of Transit Providers Policy (MT), the Minimum Free Capacity Policy (MC), and by the combination of the Minimum Free Capacity Policy and the Anticipating Failure Policy (MCAF). The results for the AF Policy (AF) alone are included in the results for MCAF when the minimum free capacity is 0%.

If we look at MT, which has a parameter that can only be increased in integer steps, it can be seen that the costs increase very quickly if the minimum number of transit providers is increased. It is important to remember that the reference costs (100%) refer to the total interconnection costs, and not only the transit costs; they are therefore very high in absolute terms. The cost increases of the MC and MCAF policies are much smoother and more controlled.

If we analyse the reliability measures, the robustness increases quickly for MT and more slowly for MC; MCAF automatically leads to full robustness because of the AF constraints. The free capacity explodes for the MT policy while it is obviously more controlled with the MC policy, because the minimal free capacity is a parameter of that

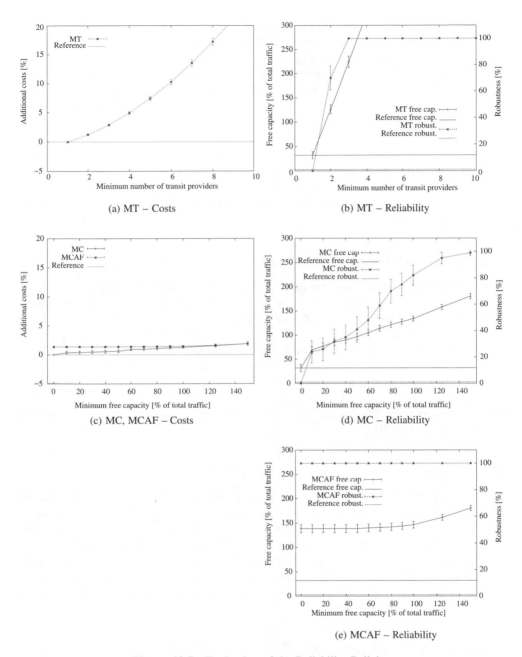

Figure 10.5 Evaluation of the Reliability Policies

policy. Because of the AF constraints in MCAF, the free capacity does not decrease for lower values of the minimum free capacity parameter.

The MT policy represents an easy rule of thumb (using a minimum number of transit providers) as it can be expected to be used by some INSPs. It can be used to increase the reliability. However, costs can explode, especially as the parameter (the number of transit providers) cannot be increased smoothly but only in steps of one. Since the policy parameter only indirectly influences reliability metrics such as free capacity and robustness, the MT policy cannot be recommended. The more sophisticated approaches developed here (MC and MCAF) perform better compared to MT. Contrary to MT, which increases free capacity by adding new transit providers, MC and MCAF can also increase the free capacity by choosing more peering providers and *bigger* providers instead of just *more* transit providers.

The MCAF policy is clearly the best choice: it offers full robustness and full control over the free capacity. Its parameter is the minimal free capacity, which can be easily estimated by the decision maker. If the failure of the biggest provider is unlikely, MC can also be used. It can save some costs compared to MCAF, but only for low values of free capacity.

10.3 Quality of Service

One of the typical parameters that an INSP would like to optimise is the QoS achievable with its interconnections. In the context of interconnections and with the information available at that abstraction level[5], the QoS can be mainly influenced by selecting interconnections so that the length of routes in terms of AS (autonomous systems) hops is kept low. In addition, peering or transit providers could be rated in some fashion with respect to the QoS they usually offer, and the solution could take those ratings into account. We will focus on the more objective measure of route lengths and show several possibilities for extending the basic model (Model 10.1) to include the QoS that is achieved by the interconnection policy chosen. These extensions can be directly combined with those discussed in Section 10.2.

10.3.1 Policies

The typical QoS metric used on the timescale of interconnections is the average number of AS hops for a route from the provider's network to the end-point. A lower number of hops correlate with lower delay and a lower loss probability for the packets, and thus a higher utility for the end-user. This is especially important for routes carrying traffic from real-time multimedia applications and network games. Peering interconnections usually offer a lower hop count than transit interconnections, because the traffic ends in the peering network. This is, in fact, the main reason why some larger INSPs accept peering with significantly smaller INSPs.

10.3.1.1 Peering Bonus (PB)

The easiest way of taking the lower hop count of peering providers into account is giving peering providers with QoS sensitive routes a bonus b_i that reduces their fixed peering costs and thus makes peering with them more attractive; see Model 10.5.

---

[5] e.g. obtained from BGP data.

Model 10.5 Peering Bonus (PB)
The parameter l_i^P of the basic Model 10.1 is replaced with the new parameter $\tilde{l}_i^P = l_i^P - b_i$.

The advantage of this approach is its ease of use; the disadvantage is that the parameter b_i can be difficult to estimate, and only indirectly influences the QoS.

10.3.1.2 Hop Constraint (HC)

Another approach that gives the decision maker more control of the QoS parameter hop count is to add an additional constraint for the average hop count of the traffic; see Model 10.6.

Parameter q_r is used as a weight when determining the average hop count of the traffic. Routes known to carry delay-sensitive traffic (e.g. to gaming sites) should obtain a higher than average q_r. Using the parameters of Model 10.6, the weighted average AS hop count is

$$H = \frac{1}{\sum_r q_r \hat{x}_r} \left(\sum_i h_i^P \cdot \sum_{r \in \Re_i} q_r x_{ir}^P + \sum_j h_j^T \cdot \sum_r q_r x_{jr}^T \right) \tag{10.22}$$

It is limited to $\overline{H}$ by Constraint (10.23) of Model 10.6. Not only the AS hop count, but also any other QoS metric can be modelled with this approach. Instead of looking at the entire traffic, this approach can be easily modified to take into account only a subset of the routes. For a more fine-grained prediction of the hop count for transit providers, h_j^T could be replaced by a route-dependent prediction h_{jr}^T for route r through the network of transit provider j.

The advantage of the Hop Constraint (HC) approach is that it gives the decision maker a finer control and, with the maximum hop count, an easy to understand design parameter. The disadvantages are the higher number of parameters and the slightly higher complexity of the optimisation model with the additional constraint.

10.3.1.3 Hop Count Penalty Costs Policy (HP)

Decreasing the hop count can quickly lead to increasing costs (as shown below). The HC policy enforces a maximal hop count without constraining the cost-increase. The hop count penalty costs policy (HP, Model 10.7) is similar but does not enforce a maximum hop count with a constraint. Instead, it adds the hop count, weighted with some Penalty Costs (PC), to the target function. This allows a trade-off between decreasing the hop count (which typically leads to increasing costs, as we will see in the evaluation below) and decreasing the costs.

10.3.2 Evaluation

In order to evaluate the QoS approaches, we use simulations based on scenario 0 again; the results observed for the other scenarios were not fundamentally different. The hop

---

Model 10.6 Hop Constraint (HC)

The following parameters and constraint are added to Model 10.1:

Parameters

h_i^P Average hop count for traffic through peering

provider i (typical value is one)

h_j^T Estimation of the expected hop count for traffic

through transit provider j

q_r Delay sensitivity of the traffic on route r

used as weight for the average hop count

$\overline{H}$ Allowed maximal average hop count allowed

Constraints

$$\sum_i h_i^P \cdot \sum_{r \in \Re_i} q_r x_{ir}^P + \sum_j h_j^T \cdot \sum_r q_r x_{jr}^T \leq \overline{H} \cdot \sum_r q_r \hat{x}_r \qquad (10.23)$$

Model 10.7 Hop Count Penalty Costs Policy (HP)

Using the parameters h_i^P, h_j^T, q_r from Model 10.6, (10.22) is added to the target function (10.1) of the otherwise unchanged Model 10.1 weighted with penalty costs ψ:

$$\text{Minimise (10.1)} + \frac{\psi}{\sum_r q_r \hat{x}_r} \left(\sum_i h_i^P \cdot \sum_{r \in \Re_i} q_r x_{ir}^P + \sum_j h_j^T \cdot \sum_r q_r x_{jr}^T \right) \qquad (10.24)$$

---

count for peering providers is set to 1, and for the transit providers, it is drawn uniformly distributed from the interval [3.0, 6.0].

The averages of $n = 100$ problem instances and the 95% confidence intervals are shown for the Peering Bonus (PB), HC, and HP policies in Figure 10.6. For reference purposes, the costs and the hop count from the plain Model 10.1, without any QoS features, are also depicted (labelled 'reference').

Figure 10.6 shows that all policies can decrease the hop count. At the same time, the costs increase. The costs for a low hop count are higher when using the PB policy than either of the other two. This occurs because the PB policy distorts the costs of the providers with the PB and minimises the distorted rather than the real costs. Also, the decision maker cannot easily guess the optimal parameter of the PB policy, therefore, this policy cannot be recommended.

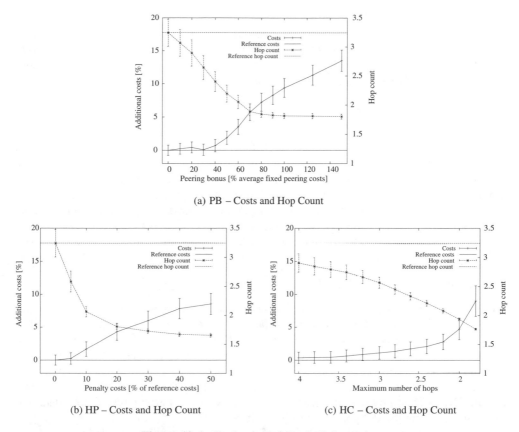

(a) PB – Costs and Hop Count

(b) HP – Costs and Hop Count   (c) HC – Costs and Hop Count

Figure 10.6   Evaluation of the QoS Strategies

The HC offers direct control over the hop count, which the other policies do not. Its parameter is therefore easy to set – if the decision maker has an indication of what the hop count should be. The costs increase more quickly for lower hop counts. The HP policy does not enforce a certain hop count but, instead, evaluates the value of the decreased hop count (expressed by the PC) against the hop count. This avoids the danger of exploding costs in case the maximum hop count of the HC has been set too low.

If a certain maximum hop count is absolutely necessary, the HC policy has to be used. Otherwise, if there is flexibility on the hop count, HP offers the best way of modelling this trade-off. HC and HP can also be combined: HC could be used to ensure that a certain (higher) hop count is not exceeded, while HP could be used to further decrease the hop count, without ignoring the cost increase.

10.4 Environment Changes

The models of Sections 10.1 to 10.3 can be used to calculate the optimal set of peering and transit providers for one INSP at one point in time. This is useful for a new INSP entering the market. An INSP that already has interconnections with a number of peering and transit providers faces a slightly different problem: *is the current set of peering and transit providers still optimal or is it worth changing interconnections, considering the technical and administrative costs for establishing a new or cancelling an existing interconnection?*

We call this the *dynamic* problem and now show that the previous models can be easily extended for the dynamic case. Again, the models are evaluated by simulations.

10.4.1 Adjusting the Basic Models

For the dynamic case, we now assume that there are interconnections to a set Θ of the I peering providers, to a set θ of the J transit providers and to a set ϑ of the N IXPs. As the traffic requirements and the cost functions of the providers change, the dynamic problem is solved every period in order to find the new optimal set of providers.

Typically some technical and administrative effort is necessary to establish a new interconnection that can be expressed by a cost term (transaction costs). Cancelling an existing interconnection also typically involves some effort that can be expressed by a cost term.

10.4.1.1 Penalty Costs Policy (PC)

The costs for establishing a new interconnection can be expressed as penalty costs per period by dividing them by the number of periods an interconnection is expected to last, or by a typical amortisation or planning horizon. These penalty costs can be added to the fixed costs of the providers that are not in set Θ or θ. Similarly, the costs for cancelling an existing interconnection can be transformed into bonus costs per period that are subtracted from the fixed costs for the providers in set Θ and θ respectively. The same can be done for the IXPs. This provides an incentive to stick with the current set of providers and IXPs; we call this the PC policy; see Model 10.8.

The advantage of this policy is that the basic models are easily extended this way and the cost terms involved can typically be estimated quickly and easily.

10.4.1.2 Limiting Change Policy (LC)

Another policy for dealing with the dynamic problem would be to limit the amount of change (new interconnections and cancelled interconnections) per period, reflecting the limited technical capacities for these changes in a period, or the risk of change the provider is ready to take. We call this policy Limiting Change (LC) policy; see Model 10.9.

Constraint (10.28) limits the permissible number of changes. The left-hand side of constraint (10.28) counts the binary y-variables that are 1 if an interconnection to provider i/j is made for all providers i/j with which no previous interconnection agreement existed. It adds all cancellations of interconnection agreements by counting the zeroes in the binary y-variables of the providers i/j with which an interconnection agreement existed during the last period.

10.4.2 Evaluation

For the simulative evaluation, we create $n = 25$ problem instances. To simulate the dynamic environment we simulate p periods per instance. At the beginning of each period,

---

Model 10.8 Penalty Costs Policy (PC)

Model 10.1 is extended as follows:

Parameters

Θ Set of peering providers that an interconnection exists with at the beginning of the current period

$s_i^P \ \forall i \notin \Theta$ (Per period) penalty costs for establishing a new interconnection with peering provider i

$b_i^P \ \forall i \in \Theta$ (Per period) bonus for not cancelling an existing interconnection with peering provider i

θ Set of transit providers that an interconnection exists with at the beginning of the current period

$s_j^T \ \forall j \notin \theta$ (Per period) penalty costs for establishing a new interconnection with transit provider j

$b_j^T \ \forall j \in \theta$ (Per period) bonus for not cancelling an existing interconnection with transit provider j

ϑ Set of IXPs connected with at the beginning of the current period

$s_n^{IXP} \ \forall n \notin \vartheta$ (Per period) penalty costs for connecting to the new IXP n

$b_n^{IXP} \ \forall n \in \vartheta$ (Per period) bonus for disconnecting from IXP n

Parameters l_i^P, l_j^T and l_{ns}^{IXP} are replaced by $\tilde{l}_i^P$, $\tilde{l}_j^T$ and $\tilde{l}_{ns}^{IXP}$ which are defined as follows:

$$\tilde{l}_i^P = l_i^P + s_i^P \ \forall i \notin \Theta \text{ and } \tilde{l}_i^P = l_i^P - b_i^P \ \forall i \in \Theta \tag{10.25}$$

$$\tilde{l}_j^T = l_j^T + s_j^T \ \forall j \notin \theta \text{ and } \tilde{l}_j^T = l_j^T - b_j^T \ \forall j \in \theta \tag{10.26}$$

$$\tilde{l}_{ns}^{IXP} = l_{ns}^{IXP} + s_n^{IXP} \ \forall s \ \forall n \notin \vartheta \text{ and } \tilde{l}_{ns}^{IXP} = l_{ns}^{IXP} - b_n^{IXP} \ \forall s \ \forall n \in \Theta \tag{10.27}$$

---

the amount of traffic, the capacity of the providers and the fixed and variable costs vary. The range of the changes is shown in Tables 10.3 and 10.4. As in Section 10.1, we analyse different scenarios where either the first or second option from Table 10.4 is used. If option *All Providers Available at Beginning* is used, all the providers are available for an interconnection agreement at period 0; the only change in this simulation is the

---

Model 10.9 Limiting Change Policy (LC)

The following parameters and constraints are added to Model 10.1:

Parameters

Θ, θ See above

W Maximum allowed number of new and

cancelled interconnections per period

Additional Constraint

$$\sum_{j \in \theta}(1 - y_{j1}^T) + \sum_{i \in \Theta}(1 - y_i^P) + \sum_{j \notin \theta} y_{j1}^T + \sum_{i \notin \Theta} y_i^P \leq W \qquad (10.28)$$

Table 10.3 Constant Parameters

Parameter Description	Interval
Growth of traffic per route per period	[15%, 25%]
Growth of capacity per period	[15%, 25%]

Table 10.4 Scenario-dependent Parameters

Bit	Parameter Description	Bit = 0	Bit = 1
1	Number of periods p	20	10
2	Change of each of the following cost terms per period: Fixed peering costs, fixed transit costs, variable costs, IXP costs	[−20%, +5%]	[−10%, 0%]
3	All providers available at beginning	Yes	No

traffic, capacity and cost change. If this option is not chosen, 25% of the providers are not available in period 0 and become available in a random period of the simulation (each period having the same probability). We now first evaluate the dependency of the results of each policy on the parameters of the policy for scenario 7 and then compare all of the policies for each scenario.

10.4.2.1 Dependency on Policy Parameters

We start by analysing the average number of changed interconnections and the probability of a period without any changes. These change metrics are depicted in Figure 10.7 (b)

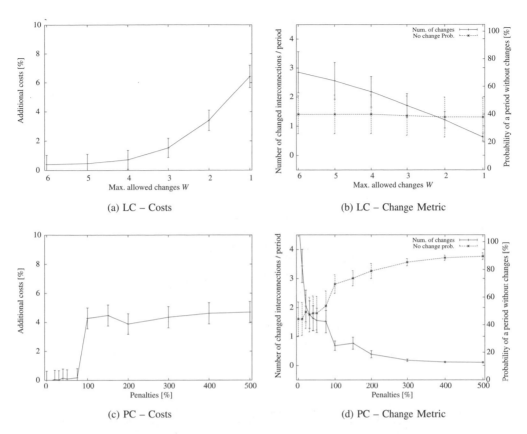

(a) LC – Costs

(b) LC – Change Metric

(c) PC – Costs

(d) PC – Change Metric

Figure 10.7 Evaluation of the Dynamic Strategies, Dependency on the Policy Parameters

for different parameters W that limit the number of permissible changes per period for the *LC policy*. The average of $n = 25$ problem instances and the 95% confidence interval are shown. We can see that the probability that no change occurs in a period remains low, independent of W. The LC policy allows a number of changes in each period and thus uniformly distributes the amount of change over all the periods. This leads to the low probability evident in the figure. The number of changed interconnections per period obviously decreases with W. The additional costs of LC relative to the results for $W = \infty$ are shown in Figure 10.7 (a). They increase by only 6% if W is decreased from 6 to 1.

For the *PC policy*, the PC were calculated as a constant percentage of the fixed peering: transit costs for establishing a new or cancelling an existing interconnection. The same procedure was followed for the IXPs. For PC of up to 100%, the probability that no change occurs increases while, at the same time, the number of changes per period decreases. At the same time, the costs increase slightly. This is a nice result; the PC policy can influence the amount of change better than the LC. However, for very high PC above 100% the amount of change is only slightly decreased. This is reasonable since the amount of change seen for very high PC is the change that is absolutely necessary, such as choosing a new transit provider because traffic demand exceeds the capacity of the existing interconnections. The PC approach is reasonable and practical. If there are

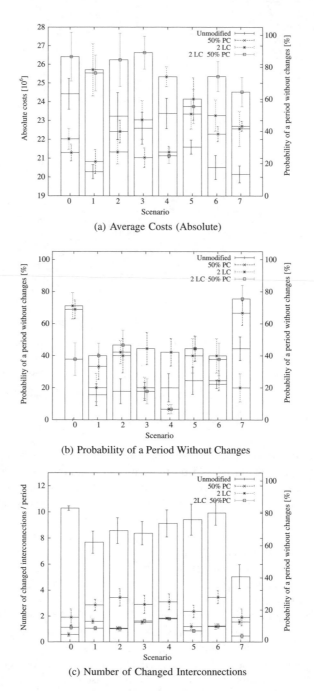

(a) Average Costs (Absolute)

(b) Probability of a Period Without Changes

(c) Number of Changed Interconnections

Figure 10.8   Evaluation of the Dynamic Strategies, Dependency on the Scenario

technical limitations restricting the amount of change, LC can be used in addition to PC. We now compare the approaches for the range of scenarios.

10.4.2.2 Evaluation of the Different Scenarios

The results for the different scenarios are shown in Figure 10.8 for the unmodified OPT Model 10.1, the PC policy with 50% penalty costs, the LC policy with $W = 2$ and the combination of PC and LC. The costs can differ by up to 32% between the policies. No one policy leads to clearly lower or higher costs than any other in all scenarios. The combined policy and LC lead to the fewest number of changes. The unmodified algorithm does not control change and thus leads to the highest change rate. PC and the combined policies lead to the highest probability of not having to change an interconnection in one period. To conclude, we can recommend using the combination of PC and LC, as it provides the most robust policy. However, the results also show that the policy and the parameter of the chosen policy strongly depend on the scenario.

10.5 Summary and Conclusions

The interconnection-related costs form one of the highest cost factors for an INSP and are therefore highly important for the efficiency of a network. In this chapter, the optimisation potential with respect to interconnection-related costs was evaluated. Several optimisation models for interconnections between providers were presented. We have shown how to find the most efficient set of peering and transit partners for a provider. Simulations show that our approach is superior to typical real-world heuristic approaches. They also show that, of the heuristics, the one performs best that connects to all possible peering partners at an IXP and chooses the optimal set of IXPs. However, the presented exact algorithm can still save 5% more total interconnection costs. Furthermore, it can be easily extended to take other issues like reliability and QoS into account.

Besides efficiency, the interconnection mix can significantly influence the achievable QoS, for example, via the AS hop count. We derived and analysed strategies for optimising both QoS and efficiency. We have also presented and discussed several ways of extending the basic strategies to take reliability issues into account. In the last part of the chapter, we have shown how to extend the basic models to the dynamic problem situation by evaluating whether a given set of peering and transit partners is still optimal, considering changes in the traffic mix or cost structure of the involved providers. We have also considered the administrative costs of changing peering and transit partners and evaluated different approaches in simulations.

Part IV

Traffic and Network Engineering

11

Traffic and Network Engineering Overview

So far in this book, the network architecture and the interconnections were analysed. Apart from choosing and tuning a network architecture and connecting its network with selected other Internet Network Service Providers (INSPs), an INSP also has to engineer its network: Nodes (Points of Presences (POPs) and routers) have to be connected by links, and these links have to be dimensioned and upgraded at regular intervals. Furthermore, with traffic engineering, the routing of traffic flows through the network can be influenced to increase the performance of the network. In this part, we investigate these engineering measures. We call the long-term engineering measures that influence the topology – for example, the link bandwidth – *network engineering*. State of the art in network engineering is discussed in Section 11.1. The medium-term engineering measures that assume the topology to be fixed and instead influence the routing of the flows through the network are called *traffic engineering* and discussed in Section 11.2. Traffic engineering and network engineering algorithms use traffic predictions between node pairs as input in the form of a traffic matrix. Because of their relevance to both topics, traffic matrices are discussed separately in Section 11.3.

11.1 Network Design and Network Engineering

In works related to network and traffic engineering, the term commodity is often found in the literature. With respect to IP networks, a *commodity* is a traffic flow between a specific pair of nodes. To be consistent in terminology with the rest of this book, the term *flow* is preferred to *commodity* here. The size of that flow is normally given as an entry in a traffic matrix.

With respect to routing, these works typically distinguish between non-bifurcated and bifurcated routing. Non-bifurcated routing (also called *singlepath routing)* implies that a flow, or commodity, is routed over a single path and cannot be split up to be routed over multiple paths. The latter is allowed if bifurcated or *multipath routing* is used.

The Competitive Internet Service Provider: Network Architecture, Interconnection, Traffic Engineering and Network Design
Oliver Heckmann © 2006 John Wiley & Sons, Ltd

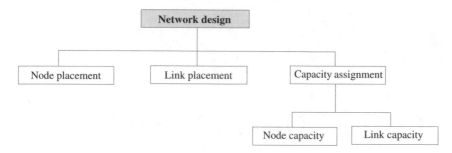

Figure 11.1   Network Design

11.1.1 Network Design

We distinguish between network design and network engineering. *Network design* (see Figure 11.1) is concerned with synthesising a new network topology. Network design consists of three parts: Node placement, link placement and capacity assignment (to nodes and links). The *node placement* sub-problem is about geographically placing the nodes of the topology that resemble the POPs of the INSP. The *link placement* sub-problem deals with connecting the nodes with each other while the *capacity assignment* problem assigns capacities (bandwidth, buffer, etc.) to the nodes and links. For designing a completely new topology, all three of these sub-problems have to be solved. Existing works often treat only a subset of these optimisation problems. The node placement especially is often assumed to be given and fixed.

Bley *et al.* (2004) describe how the GWiN backbone of the German Research Network Deutsches Forschungsnetz DFN (see Figure A.1 in the Appendix) was designed with a 2-level hierarchical approach. The set of nodes is given, so no node placement problem has to be solved. Each node becomes either an access node that is connected to a single backbone node or it becomes a backbone node that can be connected to any other node. For the capacity assignment process, a discrete list of nodes and link configurations are used.

Plain (non-bifurcated) shortest-path routing is used in that network. This fact can be exploited to simplify the mathematical programming model that results from link placement and capacity assignment. Using a Lagrangian relaxation, the optimisation problem can be split into two sub-problems: One is finding a valid network structure and hardware installation; it can be formulated as a mixed integer programming problem and solved with standard methods. The second one is the routing problem that can be solved efficiently by any shortest-path algorithm. With this approach, the design optimisation problem was solved for the size of the GWiN topology in 15 minutes on a standard PC with an optimality gap of less than 0.6%.

A classic network design paper is that of Gavish (1992). That work contains a general overview of network design. In addition, an approach to simultaneously solve all three network design sub-problems for given end-user locations and a given traffic matrix is presented. For node placement, a set of possible candidates for the backbone node locations is assumed to be given; the chosen nodes are connected by links that lead to fixed and traffic dependent costs. Besides that, a static singlepath routing scheme for the flows is derived. Quality of Service (QoS) is accounted for by including the delay between end-user nodes into the objective function. This, however, makes the resulting combinatorial optimisation problem non-linear. With a Lagrangian approach, the optimisation problem can be split into sub-problems that can be solved more easily and lead to a lower bound of the overall optimisation problem. This bound can be used to calculate the optimality gap for feasible solutions, which can be obtained by the simple heuristics described in the paper.

There is a vast amount of other works regarding network design. Han *et al.* (2000) present an approach with a more realistic cost function for the links; that function is sequence of steps as function of the link capacity (similar to the Internet Exchange Point (IXP) cost function used for the interconnection optimisation problem in Chapter 10; see Figure 10.1). The authors show that the optimisation problem can be reformulated into a simpler optimisation problem where each link is replaced by multiple links with constant costs and a capacity limit. Genetic algorithms have been successfully used to solve network design problems for example, in Berry *et al.* (1998) and Palmer and Kershenbaum (1995); a combined genetic algorithm and linear programming approach is presented in Berry *et al.* (1999). Simulated annealing is used in Randall *et al.* (2000) and tabu search in Glover and Laguna (1998). For more works, we refer to the related work cited in the references above and to the standard network design book Kershenbaum (1993).

11.1.2 Network Engineering

Contrary to network design problems that are about the synthesis of a *new* topology, network engineering is about improving an *existing* network topology either by changing nodes and/or links (*structural engineering*) or by expanding the capacity of an existing and otherwise unchanged network (*capacity expansion*).

New networks have to be designed only rarely as practically all INSPs already have existing networks. Therefore, network engineering is a more frequent and important challenge for INSPs. Traffic volumes are growing by 70–150% per year; see Odlyzko (2003). The bandwidth of a network has to be doubled roughly every year to keep pace with these rates. This leads to the conclusion that capacity expansion – especially link capacity expansion – is the most important of all network engineering challenges. Later in this part, we will therefore place the focus on link capacity expansion.

Hasslinger and Schnitter (2004) investigate link capacity expansion and traffic engineering for IP networks. On the basis of their experience with the IP backbone to Deutsche Telekom, they report capacity increase factors ranging to beyond a factor of 2 per year. They present a capacity expansion heuristic that takes into account the influence of traffic engineering on the network utilisation. Their work is similar to our experiments in Section 13.2 and discussed in that context.

Optimally expanding telecommunication network facilities have been studied in a number of works; for an overview see Chang and Gavish (1993, 1995) and Dutta and Lim (1992). Chang and Gavish (1993, 1995) present a Lagrangian decomposition approach for a rather complex network engineering problem for telecommunication networks. The approach is well suited to derive a development plan towards a given target network in a certain number of periods. The solved optimisation problem is a combined structural engineering and capacity expansion problem; nodes are considered to be given and fixed but links can be placed and upgraded. The objective is to minimise the net present worth

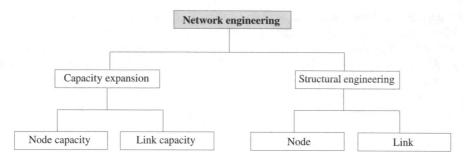

Figure 11.2   Network Engineering

of total invested costs for the given number of periods. This is contrary to our approach in Section 13.2, where the interest costs for the capacity expansion and fictive congestion costs are evaluated.

Chang and Gavish (1993, 1995) considered the fixed costs for installing a conduit for a bidirectional link between two nodes, fixed costs for upgrading the capacity of a link, and the capacity costs themselves. This leads to possible cost savings by installing excess capacity in a current period to avoid the fixed costs of later periods. Capacity is modelled by a continuous variable. The cost model and the continuous capacities are tailored for telecommunication providers and carriers but they are not suited for INSPs that typically lease the lines for their links from carriers at discrete capacities.

Dutta and Lim (1992) studied the installation of transmission capacity over time in a communication network where the nodes and the possible links are given; new nodes can be added over time but the decision of which nodes to add is not modelled. The optimisation problem is thus a combined link structural engineering and link capacity expansion problem (see Figure 11.2). Considered costs are the one-time installation costs and the per-period operation costs for links. The latter cost terms are assumed to exhibit economies of scale. The objective is to minimise the net present worth of total costs. Discrete capacities are modelled. A performance constraint based on the delay of a M/M/1 queue is also included in the model. The model is finally solved with a Lagrangian approach.

An interesting comparison of the bandwidth market and the financial market is made in d' Halluin *et al.* (2002). In that paper, capacity expansion under demand uncertainty is studied with modern financial option pricing methods. The perspective is that of a carrier that faces extremely volatile future revenues. The paper can help in explaining the current overcapacity in available bandwidth but cannot directly be transferred to the capacity expansion of INSPs that typically go to satisfy a relative constant increase in traffic volumes.

11.2 Traffic Engineering

The IETF Traffic Engineering Working Group gives the following definition of traffic engineering[1]: *Internet traffic engineering is defined as that aspect of Internet network*

---
[1] See http://www.ietf.org/html.charters/tewg-charter.html.

engineering concerned with the performance optimisation of traffic handling in operational networks, with the focus of the optimisation being minimised over-utilisation of capacity when the other capacity is available in the network.

Traffic engineering influences the forwarding decision of the routers with a specific goal in mind; it could for example, re-route flows so that they avoid a known bottleneck. Traffic engineering is basically an optimisation problem; the traffic engineering goal reflects itself in the objective function. In Chapter 12 of this book, different traffic engineering algorithms and different objective functions are discussed and evaluated.

If a plain IP forwarding architecture (see Section 6.3.1) is used, traffic engineering can be done by influencing the link weights of the routing protocol (see Section 6.4.1). Multiprotocol Label Switching (MPLS) as forwarding architecture (see Section 6.3.2) directly supports traffic engineering because it allows the creation of label switched paths independent of the routing protocol. It is therefore the preferred choice for traffic engineering.

The most straightforward *online algorithms* for routing traffic flows are based on the shortest-path algorithms such as Dijkstra (1959). The routing of flows is determined sequentially for all flows; a flow is routed on its shortest path where only links that have sufficient residual (remaining) capacity are considered. This type of algorithm can create bottlenecks and lead to underutilisation; see Suri *et al.* (2003).

A variant of the shortest-path algorithm called widest-shortest path is presented in Guérin *et al.* (1997). Here, the smallest residual link capacity of a path is maximised. The impact on other flows is neglected and still, bottlenecks can occur as shown in Suri *et al.* (2003).

The minimum interference routing algorithm of Kodialam and Lakshman (2000) takes the impact that the decision to route a flow on a certain path has on the maximum flow routable between other node pairs into account; this is called *interference*. An advanced version of this algorithm is presented in Suri *et al.* (2003); they use the solution of an off-line multicommodity flow problem (see the following text) based on an estimated traffic matrix as guidance for the on-line routing algorithm.

The term *multicommodity flow problem* designates a class of optimisation problems that is used as the basis for many off-line traffic engineering algorithms; see for example, Ahuja *et al.* (1993); Gondran and Minoux (1984); Leighton *et al.* (1995); McBride (1998) and Stein (1992). In a capacitated graph, multiple commodities (demand, traffic flows) have to be routed. A commodity is defined by a source and destination node, a size and in some cases a revenue. The multicommodity flow problem is therefore a generalisation of the well-known maximum flow problem as described by Ford and Fulkerson (1956).

For the typical type of multicommodity flow problems, the objective is to find the routing for a subset of all commodities that conforms to the capacity of the network and maximises the revenue obtained from the routed commodities. Solution algorithms thus route traffic and impose admission control on the flows. We call this class of problems the *revenue maximising multicommodity flow problems* or traditional multicommodity flow problems. However, INSPs will often have a network of sufficient capacity or will not have the possibility of admission control, for example, because they use an overprovisioned best-effort network. In that case, the optimisation problem is different: Route the traffic flows through the network so that the general QoS is maximised. We name this type of problem the *QoS maximising multicommodity flow problem*.

In the *path selection formulation* of multicommodity flow problems, the possible paths for a commodity/flow are given; typically they are determined in a preprocessing step before the actual optimisation. In a multiservice network, for flows with strict delay requirements typically only very short paths are considered while for flows with less strict delay more and longer paths can also be considered. In the *explicit routing formulation*, sometimes also called link-based formulation, no paths are precalculated, they are calculated during the optimisation process. We present an experimental evaluation of these two formulations in Section 12.4.

In the *singlepath formulation* of a multicommodity flow problem, a commodity/flow must be routed along a single path (non-bifurcated singlepath routing); this formulation is also often called integer formulation because as a combinatorial optimisation problem it can be modelled as a MIP (mixed integer program). Another formulation is the *multipath formulation*; here, a commodity can be split up along multiple paths (bifurcated multipath routing). This problem can be formulated as a linear programming model and can thus be solved in polynomial time.

In Chapter 12, the range of QoS maximising multicommodity flow problems are modelled and evaluated as LP/MIP optimisation problems in path selection and explicit routing formulation as well as singlepath and multipath formulation.

The work of Mitra and Ramakrishnan (2001) is based on the *revenue maximising multicommodity flow problem*. Two service classes are considered (QoS and best-effort). The revenue is modelled linear to the amount of carried data. For the QoS traffic, possible paths are precalculated while the best-effort flows can be routed freely through the network. The QoS traffic is routed through the network first, the best-effort traffic is then routed based on the remaining bandwidth. This complex, combined, optimisation problem is decomposed into three layered sub-problems and a scalable solution algorithm is presented in Mitra and Ramakrishnan (2001).

Bessler (2002) extends the multicommodity flow problem to multiple periods by considering the changes to the existing LSPs of the previous period. The idea is to reduce the number of changes by penalising them in the objective function as they lead to signalling overhead and a risk of service disruptions. Another work considering the trade-off between the network utilisation and the signalling/processing overhead is by Scoglio *et al.* (2001).

The multicommodity flow problem is extended with an auction-based mechanism in Bessler and Reichl (2003). Here, a bid is associated with each commodity/flow and bandwidth is distributed according to the bid order.

For the *QoS maximising multicommodity flow problems*, there are different approaches to formulate the objective function. A typical approach is to minimise the bottleneck utilisation of the network, see for example, Hasslinger and Schnitter (2002a,b); Lin and Wang (1993); Poppe *et al.* (2000) and Roughan *et al.* (2003). The motivation behind that is the fact that the QoS a flow receives is mostly influenced by the bottleneck it passes through; the utilisation of the bottleneck of a network thus determines the worst performance that flows can receive. In the next chapter, we evaluate different objective functions for the traffic routing problem and show that a congestion cost function should be preferred as objective function; for OSPF routing such a congestion cost function is used in Fortz and Thorup (2002).

Hasslinger and Schnitter (2002a,b) investigate the QoS maximising multicommodity flow problem, minimising the bottleneck utilisation. Besides solving the optimisation problem with LP/MIP methods or applying the max-flow-min-cut principle, the authors present a heuristic for the singlepath formulation of the problem based on simulated annealing. In simulations, the authors show that the maximum utilisation can be decreased by up to 42.4% compared to the utilisation with the shortest path routing. Also, the simulations indicate that the benefit of multipath routing over singlepath routing is rather low; this can also be observed in our experiments in the next chapter.

In Poppe *et al.* (2000), traffic engineering for a network with Diffserv Expedited Forwarding (EF) traffic and best-effort traffic is studied. Two traffic engineering optimisation problems are solved for the different traffic classes. For the EF traffic, the maximum utilisation is minimised as primary objective and the average load as secondary. An equivalent traffic model is also included in our experiments in the next chapter. For the best-effort traffic, fairness is maximised as the primary objective and the throughput as the secondary one. It can be argued how important fairness is for a profit-maximising INSP.

The results of the paper are that traffic engineering can significantly improve the traffic handling capabilities of a network. The findings of that paper also show that the results improve only a little if multipath routing is used instead of singlepath routing and that in the multipath case only very few flows are actually split up and routed among multiple paths. These results are consistent with our results that will be presented in the next chapter.

The off-line traffic engineering methods use a traffic matrix as input to determine the routing. In some cases, the traffic matrix can be known exactly; for example, in Diffserv networks, when only flows are considered for which SLAs exists (see Section 6.2.4.2) or in networks with reservation in advance. Normally, however, the traffic matrix is not known exactly and has to be estimated based on measurements. Traffic matrix estimation is a challenge for INSPs and discussed in detail in the next section. Roughan *et al.* (2003) asks the important question: *If traffic engineering is done based on the estimated traffic matrix, how well does it perform on the real traffic matrix?* They use the maximum utilisation as objective function for the traffic engineering algorithm, optimise the routing based on an estimated traffic matrix and verify the performance based on the real traffic matrix. The results indicate that OSPF weight optimisation combined with tomographic traffic matrix estimation (see below) performs very well, mainly because OSPF optimisation was robust to the errors found in the traffic matrix estimation. The MPLS style optimisation can determine better routing schemes but is also less robust according to Roughan *et al.* (2003).

11.3 Traffic Matrix Estimation

A traffic matrix M describes the average rate r_{ij} for a given time interval between the ingress nodes i and egress nodes j of a network.

$$M = \begin{bmatrix} \cdots & \cdots & \cdots & \cdots \\ \cdots & r_{i\,j-1} & \cdots & \cdots \\ \cdots & r_{ij} & r_{i+1\,j} & \cdots \\ \cdots & \cdots & \cdots & \cdots \end{bmatrix}$$

Traffic matrices form the input for network design and traffic engineering optimisation problems. Therefore, it is important to determine traffic matrices in real networks. However, measuring a traffic matrix is not a trivial task. Benameur and Roberts (2002) give an overview over the two distinct approaches to measure a traffic matrix: The *direct measurement* approach as advocated by Feldmann *et al.* (2000) uses NetFlow (2004) to collect flow information. This information is evaluated off-line to derive the traffic matrix using the routing tables active at the measurement time that also have to be recorded. This approach is storage space and router-Central Processing Unit (CPU) intensive and requires all routers to support NetFlow (or a similar product) but contrary to other approaches allows them to derive the point-to-multipoint traffic matrix. A point-to-point traffic matrix M models the traffic between ingress node i and egress j while the point-to-multipoint traffic matrix $\tilde{M}$ models the traffic and ingress node i and captures the fact that this traffic can exit at more than one egress j.

Another direct measurement approach that is less resource intensive is described by Schnitter and Horneffer (2004); it works for networks that employ label switching (MPLS). Every LSP has a byte-counter measuring the traffic using this LSP. Thus, if an MPLS network is built as a full mesh of LSPs, the traffic matrix can be measured directly. However, due to scalability and load balancing reasons, a full mesh of LSPs is not often used. The technique introduced in Schnitter and Horneffer (2004) can measure the traffic matrix directly if the router has a byte counter for each Forwarding Equivalence Class FEC[2]. It does not depend on the routing method (explicit LSPs with traffic engineering or plain shortest-path routing).

Most of the other works favour *deriving the traffic matrix from link measurements*, as they are more readily available for all router interfaces via SNMP (simple network management protocol) in production networks. The problem with this approach is that estimating the traffic matrix is an ill-posed inverse linear problem: In a network with N ingress/egress nodes, the traffic matrix size is $O(N^2)$ as it contains entries for each node pair. However, there are only $O(N)$ link measurements as the number of links is the average node degree times the number of nodes. Therefore, the problem becomes massively under-constrained for large N as the number of variables then exceeds the number of equations (if the problem is formulated as a linear equation system). To solve this problem, additional assumptions for example, about the traffic and the routing have to be made. Approaches to this problem can be classified into statistical tomographic methods, optimisation-based tomographic methods and other methods:

- The *Statistical tomographic methods* use higher order statistics of the link load data like the covariance between two loads to create additional constraints. Examples are Cao *et al.* (2000); Tebaldi and West (1998) and Vardi (1996). Vardi (1996) and Tebaldi and West (1998) assume a Poisson traffic model; Cao *et al.* (2000) assume a Gaussian traffic model.

---

[2] The exact requirements are that the statistics include each LSP through a router, incoming and outgoing labels, the FEC, the outgoing interface, and the byte counter. These requirements are fulfilled by the most common router operation systems like Cisco's Internet Operating System (IOS) and Juniper Network Operating System (JUNOS) (see Schnitter and Horneffer (2004)).

- The *Optimisation-based tomographic methods* select a solution out of the solution space of the under-constrained problem that optimises a certain objective function using methods like linear or quadratic programming. Goldschmidt (2000) is a simple example for this approach.

- Classified as *other methods* are approaches that combine the tomographic methods with other methods like gravity or choice models. Medina *et al.* (2002) use a logit choice model that captures the choices of users (where to download from) and network designers (how to interconnect the POPs). The decision process is modelled as a utility maximisation problem.

    Zhang *et al.* (2003a) combine a optimisation-based tomographic methods with a generalised gravity model. A gravity model can, for example, be used to estimate the traffic between edge links by assuming that the traffic between i and j is proportional to the total traffic entering at i multiplied with the total traffic exiting at j.

    Zhang *et al.* (2003b) uses an information theoretic approach that chooses the traffic matrix consistent with the measured data so that it is as close as possible to a model in which the source and destination pairs are independent and therefore the conditional probability $p(j|i)$ that source i sends traffic to j is equal to the probability $p(j)$ that the whole network sends traffic to j.

11.4 Summary and Conclusions

In this chapter, network design and network engineering were introduced. Network design is concerned with building and network engineering with upgrading a network. Then, traffic engineering was presented. It influences the routing of traffic flows through the network to increase the performance of the network. Traffic engineering and network engineering algorithms use traffic predictions between node pairs as input; we call this a traffic matrix. Ways for measuring and predicting the traffic matrix were discussed in the previous part of this chapter.

The rest of this part is structured as follows. In Chapter 12, the influence of traffic engineering on the QoS of a network and the costs and efficiency of that network are analysed. Traffic engineering strategies and performance metrics are discussed and then evaluated in a series of simulation experiments.

In Chapter 13, network engineering is discussed. The focus of that chapter lies on capacity expansion because providers have to expand the capacity of their network regularly – the Internet traffic has been growing exponentially in the last years (see Odlyzko (2003)) and there is no indication why this should not continue for the future. The capacity expansion problem therefore has to be solved much more often than the general design of a new network topology. According to the system-oriented approach of this book, we start by showing how network engineering and the network architecture in the form of QoS systems interact. Then we present and evaluate different strategies for capacity expansion. After that, we investigate the interaction between traffic engineering and capacity expansion strategies in further experiments. Finally, we investigate the elasticity of traffic matrices resulting from the elastic behaviour of TCP and the impact on capacity expansion in an analytical study.

12

Evaluation of Traffic Engineering

Traffic engineering is concerned with minimising the over-utilisation of capacity when other capacities are available in the network by re-routing traffic flows. A *traffic flow* in the context of this chapter is a macroflow consisting of all packets entering the network at the same ingress and exiting at the same egress node. The traffic flows of all ingress-egress node pairs are specified in a traffic matrix. Throughout this chapter, we assume that the traffic matrix is given. For Multi-Protocol Label Switching (MPLS) networks, the traffic matrix can be measured online and exactly by the method described in Schnitter and Horneffer (2004); for a general discussion of traffic matrix estimation techniques we refer to Section 11.3.

In this chapter, we investigate how traffic engineering influences the efficiency and the Quality of Service (QoS) of a network and explore the optimisation potential with an evaluation of different traffic engineering strategies.

We assume MPLS or an equivalent forwarding architecture (see Section 6.3) is used which allows us to explicitly establish the path on which a flow is routed through the network. Different traffic engineering strategies that differ in their objective function, their constraints and whether they can split up a macroflow to be routed along multiple paths (multipath routing) or not are investigated. Their performance with respect to different performance metrics is evaluated. In addition, the performance gain of the best traffic engineering strategies compared to a plain shortest path routing solution is evaluated. It measures the additional QoS achievable with traffic engineering and is a measure of the possible efficiency gain of traffic engineering.

As the currently dominant QoS architecture is an over-provisioned best-effort architecture, this architecture is assumed for the experiments in this work. Most of the results here, however, are also helpful for other architectures, e.g. Diffserv. One straightforward approach is to employ traffic engineering techniques sequentially for all traffic classes, starting with the highest priority traffic. The traffic of the next highest priority is then traffic engineered on the network that has capacity left that is not used by the higher priority traffic and so on. More sophisticated approaches for traffic engineering in the context of other QoS architectures, are discussed in Section 11.2.

The Competitive Internet Service Provider: Network Architecture, Interconnection, Traffic Engineering and Network Design
Oliver Heckmann © 2006 John Wiley & Sons, Ltd

For an evaluation of the traffic engineering performance of a network for a given traffic matrix, the routing determined by the traffic engineering algorithm has to be measured. The *routing* in this context consists of the paths chosen for the different macroflows. Therefore, we start by discussing different performance metrics for evaluating the routing. The average path length is an obvious performance metric. Besides it, several other performance metrics are possible, for example, the bottleneck utilisation. They are discussed in Section 12.1. In Section 12.2, different strategies for solving the QoS maximising multi-commodity flow problem are introduced; they are evaluated in a series of experiments in the rest of the chapter. The experiment set-up is described in Section 12.3. After that, the experiment results are presented and discussed:

- In the first experiment (Section 12.4), we compare the path selection and explicit routing formulation of the optimisation problem.
- A general performance evaluation of a large number of traffic engineering strategies are presented in Section 12.5. We start with a detailed evaluation of the strategies and all performance metrics in a basic experiment and then vary several parameters of the basic experiment – e.g. the used topology – to evaluate their influence.
- In Section 12.6, the performance loss of singlepath algorithms compared to multipath algorithms is evaluated.
- The most successful strategies need a number of precalculated paths. In Section 12.7, the influence of the precalculated paths on the performance of these strategies is evaluated.

Finally in Section 12.8, the conclusions are drawn and we give recommendations whether, and how to use traffic engineering.

12.1 Traffic Engineering Performance Metrics

For the evaluation of traffic engineering, the performance of the traffic flows routed through the network has to be evaluated. In this section, we discuss several metrics that can be used to evaluate the performance of the routing.

12.1.1 Path Length

Minimising the average path length between two nodes is an obvious and straightforward traffic engineering goal: With respect to different length metrics, minimising the path length is the objective of most standard interior routing protocols like Open Shortest Path First (OSPF) (see Section 6.4.1). The motivation behind the path length as performance metric is that the longer the path becomes, the more network resources are consumed and the higher the propagation delay becomes. As in a congested network the queuing delay can easily exceed the propagation delay of a hop; re-routing a flow so that it takes more hops through the network can still lead to improved overall delay besides a reduced loss probability. This observation is the basic motivation for doing traffic engineering instead of plain shortest path routing. Nevertheless, the path length remains an important performance metric for traffic engineering solutions.

12.1.2 Maximal Bottleneck Utilisation

The utilisation u_l of a link l is defined as the load l_l per capacity (bandwidth) c_l

$$u_l = \frac{l_l}{c_l} \tag{12.1}$$

The utilisation of a link is an average over a certain period[1], typical utilisation metrics measured in Internet Protocol (IP) networks are based on 5 minute, 15 minute, 2 hour and 24 hour averages.

The maximum utilisation $\max_l\{u_l\}$ describes how loaded the bottleneck link of the topology is. QoS parameters such as delay and loss are a (non-linear) function of the utilisation of a link. Because of the bursty nature of network traffic (see Section 5.1), losses occur long before an average utilisation of 100% is reached. Minimising the maximum utilisation therefore indirectly improves the QoS on the bottleneck link – the most critical link – and creates a zone of security against unpredicted traffic increases. Therefore, minimising this metric is often the dominating traffic engineering goal in related works, for example, see Hasslinger and Schnitter (2002a,b); Lin and Wang (1993); Poppe *et al.* (2000); Roughan *et al.* (2003).

One disadvantage of this metric is that it focuses exclusively on the bottleneck links, while ignoring the other links.

12.1.3 Average Utilisation

Instead of evaluating the *maximum* utilisation – that is, the utilisation of the bottleneck link – one could also evaluate the *average* utilisation of the network. This has the advantage that no link is ignored. However, a network with some highly loaded and some lightly loaded links could show the same average utilisation as a network with only medium loaded links. As the QoS flow experiences – for example, the loss probability – is largely determined by the most utilised link on its path and not by the average utilisation along its path this metric can be misleading. This is shown in some of the experiments below.

12.1.4 Average Load

The average utilisation metric does not take into account that there might be large differences in the capacity c_l of the links in the topology. The average utilisation is influenced by low capacity links the same way as by high capacity links. High capacity links, however, typically carry more traffic flows and can therefore be expected to influence more flows (or users) than smaller links. If the utilisation metric is weighted with the link capacity, the average load can be calculated.

This metric has the same disadvantages as the previous one. It is used in Poppe *et al.* (2000) as a secondary objective, for example.

---

[1] On a very short timescale a link is either 100% utilised (data is currently being transmitted) or 0% (no data is currently being transmitted).

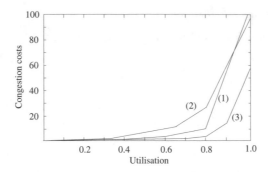

Figure 12.1 Congestion Functions

12.1.5 Congestion Costs

The high-level primary goal of traffic engineering should be to maximise the overall utility of the customers given the available network resources. This is a special form of the network efficiency we used throughout this book. The utility depends on the application, on the traffic mix and on network parameters like the loss or the queuing delay (see Chapter 8). On the timescale of traffic engineering, it is the average of the network parameters like loss and delay that can be influenced. The network parameters are a non-linear function of the utilisation or the load situation on a link. Assume for example, a M/M/1/B queue (see Section 3.1.5). The M/M/1/B queue is not the most realistic representation of an Internet link but is a commonly used one because it can be mathematically handled very well. For more realistic queuing models, see the queuing models of Appendix D. They, however, show a similar basic behaviour.

Loss and queuing delay of the M/M/1/B are depicted in Figure 3.4. As one can see there, the loss and delay are non-linear convex functions of the utilisation. In Section 13.1, we present a more detailed analysis based on the utility for various QoS systems using packet-level simulations instead of the M/M/1/B formulas. The results there also point out a convex relationship.

The convex relationship between the utilisation and network congestion indicators like loss and delay has an important implication for traffic engineering. If the load of one highly utilised link is reduced by a certain amount due to a routing change, the overall performance can improve even if the load on multiple other (but not so highly utilised) links is increased because of the routing change. The lower the utilisation becomes, the less can be gained by re-routing. This behaviour is not correctly expressed by any of the above-mentioned metrics. Therefore, we propose to use the following metric called *congestion costs*[2] that captures this non-linear behaviour. Figure 12.1 presents three different stepwise linear convex congestion cost functions $p^x(u)$ that we use throughout this chapter to model how the congestion situation of a link depends on the utilisation of the link. Fortz and Thorup (2002) use a very similar metric to evaluate OSPF-based traffic engineering. The parameters of the congestion cost functions are arbitrarily chosen but roughly oriented to

---

[2] The reason for calling this congestion measure "costs" becomes more visible in Chapter 13 where it has to be added to the costs for expanding the capacity of the network and therefore has to have the same unit as true monetary costs. For the experiments of this chapter, the scale and unit of this congestion measure do not matter and do not influence the results.

Figure 3.4 and the results of Section 13.1. The default congestion cost function is labelled (1) and used in the following experiments by default if nothing else is mentioned; it is varied in Section 12.5.2.

The congestion costs are calculated for every link and can be summed up for the complete topology in the following two ways:

- Weighted congestion costs: $\sum_l l_l \cdot p^x(u_l)$
  The motivation behind weighting the congestion costs with the load l_l follows the same argument as for the average load versus average utilisation metric above: Links with a high load are likely to affect more customers and can therefore be judged more important than links with a lower load.
- Unweighted congestion level: $\sum_l p^x(u_l)$
  For comparison reasons, we will also investigate the unweighted congestion costs metric in this chapter.

12.2 Traffic Engineering Strategies

We use optimisation models to describe different traffic engineering strategies mathematically, using the following notation:

A network (η, ζ) consists of a set of nodes η and a set of directed links $l_{ij} \in \zeta$ with link l_{ij} connecting node i to j. A link l_{ij} has a capacity c_{ij}.

A subset η_e of the nodes is marked as *edge nodes*. Customers and interconnection partners are connected to these nodes; therefore the edge nodes are potential sources and sinks for the traffic flows while the other nodes $n \in (\eta \backslash \eta_e)$ only forward traffic (*core nodes*).

There are F traffic flows f that have to be routed through the network. A traffic flow f is characterised by its ingress node $i_f \in \eta_e$ and egress node $e_f \in \eta_e$ and its size r_f; the size of the flow is its traffic volume or – if we assume time periods of a fixed duration as a basis – its average transmission rate.

The ingress and egress nodes (i_f, e_f) of flow f are connected by a set of different paths ρ_f. Each path $p \in \rho_f$ is an ordered set of links $\phi_p = \{l_{i_f j_1}, l_{j_1 j_2}, \ldots, l_{j_k e_f}\}$ from the ingress i_f to the egress node e_f. For our analysis, we assume that the length l_p of a path p is the number of links it contains; for a real network other factors such as path length metric could also be taken into account, for example, the propagation delay.

12.2.1 Traffic Engineering Objectives

The overall goal of traffic engineering is to optimise the routing of flows through a network of given and fixed capacity; traffic engineering is thus an optimisation problem. Several specific objectives can be formulated as an objective function of the traffic engineering problem. As several objectives can be optimised at the same time, the optimisation problem can be a multi-objective optimisation problem[3]. The different objective functions can be combined, either as prioritised objectives (multilevel programming) or as weighted summed objectives. In the first case, the problem is first optimised with the primary objective function only in mind and among all the solutions that optimise

---

[3] For multi-objective optimisation see Eschenauer *et al.* (1990); Statnikov and Matusov (1995).

the primary objective function the one that optimises the secondary objective function is selected. In the latter case, both objective functions are added with certain weights to a single objective function and the resulting problem is then optimised for the aggregate objective function. Prioritised objective functions can be approximated with the weighted ones by giving the primary objective function a much larger weight than the secondary one; because of that, we restrict ourselves to the second approach with weighted objective functions in the optimisation problems that we present and discuss below.

In Section 12.1, metrics for evaluating the performance of traffic engineering strategies were presented and discussed. Obviously, they can also be used as objective functions for the traffic engineering problems. We do so by integrating them into the more sophisticated traffic engineering strategies below.

12.2.2 Shortest Path Routing

The shortest path routing strategy is straightforward: Each traffic flow f is routed along its shortest path p^* with $l_{p^*} = \min_{p \in \rho_f} \{l_p\}$. The shortest path can, for example, be determined with the Dijkstra algorithm (see Dijkstra (1959)). Each flow is routed along a single path only, multipath routing is never used. The shortest path routing algorithm minimises the average path length metric only, other target functions are not considered. This strategy is used as a reference because it is the default strategy of a network with a standard routing protocol and no traffic engineering functionality.

12.2.3 Equal Cost Multipath

As another reference strategy, we include an equal cost multipath algorithm. It splits a flow evenly among a given number of paths. The equal cost multipath algorithm we use has two parameters n and Δl. n denotes the maximum number of paths considered. For a flow f, the n shortest paths are determined with a modified Dijkstra algorithm. The shortest of these paths is denoted as p^*. All paths that are more than Δl hops longer than p^* are discarded. If there are more than n shortest paths left within Δl hops, those that have the most overlapping (same links) with the shortest path are discarded until only n paths are left. The traffic is split up evenly among the remaining paths. This algorithm does not directly minimise any of the metrics of Section 12.1; it is included for reference purposes only.

12.2.4 Explicit Routing

The explicit routing strategy is based on the explicit routing form of the multi-commodity flow problem (see Section 11.2). The network's topology is modelled by the set I_n and O_n that contains the ingoing and outgoing links l of node n. The explicit routing optimisation problem is given with Model 12.1 as a singlepath model and Model 12.2 as a multipath model, both with the weighted maximum utilisation and average utilisation criteria as objective function (12.2).

Variable a_{lf} describes which proportion of flow f is routed via link l. Constraint (12.3) is the flow conservation constraint: For all nodes that are not the ingress or egress node of flow f, the amount of traffic from flow f that flows into node n also

---

Model 12.1 Explicit Routing (Singlepath)

Indices

$f = 1, \ldots, F$	Flow f
$n = 1, \ldots, N$	Node n
$l = 1, \ldots, L$	Link l

Parameters

r_f	Size of flow f
I_n	Set of incoming links of node n
O_n	Set of outgoing links of node n
i_f	Ingress (start) node of flow f
e_f	Egress (end) node of flow f
w^ξ	Weight for the maximum utilisation objective
w^u	Weight for the average utilisation objective
c_l	Capacity of link l

Variables

ξ	Maximal link utilisation
a_{lf}	Routing variable, flow f is routed by this proportion on link l

$$\text{Minimise } w^\xi \xi + w^u \frac{1}{L} \sum_l \sum_f \frac{r_f a_{lf}}{c_l} \tag{12.2}$$

subject to

$$\sum_{l \in O_n} a_{lf} = \sum_{l \in I_n} a_{lf} \qquad \forall f \; \forall n \neq i_f, e_f \tag{12.3}$$

$$\sum_{l \in O_{i_f}} a_{lf} = 1 + \sum_{l \in I_{i_f}} a_{lf} \qquad \forall f \tag{12.4}$$

$$\sum_f r_f a_{lf} \leq c_l \xi \qquad \forall l \tag{12.5}$$

$$0 \leq \xi \leq 1 \tag{12.6}$$

$$a_{lf} \in \{0, 1\} \qquad \forall l \; \forall f \tag{12.7}$$

Model 12.2 Explicit Routing (Multipath)

Constraint 12.7 is replaced with the following constraint in the otherwise unchanged Model 12.1:

$$0 \leq a_{lf} \leq 1 \qquad \forall l \, \forall f \tag{12.8}$$

---

has to leave node n. Constraint (12.4) specifies that 100% of a flow f is inserted into the network at the ingress node i_f. Because of (12.3) and (12.4) no extra constraint for the egress node e_f is necessary. (12.5) forces variable ξ to the maximum utilisation of all links l and in combination with (12.6) ensures that the capacity c_l of a link l is not exceeded.

The multipath explicit routing problem can be solved with the simplex algorithm, for example (see Section 3.3), the singlepath version is harder to solve because of the binary constraint (12.7). It has to be solved with Mixed Integer Programming (MIP) solving techniques like, Branch & Bound with LP relaxation as discussed in Section 3.3. Let F denote the number of flows, N the number of nodes and L the number of links. As $O(L) = O(N)$ and $O(F) = O(N^2)$ the number of constraints and therefore the complexity of the explicit routing LP/MIP models is $O(N^3)$. The number of (computationally expensive) binary variables in the singlepath model are $O(N^3)$. As this is a rather high complexity, we next present a more efficient model for traffic engineering.

12.2.5 Path Selection

As mentioned above, the explicit routing model is of high complexity. The main reason for this is that it explicitly models the topology and thus the solution algorithm searches for paths through the network at the same time as assigning the flows to these paths so that the traffic engineering goals are optimised. For computing paths through the network, especially, there exist efficient specialised algorithms like the Dijkstra algorithm rather than the general LP/MIP solving algorithms.

Therefore, the optimisation problem can be simplified by precomputing the possible paths for all flows in a first step. Then in a second step, the path(s) for each flow are selected in a way that optimises the objective function. Precomputing the paths can, for example, be done with a (modified) Dijkstra algorithm in polynomial time. The optimisation models for selecting one or more paths for each flow among the precomputed ones are discussed below and called *path selection models*.

If all possible paths for all flows f are precomputed and used as input in the path selection models, the path selection models yield the same optimal solution as the explicit routing model. However, as for a large topology the number of possible paths is extremely high, only the shortest n paths for each flow can typically be considered in the path selection model, making the solution space of the path selection smaller than that of the explicit routing problem. In that case, it is possible that the path selection model does not find the globally optimal solution. We investigate this experimentally in Section 12.4. At first glance, this might seem a drawback, but, in actuality, the fact that the path selection models use precomputed paths gives the decision maker more control over the possible paths. The explicit routing models could route a flow over a path that is much longer than the shortest path. For the path selection models, the decision maker can limit the paths, for example, so that they do not have more than Δl additional hops than the shortest path between two nodes.

The basic path selection model is mathematically specified as a mixed integer programming (MIP) model in Model 12.3. It is a singlepath model. The multipath version of Model 12.3 is given by Model 12.4.

Model 12.3 accounts for four of the five traffic engineering goals discussed in Section 12.1. To account for the congestion costs, additional parameters and variables are necessary. Model 12.5 is an extension of Model 12.3 that also accounts for the congestion costs in the objective function.

The path selection models can be solved with the same methods as the explicit routing models. Their complexity is reduced to $O(N^2)$.

The objective function (12.9) of Model 12.3 minimises the maximum utilisation, the average utilisation, the average load, and the average path length. Each of these criteria is weighted with a special parameter w, if a parameter w is set to zero, the according criterion is ignored when searching for the optimal solution.

Constraint (12.10) is the routing constraint and makes sure that every flow is routed along *one* path. Please note that in the basic model variable a_{fp} is a binary variable. If the binary condition (12.15) is relaxed towards (12.16) in Model 12.3, multipath routing is allowed and a flow can be split up.

Constraint (12.11) sets the utilisation u_l of a link l in relation to the amount of traffic routed through that link and its capacity. Constraint (12.12) forces ξ to the maximum utilisation. (12.13) to (12.15) form the non-negative binary constraints of Model 12.3.

In Model 12.5, the congestion costs are additionally added to the objective function (12.17). They are measured with variable x_{sl} that is set in (12.18) to the value by which the lower threshold of step s of the congestion cost function is exceeded on link l. The congestion with added weighted high capacity links are likely to be used by more users than low capacity links; therefore, they should be weighted higher. The unweighted congestion costs (the last term in the objective function) are included for reference only.

Please note that any algorithm could be used to calculate the paths that are used as input for the path selection models. Throughout our experiments we use the same method described above in Section 12.2.3 to the n shortest paths with minimal overlappings that have no more than Δl additional hops than the shortest path. How to choose the parameters n and Δl is discussed in Section 12.7.

12.3 Experiment Setup

In the rest of the chapter, the above presented traffic engineering strategies are evaluated in a number of experiments. Each experiment is repeated N times. The average of the performance metrics of Section 12.1 and the 95% confidence intervals are derived from the results. They are presented and discussed in the following sections.

For each experiment, a topology is selected; we use the German Research Network (DFN) topology as the default topology for all the experiments. For some experiments,

---

Model 12.3 Path Selection (Singlepath)

Indices

$f = 1, \ldots, F$ Flow f

$p \in \rho_f$ Path p for flow f

$l = 1, \ldots, L$ Link l

Parameters

r_f Size of flow f

ρ_f Set of available paths for flow f

l_p Length of path p

c_l Capacity of link l

ϕ_p Set of links belonging to path p

w^ξ Weight for the maximum utilisation objective

w^u Weight for the average utilisation objective

w^l Weight for the average load objective

w^p Weight for the average pathlength objective

Variables

ξ Maximal link utilisation

u_l Utilisation of link l

a_{fp} Routing variable, flow f is routed via path p by the amount denoted with a_{fp}

$$\text{Minimise } w^\xi \xi + w^u \frac{1}{L} \sum_l u_l + w^l \frac{1}{L} \sum_l c_l u_l + w^p \frac{1}{F} \sum_f \sum_{p \in \rho_f} l_p a_{fp} \qquad (12.9)$$

subject to

$$\sum_{p \in \rho_f} a_{fp} = 1 \qquad \forall f \qquad (12.10)$$

$$\sum_f \sum_{p \mid l \in \phi_p} r_f a_{fp} = c_l u_l \qquad \forall l \qquad (12.11)$$

$$u_l \leq \xi \qquad \forall l \qquad (12.12)$$

$$\xi \geq 0 \qquad (12.13)$$

$$0 \leq u_l \leq 1 \qquad \forall l \qquad (12.14)$$

$$a_{fp} \in \{0, 1\} \qquad \forall f \, \forall p \in \rho_f \qquad (12.15)$$

Model 12.4 Path Selection (Multipath)

Constraint (12.15) of the otherwise unchanged Model 12.3 is replaced with

$$0 \leq a_{fp} \leq 1 \qquad \forall f \, \forall p \in \rho_f \qquad (12.16)$$

Model 12.5 Path Selection with Congestion Costs (Singlepath)

Model 12.3 is extended as follows:

Index

$s = 1, \ldots, S$ Step s of the congestion costs function, see Figure 12.1

Parameters

p_s^x Additional congestion costs in step s of the congestion costs function

q_s Lower threshold of step s of the congestion costs function

w^x Weight for the congestion costs objective (weighted with capacity)

$\tilde{w}^x$ Weight for the congestion costs objective (not weighted)

Variables

x_{sl} Congestion costs variable, denotes by how much the threshold of

step s of the congestion cost function has been exceeded on link l

$$\text{Minimise } (12.9) + w^x \sum_l c_l \sum_s p_s^x x_{sl} + \tilde{w}^x \sum_l \sum_s p_s^x x_{sl} \qquad (12.17)$$

subject to (12.10)-(12.15) and

$$x_{sl} \geq u_l - q_s \qquad \forall s \, \forall l \qquad (12.18)$$

$$x_{sl} \geq 0 \qquad \forall s \, \forall l \qquad (12.19)$$

---

we also vary the topology. The topology is modelled as a directed graph, the capacity of opposing links is assumed equal in all experiments of this section.

A traffic matrix[4] is necessary to evaluate the strategies. Unfortunately, measured traffic matrices are not available as providers are reluctant to reveal information about their topology and traffic characteristics or prohibit publication. Therefore, we have to generate

---

[4] More exactly: The structural relationship between the traffic matrix and the link capacities of the topology.

artificial traffic matrices based on the information known about the characteristics of traffic matrices. We generate multiple traffic matrices per experiment and vary the generation method during the experiments – our experiments below show that the results are stable for different traffic matrices and traffic distributions.

12.3.1 Traffic Creation

Traffic flows are created between all node pairs. Bhattacharyya et al. (2001) show that traffic flows differ drastically in their size (hence often named mice and elephants) and that points of presence (POPs) nodes in a POP level topology show large differences in throughput. We model this behaviour with node weights; the node weight of the source and sink node massively influence the flow size. The node weight can be imagined to represent the size of the customer base served by this node. Prior to the traffic generation, for each node n a node weight w_n is randomly selected from the list (1, 2, 3, 4) with the probabilities (60%, 20%, 10%, 10%).

Then, the size r_f of traffic flow f between ingress node i_f and egress node e_f is drawn from a uniform distribution in interval $[0.6 \cdot w_{i_f} \cdot w_{e_f}, 3.0 \cdot w_{i_f} \cdot w_{e_f}]$.

12.3.2 Capacity Assignment

Finally, the capacities of the links have to be determined. As the link capacities very strongly influence the performance (see Section 13.1), it is very important to set them to "realistic" values. Similar to a QoS system, traffic engineering has the highest impact in times when the network is highly loaded. Therefore, for our evaluation, a high-load situation is assumed as they typically occur in the late morning or early evening hours (see Roberts (2001)) .

In a real network, traffic volumes increase over time and link capacities are upgraded at regular intervals and in discrete steps by adding new or upgraded line cards to the routers. A typical approach is to double the capacity of a link once a certain utilisation threshold is exceeded. How large this threshold is strongly depends on the timescale used for the utilisation. For our evaluation of traffic engineering, we assume that the evaluation is based on a rather short timescale and a busy period.

We use the following algorithm to set the link capacities (bandwidths) in order to reflect that the network has a history and has grown to satisfy the traffic patterns:

1. Each link is assigned an arbitrary starting bandwidth of 155. This value is motivated by the bandwidth provided by Synchronous Transfer Mode-1 (STM-1)/Optical Carrier-3 (OC-3) links, see Table 4.2 in Section 4.1.4.
2. The utilisation of all links is determined based on the assumption that the flows are routed on their shortest path through the network.
3. If the utilisation of a link exceeds 80%, the bandwidth of the link is doubled successively until the utilisation is below 80%. This represents the "history" of the network and that it has grown to accommodate the traffic.

   The drawback of this approach is that the network capacities will be optimised to a certain extent for the shortest path routing algorithm which can give it a slight edge compared to the other algorithms. In Section 12.5.4, the generation method is therefore varied and different traffic distributions are analysed.

4. As the next step, each traffic flow is increased randomly by 1% to 10% to introduce more variation and to make sure that the capacities are not fully optimised for shortest path routing. One can imagine that this represents traffic growth since the network was expanded the last time.
5. If the bandwidths of two opposite links are not equal, they are set to the maximum of the two bandwidths so that the bandwidth between two nodes is symmetrical.

12.4 Explicit Routing versus Path Selection

As mentioned above, the path selection strategies offer a reduced computational complexity over the explicit routing strategies at the costs of a reduced solutions space because the choice of paths is restricted. The reduced solution space can lead to sub-optimal results with respect to the selected objective function. To evaluate how likely sub-optimal results are, we run an experiment with $N = 50$ repetitions with the singlepath and multipath strategies for the DFN topology (see Figure A.1). For the path selection algorithm we chose two different sets of paths, one with a maximum number of $n = 5$ paths between each node pair and maximal $\Delta l = 2$ additional hops and one with the shortest $n = 10$ paths and any number of additional hops allowed ($\Delta l = \infty$). The maximum utilisation was chosen as objective function with a weight of 1000 and the average utilisation with a weight of 1. As can be seen from the results in Tables 12.1 and 12.2, the $10/\infty$ path selection and the explicit routing strategy came to the same solution for all 50 different problem incarnations. However, the explicit routing strategy needed considerably more time[5]. The 5/2 path selection strategy leads to the same results for the primary objective

Table 12.1   Explicit Routing versus Path Selection, Multipath

Strategy	Time to Solve [s]	Maximum Utilisation (%)	Average Utilisation (%)
Shortest-Path	0.289	88.906	53.7
Path Selection 5/2	1.748	82.329	52.34
Path Selection 10/∞	5.296	82.329	52.21
Explicit Routing	18.553	82.329	52.21

Table 12.2   Explicit Routing versus Path Selection, Singlepath

Strategy	Time to Solve [s]	Maximum Utilisation (%)	Average Utilisation (%)
Shortest-Path	0.289	88.906	53.7
Path Selection 5/2	9.383	85.376	53.63
Path Selection 10/∞	17.282	85.376	53.18
Explicit Routing	33.695	85.376	53.18

---

[5] The *time to solve* in Tables 12.1 and 12.2 was measured on a 2 GHz Mobile Pentium with 512 MB Random Access Memory (RAM) using the MIP solver CPlex (see ILOG CPLEX (2004)).

function and due to the reduced solution space to slightly worse results for the secondary objective function. It is, however, very fast to solve.

Because of their better computational performance, their increased flexibility, and the insignificant difference in the results, we focus on the path selection strategies in the rest of the chapter.

12.5 Performance Evaluation

In this section, the performance of different traffic engineering strategies is evaluated. The shortest path strategy is used as a reference and several path selection strategies with different objective functions are evaluated. Their parameters n and Δl are set to $n = 5$ and $\Delta l = 2$. The effect of changing these parameters is analysed in Section 12.7. The first experiments are based on the DFN topology, other topologies are evaluated in Section 12.5.3. We start with multipath routing. The discussion of the singlepath variant of the strategies will be the subject of Section 12.6. Table 12.3 lists the selected strategies and their abbreviations.

We evaluate the performance of the strategy based on all metrics discussed in Section 12.1. Our focus, however, will be on the congestion costs because it best captures the overall performance of a network. The absolute value of the congestion costs and the link load bears no deeper meaning, therefore these values are normalised relative to those yielded by the SP strategy.

12.5.1 Basic Experiment

The average and 95% confidence intervals over all $N = 20$ different randomly created problem incarnations are summarised in Table 12.4 and shown in Figure 12.2.

Table 12.3 Abbreviations of the Traffic Engineering Strategies

Strategy	Denotation
SP	Shortest path routing
CC	Path selection: Minimise (weighted) congestion costs
CC_{uw}	Path selection: Minimise unweighted congestion costs
U_{max}	Path selection: Minimise maximum utilisation
$U_{max}L_{av}$	Path selection: Minimise maximum utilisation with $w^\xi = 1000$ and average load with $w^l = l_{SP}$ (l_{SP} is the average load of the SP strategy)
$U_{max}P_{av}$	Path selection: Minimise maximum utilisation with $w^\xi = 1000$ and average path length with $w^p = 1$
$U_{max}U_{av}$	Path selection: Minimise maximum utilisation with $w^\xi = 1000$ and average utilisation with $w^u = 1$
U_{av}	Path selection: Minimise average utilisation
$U_{av}P_{av}$	Path selection: Minimise average utilisation with $w^u = 1000$ and average path length with $w^p = 1$
$P_{av}L_{av}$	Path selection: Minimise average path length with $w^p = 1000$ and average load with $w^l = 1$
L_{av}	Path selection: Minimise average load

Table 12.4 Results of the Basic Experiment

Strategy	Congestion Costs		Unweighted Congestion Costs		Maximum Utilisation		Average Utilisation		Average Path Length		Average Load	
	Average	Confidence Interval	Average	Confidence Interval	Average	Confidence Interval	Average	Confidence Interval	Average	Confidence Interval	Average	Confidence Interval
SP	1.00	–	1.00	–	0.80	0.01	0.49	0.02	2.64	0.00	1.00	–
CC	0.86	0.02	0.92	0.02	0.76	0.03	0.50	0.02	2.66	0.01	1.00	0.00
CC_{uw}	0.89	0.02	0.88	0.02	0.76	0.03	0.50	0.02	2.66	0.01	1.01	0.00
U_{max}	0.99	0.01	0.99	0.01	0.76	0.03	0.49	0.02	2.64	0.00	1.00	0.00
$U_{max}L_{av}$	0.98	0.01	0.99	0.01	0.76	0.02	0.49	0.02	2.64	0.00	1.00	0.00
$U_{max}P_{av}$	1.04	0.05	1.26	0.09	0.76	0.03	0.54	0.02	2.64	0.00	1.00	0.00
$U_{max}U_{av}$	1.27	0.09	1.09	0.05	0.76	0.03	0.47	0.01	2.75	0.03	1.05	0.01
U_{av}	2.59	0.44	1.94	0.25	0.95	0.03	0.47	0.01	2.79	0.03	1.07	0.01
$U_{av}P_{av}$	1.02	0.03	1.00	0.02	0.80	0.02	0.49	0.02	2.64	0.00	1.00	0.00
$P_{av}L_{av}$	1.83	0.26	2.76	0.52	0.94	0.03	0.56	0.02	2.64	0.00	1.00	0.00
L_{av}	0.98	0.01	1.00	0.01	0.79	0.01	0.49	0.02	2.64	0.00	1.00	0.00

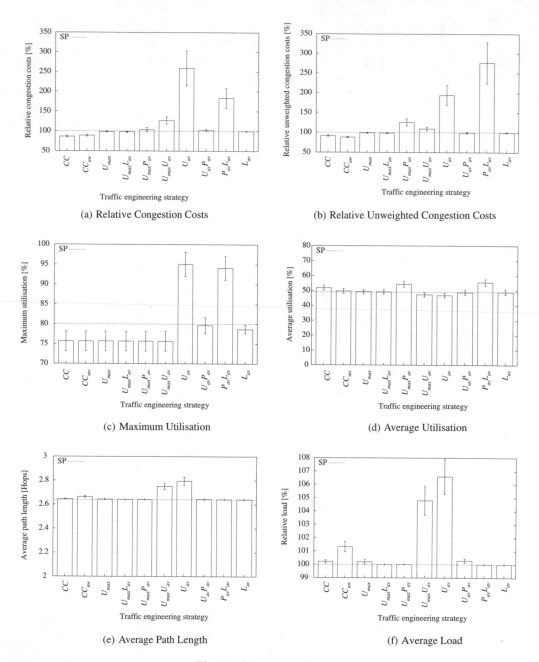

Figure 12.2   Basic Results

The *congestion costs* are evaluated first. As the CC and CC_{uw} strategies directly optimise the congestion costs, they yield the minimal weighted or unweighted congestion costs. All other strategies show a bad performance with respect to congestion. Only a few of them (U_{max}, $U_{max}L_{av}$, L_{av}) perform a little better than the shortest path (SP) reference strategy. Compared to SP, they can reduce the overall congestion of the network only by 1% or 2%. The other strategies perform worse than the shortest path strategy with respect to the congestion. Of all, the U_{av} strategy leads to the worst performance. Part of these results can be attributed to the fact that due to our method of generating the traffic and the link capacities the network capacities are relatively well adapted for the shortest path strategy. The extent of this effect is analysed in Section 12.5.4. Also, for other topologies the performance of the traffic engineering strategies compared to SP improves, see Section 12.5.3.

Comparing the results for the unweighted congestion costs with those of the weighted (default) congestion costs, some interesting effects can be observed. To explain them, one has to keep in mind that the difference between the utilisation of a link and the load of a link is the factor link capacity. The link capacity is also the difference between the weighted and unweighted congestion costs – the link capacity influences the weighted but not the unweighted congestion costs. This explains why the strategies that consider the average load ($U_{max}L_{av}$, $P_{av}L_{av}$, L_{av}) and therefore (indirectly) the link capacities, perform relatively better for the weighted congestion costs than for the unweighted ones. Vice versa, the strategies that consider the average utilisation ($U_{max}U_{av}$, U_{av}, $U_{av}P_{av}$) and therefore ignore the link capacities when calculating the average, perform relatively better for the unweighted congestion costs.

Next, the *maximum utilisation* performance metric is evaluated for all strategies. The maximum utilisation of a network shows how loaded the bottleneck links of that network are. As can be seen from Table 12.4, all U_{max} strategies lead to the lowest maximum utilisation as the maximum utilisation is their objective function. Besides these strategies, the CC and CC_{uw} strategies – despite having a different objective function – also lead to the lowest maximum utilisation. This is also not surprising, considering the convex shape of the congestion cost function that gives strong incentives to keep the utilisation low.

The U_{av} and $P_{av}L_{av}$ strategies lead to an unacceptably high maximum utilisation and thus create at least one bottleneck that is higher utilised than the bottleneck in the shortest path routed network. This behaviour should be avoided by traffic engineering strategies. These strategies cannot therefore be recommended.

Looking at the *average utilisation* as a performance metric one can notice that all strategies except $U_{max}P_{av}$ and $P_{av}L_{av}$ lead to an average utilisation very close to that of the SP reference strategy. The strategies minimising the average utilisation – especially $U_{max}U_{av}$ and U_{av} – lead to a slightly lower average utilisation. There is a trade-off between optimising average load and average utilisation. This can also be seen in the results for the *average load* performance metric. There, all strategies except $U_{max}U_{av}$ and U_{av} lead to almost the same average load[6] as the SP reference strategy while $U_{max}U_{av}$ and U_{av} lead to significantly higher average loads. There is no potential for reducing the average load compared to SP, as the average load is automatically minimised if flows are routed along their shortest path. Only if flows are routed on a path that is longer than the shortest path the average load is increased – and besides that obviously also the average path length. This is also visible for the *average path length,* only $U_{max}U_{av}$ and U_{av} show a significant increase in the average path length compared to the reference strategy, the increase of the path length for the other strategies is very small. This result shows that there is no reason

---

[6] Most differences are smaller than 10^{-2}.

to worry about the increase of the propagation delay for the traffic engineering strategies. Also, for all path selection strategies the maximum increase of the propagation delay is controlled by the parameter Δl.

Next, the performance for the individual strategies is summarised. CC performs very well for all criteria and can therefore be recommended without doubt. Also, it shows the best performance with respect to the congestion cost metric which we deem the most important metric. CC also performs significantly better than the SP strategy. It reduces the overall congestion by 14%.

The excellent performance of CC also reflects itself in the performance of the related CC_{uw} strategy. Here, the congestion costs are not weighted in the objective functions, congestion on a low bandwidth link is therefore treated the same as congestion on a high bandwidth link. As explained before, we do not recommend doing this, nevertheless, the performance of the CC_{uw} strategy is very good.

The U_{max} strategy and the derivatives of that strategy that minimise the average load, utilisation or path length as secondary objective obviously show the best performance for the maximum utilisation metric. Also they perform well for the average utilisation (except $U_{max}P_{av}$), path length (except $U_{max}U_{av}$) and load (except $U_{max}U_{av}$). However, for the congestion costs they do not perform well. $U_{max}P_{av}$ and $U_{max}U_{av}$ perform especially badly and cannot be recommended. If a U_{max} strategy has to be used, $U_{max}L_{av}$ should be used. However, the CC strategies perform significantly better and should be favoured.

U_{av} only minimises the average utilisation and cannot be recommended. The performance improves considerably if the objective function is combined with a second objective function as in $U_{av}P_{av}$. However, $U_{max}L_{av}$ and especially CC then still perform better. Similarly, $P_{av}L_{av}$ and L_{av} perform worse than these two mentioned strategies.

12.5.2 Variation of the Congestion Cost Function

We argued above that the congestion cost function is the best and most important traffic engineering performance metric. While it is clear that the congestion cost function is of a convex shape, the question remains how the exact shape of the function influences the results. In this section, we evaluate this influence by repeating the above experiments for the three different congestion cost functions of Figure 12.1. The resulting congestion costs are summarised in Table 12.5. The evaluation of other criteria like the *average utilisation*, the *average load* and the *average path length* was not affected more than 1%.

As one can see, the strategies that perform exceptionally badly with respect to congestion costs (U_{av}, $P_{av}L_{av}$) are influenced to a great extent by the exact shape of the congestion cost function. Nevertheless, independent of the shape, they remain the worst strategies with respect to congestion costs.

The other strategies are only slightly influenced by the congestion cost function. The exact shape of the congestion function does not influence the ranking of the strategies. However, the advantage of the CC strategies compared to the SP strategy depends on the shape of the congestion cost function. In the experiment, this advantage varies between 5% and 14%. The relatively small advantage for the congestion cost function (3) can be explained by the relatively small steepness of the function for high values of utilisation. By re-routing flows, highly utilised links are relieved by the CC strategy. The higher the steepness of the function, the higher the lowered utilisation reflects itself in the results.

Table 12.5 Congestion Cost Metric for Different Strategies and Congestion Cost Functions

	Original (1)		Function (2)		Function (3)	
	Average	Confidence Interval	Average	Confidence Interval	Average	Confidence Interval
SP	1.00	–	1.00	–	1.00	–
CC	0.86	0.02	0.91	0.02	0.95	0.01
CC_{uw}	0.89	0.02	0.94	0.02	0.98	0.02
U_{max}	0.99	0.01	0.99	0.01	0.98	0.01
$U_{max}L_{av}$	0.98	0.01	0.98	0.01	0.98	0.01
$U_{max}P_{av}$	1.04	0.05	1.03	0.05	1.05	0.04
$U_{max}U_{av}$	1.27	0.09	1.23	0.08	1.16	0.07
U_{av}	2.59	0.44	2.03	0.27	3.40	0.84
$U_{av}P_{av}$	1.02	0.03	1.01	0.02	1.02	0.03
$P_{av}L_{av}$	1.83	0.26	1.49	0.15	2.52	0.49
L_{av}	0.98	0.01	1.00	0.00	1.00	0.00

See Figure 12.1 for the shape of the congestion cost functions.

The results of this experiment show that for the choice of the strategy the exact shape of the congestion cost function is not important. This is important for the application of the congestion cost strategies because it cannot be expected that a single function can be specified for a network that exactly represents the influence of the link utilisation on the congestion for all traffic types and users (see also Section 13.1). Congestion cost functions will always be approximations and estimates. Due to the relatively small influence of the exact shape, however, this does not matter much.

12.5.3 Influence of the Topologies

The previous experiments were based on the DFN topology. In this section, the influence of the topology network graph on the performance of the traffic engineering strategies is evaluated. The different analysed topologies and their basic connectivity properties like the diameter and the out-degree distribution are presented in Appendix A.

Because of the little influence of the other metrics in the previous experiments, the evaluation is restricted here to the congestion cost metric (Table 12.6) and the maximum utilisation metric (Table 12.7).

As one can see from the results, the topology significantly influences the performance of all traffic engineering strategies. We first address the question of how the topology influences the ranking of the strategies and next, how the topology influences the overall benefits of traffic engineering compared to shortest path routing.

The *ranking of the strategies* depends on the topology. While most strategies show similar behaviour for all topologies, the performance and ranking of $U_{max}U_{av}$, U_{av}, and $P_{av}L_{av}$ with respect to congestion costs depend strongly on the topology. $U_{max}U_{av}$ becomes the best strategy of all U_{max} based strategies for topologies like Colt and Artificial-2/3 and the worst of them for topologies like the DFN and C&W. The different parameters of the topologies (Table A.1) offer no clear explanation for that. U_{av} and $P_{av}L_{av}$ show the same trend for the same topologies as $U_{max}U_{av}$. Looking at the maximum utilisation,

Table 12.6   Normalised Congestion Costs for Different Topologies

	DFN		Deutsche Telekom		Colt		C&W	
	Average	Confidence Interval	Average	Confidence Interval	Average	Confidence Interval	Average	Confidence Interval
SP	1.00	–	1.00	–	1.00	–	1.00	-
CC	0.86	0.02	0.94	0.05	0.87	0.02	0.85	0.01
CC_{uw}	0.89	0.02	0.94	0.05	0.89	0.02	0.88	0.02
U_{max}	0.99	0.01	0.97	0.05	0.99	0.01	1.00	0.01
$U_{max}L_{av}$	0.98	0.01	0.95	0.05	1.00	0.01	1.00	0.02
$U_{max}P_{av}$	1.04	0.05	0.94	0.05	0.99	0.1	1.02	0.04
$U_{max}U_{av}$	1.27	0.09	0.94	0.05	0.96	0.02	1.20	0.06
U_{av}	2.59	0.04	0.99	0.06	1.09	0.05	2.14	0.19
$U_{av}P_{av}$	1.02	0.03	0.99	0.07	1.05	0.03	1.36	0.10
$P_{av}L_{av}$	1.83	0.03	0.97	0.06	1.30	0.07	1.67	0.14
L_{av}	1.00	0.00	0.96	0.06	1.14	0.06	1.11	0.07

	SWITCH		Artificial-1		Artificial-2		Artificial-3	
	Average	Confidence Interval	Average	Confidence Interval	Average	Confidence Interval	Average	Confidence Interval
SP	1.00	–	1.00	–	1.00	–	1.00	-
CC	0.91	0.02	0.91	0.02	0.79	0.02	0.78	0.02
CC_{uw}	0.94	0.02	0.94	0.02	0.82	0.02	0.79	0.02
U_{max}	1.07	0.07	1.01	0.03	0.94	0.08	0.89	0.03
$U_{max}L_{av}$	0.99	0.01	0.97	0.02	0.94	0.03	0.91	0.02
$U_{max}P_{av}$	0.98	0.01	0.96	0.02	0.96	0.02	0.83	0.02
$U_{max}U_{av}$	1.16	0.03	1.00	0.02	0.90	0.03	0.87	0.02
U_{av}	1.67	0.13	1.08	0.05	1.00	0.10	0.94	0.03
$U_{av}P_{av}$	1.07	0.07	1.01	0.03	0.94	0.08	0.89	0.03
L_{av}	1.00	0.00	1.23	0.08	1.37	0.10	1.07	0.06
$P_{av}L_{av}$	1.05	0.03	1.31	0.07	1.47	0.11	1.06	0.04

the Deutsche Telekom topology shows a very low overall utilisation because its very small size (see Table A.1)leads to sufficient bandwidth on most links in the first step of the bandwidth assignment, see Section 12.3. This stresses that – as in every experiment based on randomly generated traffic – it is important to vary the generation method. We do so in the next section.

Besides that, the maximum utilisation results also show that different topologies have different potentials for optimisations. The SP strategy has a maximum utilisation close to 80% in all topologies (except Deutsche Telekom). The U_{max} strategies can reduce the maximum utilisation by 2% to 7% depending on the topology.

CC remains the best overall strategy for all topologies, it reduces congestion by 6% to 22%. For some of the topologies, it also leads to the optimal maximal utilisations and in that respect is always better than SP.

Table 12.7 Maximum Utilisation for Different Topologies

	DFN		Deutsche Telekom		Colt		C&W	
	Average	Confidence Interval	Average	Confidence Interval	Average	Confidence Interval	Average	Confidence Interval
SP	0.80	0.02	0.53	0.07	0.80	0.01	0.80	0.01
CC	0.76	0.03	0.40	0.05	0.79	0.01	0.78	0.02
CC_{uw}	0.76	0.03	0.41	0.05	0.78	0.01	0.78	0.02
U_{max}	0.76	0.02	0.42	0.07	0.79	0.01	0.77	0.02
$U_{max}L_{av}$	0.76	0.02	0.39	0.07	0.77	0.01	0.76	0.02
$U_{max}P_{av}$	0.76	0.03	0.39	0.05	0.75	0.02	0.76	0.02
$U_{max}U_{av}$	0.76	0.03	0.39	0.05	0.75	0.02	0.76	0.02
U_{av}	0.95	0.03	0.44	0.08	0.89	0.03	0.98	0.02
$U_{av}P_{av}$	0.80	0.02	0.45	0.08	0.82	0.01	0.93	0.03
$P_{av}L_{av}$	0.94	0.03	0.44	0.07	0.98	0.01	0.97	0.02
L_{av}	0.94	0.03	0.44	0.07	0.98	0.01	0.97	0.02

	SWITCH		Artificial-1		Artificial-2		Artificial-3	
	Average	Confidence Interval	Average	Confidence Interval	Average	Confidence Interval	Average	Confidence Interval
SP	0.79	0.01	0.79	0.01	0.80	0.01	0.79	0.01
CC	0.78	0.01	0.77	0.02	0.76	0.02	0.73	0.03
CC_{uw}	0.78	0.01	0.77	0.02	0.76	0.02	0.73	0.03
U_{max}	0.78	0.02	0.76	0.02	0.76	0.02	0.73	0.03
$U_{max}L_{av}$	0.78	0.02	0.76	0.02	0.76	0.02	0.73	0.03
$U_{max}P_{av}$	0.78	0.02	0.76	0.02	0.76	0.02	0.73	0.03
$U_{max}U_{av}$	0.78	0.02	0.76	0.02	0.76	0.02	0.73	0.03
U_{av}	0.96	0.02	0.85	0.04	0.81	0.04	0.83	0.04
$U_{av}P_{av}$	0.82	0.03	0.83	0.04	0.80	0.03	0.81	0.04
$P_{av}L_{av}$	0.87	0.04	0.99	0.01	0.97	0.01	0.93	0.03
L_{av}	0.79	0.01	0.96	0.03	0.95	0.02	0.92	0.04

12.5.4 Variation of the Traffic Distribution

As has been pointed out before, the influence of the traffic distribution also has to be evaluated. The following variations to the procedure described in Section 12.3 were evaluated for a subset of all traffic engineering strategies:

1. Assignment of equal node weights for all nodes in the network.
   If equal node weights are assigned to all nodes, the traffic is spread more evenly among the topology than in the basic set-up.
   Table 12.8 depicts the results (Experiment Setup 1). The benefit of traffic engineering improves a lot if the traffic is spread more evenly among the topology. In that case, all strategies show far better performance than shortest path routing. The maximum utilisation is now almost half of that of the shortest path routing.

Table 12.8   Variation of the Traffic Distribution

| | Congestion Costs | | | | | | | |
| | Default | | 1 | | 2 | | 3 | |
Experiment Set-up Strategy	Average	Confidence Interval	Average	Confidence Interval	Average	Confidence Interval	Average	Confidence Interval
SP	1.00	–	1.00	–	1.00	–	1.00	–
CC	0.86	0.02	0.81	0.01	0.90	0.04	0.90	0.04
CC_{uw}	0.89	0.02	0.81	0.02	0.90	0.04	0.90	0.04
U_{max}	0.99	0.01	0.84	0.01	0.97	0.03	0.98	0.01
$U_{max}L_{av}$	0.98	0.01	0.81	0.01	0.91	0.03	0.92	0.03
$U_{max}P_{av}$	1.04	0.05	0.81	0.01	0.91	0.04	0.90	0.04
$U_{max}U_{av}$	1.27	0.09	0.81	0.01	0.92	0.04	0.92	0.03
$U_{av}P_{av}$	1.02	0.03	0.81	0.01	0.91	0.04	0.91	0.04

| | Maximum Utilisation | | | | | | | |
| | Default | | 1 | | 2 | | 3 | |
Experiment Set-up Strategy	Average	Confidence Interval	Average	Confidence Interval	Average	Confidence Interval	Average	Confidence Interval
SP	0.80	0.01	0.75	0.01	0.70	0.05	0.68	0.06
CC	0.76	0.03	0.42	0.01	0.65	0.06	0.62	0.08
CC_{uw}	0.76	0.03	0.42	0.01	0.65	0.06	0.62	0.08
U_{max}	0.76	0.03	0.42	0.01	0.65	0.05	0.62	0.08
$U_{max}L_{av}$	0.76	0.02	0.42	0.01	0.65	0.06	0.62	0.08
$U_{max}P_{av}$	0.76	0.03	0.42	0.01	0.65	0.06	0.62	0.08
$U_{max}U_{av}$	0.76	0.03	0.42	0.01	0.65	0.06	0.62	0.08
$U_{av}P_{av}$	0.80	0.02	0.49	0.01	0.65	0.05	0.63	0.07

The behaviour is explained by the fact that if node weights differ, the flow between two different node pairs can differ by a great amount. If that is the case, the bandwidths of the links of the network are also likely to differ to some extent as we assumed the network to be roughly adapted to the traffic. The differing flows and link bandwidths limit the re-routing of flows as large flows can only be re-routed to a great extent on other high-bandwidth links. This limits the traffic engineering potential in the case of different node weights and explains the observed behaviour.

2. Assignment of equal bandwidth to all links.

In a different set-up, we assign all links equal bandwidth. This removes possible advantages for the SP strategy because the bandwidth assignment process in the basic set-up used the shortest paths to derive reasonable bandwidth settings.

The results are shown in Table 12.8 (Experiment Set-up 2). All traffic engineering strategies now show very similar performances, the congestion can be reduced by 10%, the maximum utilisation by 5%. The now smaller advantages of the CC strategies compared to the others with respect to the congestion is explained by the fact that due to the different setting of bandwidth the network is now less utilised on average. This is also visible from the maximum utilisation values of the SP strategy. Because of the lower overall utilisation and the exponential shape of the congestion cost function CC has less advantages and the performance differences between the strategies are smaller.

3. Bandwidth assignment based on the *EQMP* (equal cost multipath) strategy instead of the SP strategy.

   A possible bias towards SP can be analysed by replacing SP in the creation process with a different strategy, in this case *EQMP*.

   The results for $EQMP$ with $n = 3$ paths are shown in Table 12.8 (Experiment Set-up 3). A behaviour similar to that in experiment set-up 1 can be observed, albeit not as extreme. The explanation is similar; flows are now assumed to be spread over the three shortest paths for the bandwidth calculation which creates a more even traffic distribution leading to the effects observed and explained above.

12.5.5 Conclusions

As a conclusion of the performance evaluation we recommend the CC strategy for traffic engineering as its overall performance is better than that of the other strategies under all evaluated circumstances. It optimises the congestion costs that we deem the most important metric. The congestion costs consider all links but – because of the convex shape – higher utilised links influence the routing decision more. Also, we recommend the use of the weighted congestion costs with the link bandwidth because high-bandwidth links are likely to be used by more users and should thus have more influence on the routing decision. Therefore, the weighted congestion costs were used in this section as default. Other strategies try minimising the maximum or average utilisation, the average load or the path length or a combination of these objectives and did not perform well in all experiments.

12.6 Singlepath versus Multipath

So far, the evaluation was focused on the multipath strategies that were allowed to split a flow in order to be routed on multiple different paths through the network. Contrary to that, singlepath strategies route one traffic flow on a single path through the network. As the solution space of the singlepath strategies is a subset of the multipath solution space, singlepath solution strategies can never show a better performance with respect to the objective function than the corresponding multipath strategy. In this section, we evaluate the performance loss for the traffic engineering strategies. We focus on the congestion costs and maximum utilisation, as the other metrics did not show a significant difference.

The relative difference in congestion costs and maximum utilisation of the singlepath variants of the previously discussed traffic engineering compared to the multipath solution is presented in Table 12.9 for different topologies.

The singlepath CC strategy shows a very small and almost negligible performance loss compared to the multipath CC strategy. The largest performance loss is 0.46%, occurring at the relatively small Telekom topology. For the larger topologies, the performance loss is below 0.06%.

The performance loss of CC_{uw} is of the same order of magnitude. For the U_{max} strategies, the maximum utilisation only increases by less than 0.01%, that performance

Table 12.9 Relative Difference in *Congestion Costs* and *Maximum Utilisation* of the Singlepath Strategy Compared to the Multipath Strategies for Different Topologies

| Strategy | DFN (%) | Congestion Costs | | | |
		Deutsche Telekom	Colt Telekom (%)	Cable & Wireless (%)	Artificial-2 (%)
CC	0.03	0.46	0.01	0.06	0.01
CC_{uw}	0.34	0.52	0.11	0.36	0.14
U_{max}	7.61	7.19	2.03	9.76	−12.39
$U_{max}L_{av}$	1.03	0.27	−0.62	−0.31	−3.51
$U_{max}P_{av}$	−6.88	2.22	−0.84	−6.14	9.80
$U_{max}U_{av}$	−1.23	1.69	−0.16	−1.39	−0.66

| Strategy | DFN (%) | Maximum Utilisation | | | |
		Deutsche Telekom (%)	Colt Telekom (%)	Cable & Wireless (%)	Artificial-2 (%)
CC	0.01	0.03	0.00	0.00	0.00
CC_{uw}	0.01	0.03	0.00	0.00	0.00
U_{max}	0.00	0.00	0.00	0.00	0.00
$U_{max}L_{av}$	0.00	0.00	0.00	0.00	0.00
$U_{max}P_{av}$	0.00	0.00	0.00	0.00	0.00
$U_{max}U_{av}$	0.00	0.00	0.00	0.00	0.00

loss is negligible. However, if the congestion costs are evaluated for these strategies, it becomes obvious that the singlepath and multipath solutions differ in their routing. The congestion costs are influenced randomly by the singlepath routing variant, because they are not optimised by the U_{max} strategies directly. Depending on the strategy and topology, they can significantly improve the congestion situation. Despite this effect, the CC strategies still always perform significantly better than the U_{max} strategies.

To summarise, the performance loss of singlepath strategies compared to multipath strategies is negligible. The only drawback of the singlepath strategies is therefore the fact that they need more time to solve (see Table 12.1 and 12.2), as the singlepath MIP models use binary variables.

12.7 Influence of the Set of Paths

The path selection strategies use a precomputed set of paths for their optimisation. In this section, the influence of this set of paths on the performance of the path selection strategies is evaluated.

Two parameters (n and Δl) are used to precompute the paths for each node pair. Parameter n is the upper bound on the number of paths that are taken into account. Parameter Δl denotes the maximum number of additional hops compared to the shortest path that are allowed for paths in the set.

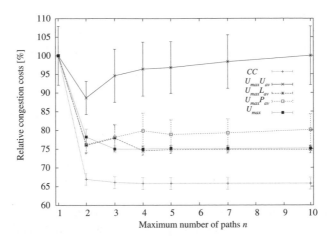

Figure 12.3   Influence of n on the Performance

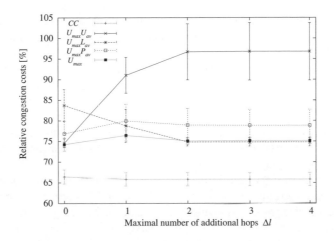

Figure 12.4   Influence of Δl on the Performance

The results are depicted in Figures 12.3 and 12.4. Figure 12.3 shows the congestion costs for the DFN topology and for different traffic engineering strategy. The maximum number of paths n is shown on the x-axis. It can be seen that the CC strategy clearly improves if n is increased. This can be expected. The largest performance increase occurs if n is increased from 1 – where all path selection strategies are effectively identical to the SP strategy – to 2. After that, the performance increase is significantly smaller.

Figure 12.4 shows the performance change if Δl is increased (for $n = 5$). The performance gain of CC is very small. This can be expected, as the previous experiments have already shown that the CC strategy does not tend to increase the average path length very much – therefore it does not make much use of the additional (longer) paths.

An important question to answer is what the optimal settings are for n and Δl. As the performance increase for values of $n > 5$ and $\Delta l > 2$ is negligible for CC, we recommend 5 and 2 for n and Δl. Higher values only lead to more computational complexity.

12.8 Summary and Conclusions

In this chapter, different traffic engineering strategies were discussed. They can be distinguished as path selection and explicit routing models. Explicit routing models show a very small performance advantage at the cost of computational complexity that prohibits their use for large networks. Path selection strategies can be computed much faster, are also more flexible, and offer the decision maker more control as they use a set of precomputed paths.

Traffic engineering strategies can also be distinguished into singlepath and multipath strategies, depending on whether they can split a flow into subflows and route them over different paths through the network. Multipath strategies have a theoretical performance advantage. In our experiments, it turned out that this advantage is extremely small for realistic topologies.

We introduced different metrics for measuring the performance of traffic engineering. Naturally, it makes sense to use these metrics as objective functions for the traffic engineering strategies. We did so for the path selection strategies. We argued that the congestion costs are the best performance metric. The strategies were evaluated in extensive simulations during which we investigated different topologies, different congestion cost functions, and traffic distributions. Throughout all these experiments, the CC strategy showed the best overall performance. Contrary to most other strategies, it performed well for practically all performance metrics. It can therefore be recommended without doubt. The other strategies showed flaws and bad performance in some or many cases and cannot therefore be recommended.

Using the correct strategy, traffic engineering can reduce the congestion of a highly loaded network and therefore directly improve the QoS. This advantage can also be used to increase the efficiency because more traffic can be served with the same capacity; correspondingly capacity expansions can be delayed and costs saved. This effect is also visible in the next chapter where capacity expansion is discussed. However, for several topologies and traffic distributions the advantages were rather small compared to the much simpler (and expectedly cheaper) solution of simply using shortest path routing. Therefore, traffic engineering cannot be recommended generally; an Internet Network Service Provider (INSP) has to carefully weigh the benefit of the increased QoS against the additional costs for the traffic management equipment, costs for staff and training, etc.

13

Network Engineering

In this chapter, the influence of network engineering on the efficiency and quality of service of a network is investigated. As argued in Section 11.1, capacity expansion is the most frequent network engineering task of an INSP. Therefore we focus on capacity expansion. We start by evaluating the influence of capacity on the performance of different QoS systems in Section 13.1. Different capacity expansion strategies are evaluated in Section 13.2. We base this analysis on the results of the previous chapter by incorporating the previously found best traffic engineering algorithms into our analyses. The mutual influence of capacity expansion and traffic engineering is also analysed in that section. Finally, in Section 13.3, we investigate the effect of elastic traffic on traffic matrices in the context of capacity expansions.

13.1 Quality of Service Systems and Network Engineering*

Capacity expansion (CE) deals with increasing the network capacity of a network. Internet traffic volumes are growing very fast. Numbers presented, for example in Odlyzko (2003) indicate that the traffic volume is increasing by 70 to 150% per year. Therefore, the capacity of a network has to be adapted regularly to the growing needs.

The effect of capacity expansion on the performance of different QoS systems is analysed by the following experiment. It is based on the packet simulations that are described in detail in Chapter 8, especially Section 8.5. We repeat the experiments of Section 8.5[1] with varying levels of capacity (bandwidth and buffer), starting with half the capacity used in Section 8.5; the capacity multiplicator is depicted on the x axis of the following graphs.

The utility of the accepted flows is used as the performance measure of the overall network performance; see Chapter 8 for details. For the four different types of traffic of Chapter 8, it is depicted in Figures 13.1 and 13.2. Please note, that the maximum possible utility is 1.0.

As one can see for all QoS systems, the overall utility obviously increases with the amount of available capacity. There are, however, great differences between the different QoS systems.

---

* Reproduced with permission from Oliver Heckmann and Ralf Steinmetz, Capacity Expansion for MPLs Networks, Proceedings of INOC 2005.
[1] DFN topology, traffic mix A.

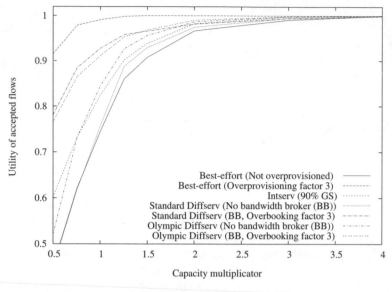

(a) Short-lived TCP Traffic

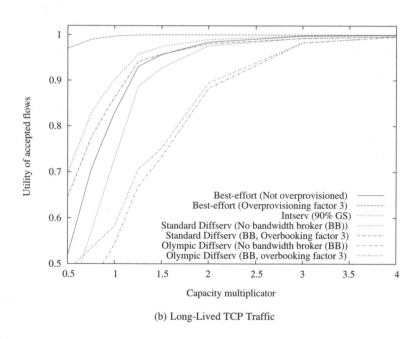

(b) Long-Lived TCP Traffic

Figure 13.1 Utility of the TCP Flows for Different QoS Systems as a Function of the Capacity

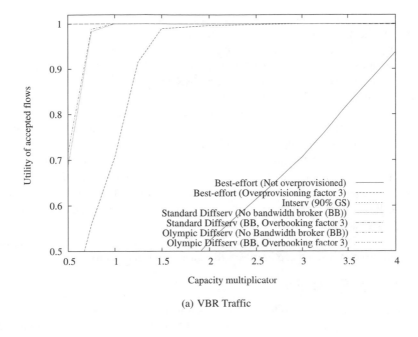

(a) VBR Traffic

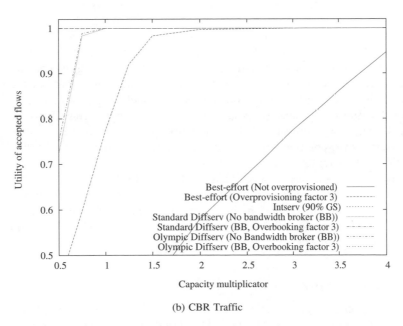

(b) CBR Traffic

Figure 13.2   Utility of the Accepted Inelastic Flows for Different QoS Systems as a Function of the Capacity

The utility of all QoS systems without admission control (Best-effort, Diffserv without Bandwidth Broker) breaks down quickly if the capacity of the network is too low. For the QoS systems with admission control, the utility of the accepted CBR and VBR flows does not break down as these flows are protected by the QoS system. However, the number of rejected customers increases: For Intserv and CBR traffic, the rejection rate is 48.1% for a capacity factor of 1 and 3.4% for one of 4.

For the experiment, the CBR and VBR flows were assumed to be of higher importance than the short-lived TCP flows. The long-lived TCP flows were assumed to resemble peer-to-peer or similar traffic with the lowest importance. The strongest differentiation between the flows is visible for the Olympic Diffserv QoS systems, where the performance of the long-lived TCP flows breaks down long before that of the CBR/VBR flows.

Figures 13.1 and 13.2 also show that in order to support CBR/VBR (multimedia) flows with a plain best-effort architecture, sufficient capacity is even more important than for the other QoS systems.

On the other side, the experiment also shows that if capacity is available in abundance, there is no significant difference between the various QoS systems.

13.2 Capacity Expansion*

Because Internet traffic is continuously increasing, capacity expansion is extremely important to maintain QoS. While QoS systems differ in their ability to maintain a high QoS in the face of scarce capacities, the performance of all systems breaks down if the capacity is too low, as was shown in the previous section. On the other hand, if capacity is expanded too early, the additional capacity remains largely unused for some time and the efficiency of the network suffers. We found that most INSPs use rules of thumb as link capacity expansion strategy in a continuous planning process. The typical rule of thumb is to trigger the expansion of a link once a certain utilisation threshold is exceeded.

In this section, the capacity expansion problem is modelled as an optimisation problem. The mutual influence of capacity expansion and traffic engineering is also considered. Different strategies are compared with the mentioned rule of thumb and some variations of it in a series of experiments in order to analyse the influence of the strategies and to identify the best strategy.

Capacity expansion is based on predictions of future traffic that are typically uncertain – contrary to the traffic engineering experiments in the previous chapter that is based on actual (measurable) traffic. Therefore, and contrary to almost all of the related works (see Section 11.1), we now also consider the uncertainty involved in predicting future traffic demand in our experiments.

13.2.1 Capacity Expansion Process

The typical capacity expansion process is depicted in Figure 13.3. Multiple periods t are investigated; the traffic changes from period to period. In every period, the traffic is routed through the network. If traffic engineering is used, the routing can change from period to period, adapting to changed capacities and flow sizes.

---

* Reproduced with permission from Oliver Heckmann and Ralf Steinmetz, Capacity Expansion for MPLs Networks, Proceedings of INOC 2005.

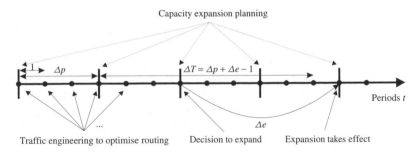

Figure 13.3 Capacity Expansion Process

An INSP decides on capacity expansions every Δp periods (e.g. once per quarter). It takes Δe periods from the decision to expand the capacity of a link until the expansion actually takes effect and the capacity is increased. Δe can be larger or smaller than Δp.

At the point in time when the decision is made, the link utilisations of the current period are known exactly as they can be measured. We assume that there are predictions available for at least the next ΔT periods. The predictions, however, are subject to uncertainty. ΔT is called the *planning horizon*. ΔT has to be at least $\Delta p + \Delta e - 1$ so that all periods are covered by the capacity expansion process.

13.2.2 Capacity Expansion Strategies

To describe the capacity expansion strategies, the same modelling parameters and simulation environment are used as for the traffic engineering strategies in the previous chapter (see Section 12.2).

The traffic volume is increasing in the long run, so the link capacities have to be expanded sooner or later. We assume that the capacity expansion of a link results in a doubling of the available link bandwidth. This is common practice at INSPs and represents adding either a second line card for one link to a router doubling the available bandwidth or – if two line cards are already present – switching to the next higher SONET/SDH data rate, which also results in effectively doubling the bandwidth (see Table 4.2).

The topology is modelled as a directed graph to be consistent with the models of the previous chapter; however, in a network the connection between two routers typically has the same bandwidth in both directions. Therefore, we assume that two opposing links between the same node pair always have the same capacity.

There are two types of *costs* involved. (a) The costs for the capacity expansion and (b) the increased congestion if capacity is expanded too late.

Assuming that the Internet traffic continues growing in the long run, the *costs for capacity expansion* are not the absolute costs for the equipment, as that equipment has to be bought anyway sooner or later. Also, the question answered by the capacity expansion strategies in the long run is not *whether* to expand but rather *when* to expand. The true costs of the capacity expansion in period t_a are the opportunity costs representing the missed earnings that could be realised if the expansion was delayed until a later period t_b. These opportunity costs consist of the interest for the invested money plus the savings if the prices for the equipment (line cards, leased lines) falls until period t_b[2]. We accumulate all these costs with the interest cost factor p^i and assume that they are proportional to the capacity.

Obviously, a capacity expansion cannot be delayed forever because the congestion would rise to an unbearable level. In the previous chapter, the congestion was modelled with a *congestion cost function* $p^x(u_l)$ that increases exponentially with the utilisation u_l of a link l (see Section 12.1 and Figure 12.1). The same approach is used in this chapter to model the fictive costs resulting from the congestion of the network. These costs result from the decreased QoS the network offers and the risk of, for example, losing profit and customers as a result of that. This cost term can be hard to quantify exactly in reality as it depends on many variables and on market conditions. The more important the network QoS is for a provider, the higher this cost factor will be.

While this second cost factor is influenced by the network QoS, the first cost factor leads directly to monetary expenses and therefore directly influences the overall network efficiency. Solving the capacity expansion problem means finding a compromise between these two goals. Therefore, we introduce a parameter c that measures how these two goals are weighted with each other. c measures the ratio between the interest cost factor and the congestion cost function. c describes where along the optimal performance boundary (see Figure 1.1) a provider wants to operate. Because the congestion costs depend on the utilisation, we arbitrarily define a reference point for a utilisation of 60% to quantify c

$$c = \frac{p^i}{p^x(60\%)} \tag{13.1}$$

In the experiments below, we evaluate the influence of c on the results.

Throughout this section, we assume that the traffic volume rates r_{tf} are influenced by the capacity expansion itself. This is the typical approach in almost all traffic engineering and network design problems (see Chapter 11). In Section 13.3, we drop this assumption and analyse the effect of elastic traffic – for example, TCP – on capacity expansion.

13.2.2.1 Threshold-based Capacity Expansion Strategy (T)

The *threshold-based capacity expansion strategy* (T_{ut}^{la} or short T) is a simple heuristic with two parameters la and ut. la is called the *look-ahead time* and ut the *utilisation threshold*.

The heuristic works as follows: t_0 is the current period; if the utilisation threshold ut of a link is reached or exceeded in period $t_0 + la$, a capacity expansion is triggered.

For $la > 0$, a prediction of the r_{tf} for future periods is necessary. For $la = 0$, the measured utilisation of the current period is used. This heuristic with $la = 0$ resembles the rules of thumb often used by INSPs. The experiments below will show if the performance can be improved by basing the decision on predicted traffic demands.

This strategy without a look-ahead time is also the basic strategy in the paper of Hasslinger and Schnitter (2004).

---

[2] Prices for line cards seem to be relatively stable and therefore their price development should not influence interest costs. Contrary to that, the price of pure transmission rates dropped significantly in the past. d' Halluin *et al.* (2002) list some numbers for OC-48 links between 1999 and 2002. The prices decline between 5 and 43% per year, which corresponds to 0.4 and 3% per period assuming that a period equals a month.

13.2.2.2 Capacity Expansion Strategy (CE)

Strategy   The basic capacity expansion strategy (CE) uses the solution of the optimisation problem that is specified in Model 13.1 in mixed integer programming (MIP) form:

The objective function (13.2) consists of the interest costs for capacity expansion and the congestion costs. The capacity doubling is modelled with constraints (13.3) to (13.6). As two opposing links have to have the same capacity constraints, (13.7) and (13.8) are necessary. To account for the congestion costs, constraint (13.9) is necessary. Finally, constraints (13.10) to (13.12) are the nonnegative binary constraints of the variables.

The optimal solution of Model 13.1 can be obtained with standard MIP solving methods (see Section 3.3) or with the faster algorithm that is presented below. It shows the optimal capacity expansion plan, the variables e_{tl} indicate the periods when the expansion of link l should be finished. The expansion of that link has to be triggered Δe periods before that. Please note that if Δe is rather long, it is possible that the optimal solution indicates that the expansion should have already been triggered before the current planning period t_0. In that case, the strategy triggers the expansion immediately in period t_0. Because of the uncertainty involved in the traffic predictions, this situation can be expected to occur for higher Δp.

Model 13.1 uses the predicted link loads v_{tl} as input. If the link loads are predicted correctly, it leads to the optimal capacity expansion plan. In a network with the shortest-path or any static routing, the link loads can be calculated directly from the predicted flow sizes of the predicted traffic matrix In a network that is using traffic engineering to optimise the routing, however, the routing can change from period to period. Flows are more likely to be routed over links that have just been expanded. Therefore, there is a mutual influence of the traffic engineering and the capacity expansion that cannot be accounted for with the above model as the capacity exact routing is not known in advance. The combined traffic engineering and capacity expansion (TMCE) strategy below extends Model 13.1 and takes this mutual influence into account by optimising the routing and the capacity expansion at the same time.

Faster Algorithm   Model 13.1 models the capacity expansion problem assuming that the load of individual links can be predicted. In the resulting problem, the links *between different node pairs* are unconnected in the objective function (13.2) and in all constraints from (13.3) to (13.12). Therefore, the problem can be split up into smaller subproblems (one for every connected node pair). They can be solved independent of each other, resulting in the same optimal solution as Model 13.1. The subproblems can be solved efficiently with the following break-even algorithm:

For links l_1 and l_2 with $(l_1, l_2) \in \Omega$, the optimal period for the capacity to be doubled is when the additional congestion costs ΔC, that would be incurred if the capacity is not expanded, exceed the interest costs ΔI that can be saved by further delaying the capacity expansion. With the congestion costs function $p^x(u)$, the additional congestion costs ΔC in period t are

$$\Delta C = p^x\left(\frac{v_{t l_1}}{c_{t l_1}}\right) + p^x\left(\frac{v_{t l_2}}{c_{t l_2}}\right) - p^x\left(\frac{v_{t l_1}}{2 \cdot c_{t l_1}}\right) - p^x\left(\frac{v_{t l_2}}{2 \cdot c_{t l_2}}\right) \tag{13.13}$$

while the saved costs of delaying the capacity expansion of one period ΔI is given by

$$\Delta I = p^i c_{t l_1} + p^i c_{t l_2} \tag{13.14}$$

Model 13.1 Capacity Expansion (CE)

Indices

$t = t_0, \ldots, (t_0 + \Delta T)$	Period t
$s = 1, \ldots, S$	Step s of the congestion costs function, see Figure 12.1
$l = 1, \ldots, L$	Link l

Parameters

t_0	Current period
ΔT	Planning horizon
v_{tl}	Prognosed load of link l in period t
$c_{(t_0-1)l}$	Initial capacity of link l
p^i	Interest costs for link capacity
p^x_s	Additional costs in step s of the congestion costs function
q_s	Lower threshold of step s of the congestion costs function
M	Sufficiently large number, $M \geq \max_l(2^{\Delta T - 1} c_{0l})$
Ω	Set of link pairs (l_1, l_2) with opposite directions

Variables

x_{stl}	Congestion costs variable, denotes by how much traffic the threshold of step s of the congestion cost function has been exceeded on link l
c_{tl}	Capacity of link l in period t
e_{tl}	Binary variable, 1 if the capacity of link l was doubled at the beginning of period t, and 1 otherwise

$$\text{Minimise} \quad \sum_t \sum_l p^i c_{tl} + \sum_t \sum_s \sum_l p^x_s x_{stl} \tag{13.2}$$

subject to

$$c_{tl} \geq c_{t-1l} \qquad \forall t \; \forall l \tag{13.3}$$

$$c_{tl} \leq 2 \cdot c_{t-1l} \qquad \forall t \; \forall l \tag{13.4}$$

$$c_{tl} \leq c_{t-1l} + M \cdot e_{tl} \qquad \forall t \; \forall l \tag{13.5}$$

$$c_{tl} \geq 2 \cdot c_{t-1l} + M \cdot (1 - e_{tl}) \qquad \forall t \; \forall l \tag{13.6}$$

$$e_{tl_1} = e_{tl_2} \qquad \forall t \; \forall (l_1, l_2) \in \Omega \tag{13.7}$$

$$c_{tl_1} = c_{tl_2} \qquad \forall t \; \forall (l_1, l_2) \in \Omega \tag{13.8}$$

$$x_{stl} + q_s c_{tl} \geq v_{tl} \qquad \forall s \; \forall t \; \forall l \tag{13.9}$$

$$c_{tl} \geq 0 \qquad \forall t \; \forall l \tag{13.10}$$

$$x_{stl} \geq 0 \qquad \forall s \; \forall t \; \forall l \tag{13.11}$$

$$e_{tl} \in \{0, 1\} \qquad \forall t \; \forall l \tag{13.12}$$

Let t^* be the smallest period with $\Delta C > \Delta I$. The capacity should be expanded in that period. As the expansion takes Δe periods, it has to be triggered in period $t^* - \Delta e$.

13.2.2.3 Combined Traffic Engineering and Capacity Expansion (TMCE)

The TMCE strategy is similar to the CE strategy except that it is based on Model 13.2. Routing and the capacity expansion are considered at the same time. The model accounts for the fact that the routing in a subsequent period can be adapted to exploit the increased capacities of the links that were upgraded. The model is a combination of the CC traffic engineering strategy[3] described by Model 12.5 and the capacity expansion strategy of Model 13.1.

The objective function (13.15) consists of the total interest costs for capacity expansion and the total congestion costs. Constraint (13.16) is the routing constraint and constraint (13.17) is used to calculate the true load based on the expanded capacities.

The capacity increase to twice the previous capacity is modelled with constraints (13.18) to (13.21); opposing links are forced to the same capacity by constraints (13.22) and (13.23). The congestion costs are accounted for by constraint (13.24). Finally, constraints (13.25) to (13.29) are the nonnegative binary constraints of the variables.

Please note that Model 13.2 cannot be divided into subproblems as Model 13.1; therefore, the fast algorithm presented for the CE strategy cannot be used here. Instead, the MIP model has to be solved directly.

Model 13.2   Combined Traffic Engineering and Capacity Expansion (TMCE)

	Indices
$t = t_0, \ldots, (t_0 + \Delta T)$	Period t
$s = 1, \ldots, S$	Step s of the congestion costs function, see Figure 12.1
$l = 1, \ldots, L$	Link l
$f = 1, \ldots, F$	Flow f
$p \in \rho_f$	Path p

	Parameters
ΔT	Planning horizon
r_{tf}	Size of flow f in period t
ρ_f	Set of paths for flow f
ϕ_p	Set of links belonging to path p
$c_{(t_0-1)l}$	Initial capacity of link l
p^i	Interest costs for link capacity
p^c	Price for new link capacity

---

[3] Any of the other strategies could also be easily used, but CC was the best traffic engineering strategy in Chapter 12.

p_s^x	Additional congestion costs in step s of the congestion costs function
q_s	Lower threshold of step s of the congestion costs function
M	Sufficiently large number, $M \geq \max_l (2^{\Delta T - 1} c_{0l})$
Ω	Set of link pairs (l_1, l_2) with opposite directions

Variables

v_{tl}	Load of link l in period t
a_{tfp}	Routing variable, flow f is routed via path p by this proportion
x_{stl}	Congestion costs variable, denotes by how much traffic the threshold of step s of the congestion cost function has been exceeded on link l
c_{tl}	Capacity of link l in period t
e_{tl}	Binary variable, 1 if the capacity of link l was doubled at the beginning of period t, and 1 otherwise

$$\text{Minimise} \sum_t \sum_l p^i c_{tl} + \sum_t \sum_s \sum_l p_s^x x_{stl} \tag{13.15}$$

$$\text{subject to}$$

$$\sum_{p \in \rho_f} a_{tfp} = 1 \qquad \forall t \, \forall f \tag{13.16}$$

$$\sum_f \sum_{p \mid l \in \phi_p} r_{tf} a_{tfp} = v_{tl} \qquad \forall t \, \forall l \tag{13.17}$$

$$c_{tl} \geq c_{t-1l} \qquad \forall t \, \forall l \tag{13.18}$$

$$c_{tl} \leq 2 \cdot c_{t-1l} \qquad \forall t \, \forall l \tag{13.19}$$

$$c_{tl} \leq c_{t-1l} + M \cdot e_{tl} \qquad \forall t \, \forall l \tag{13.20}$$

$$c_{tl} \geq 2 \cdot c_{t-1l} + M \cdot (1 - e_{tl}) \qquad \forall t \, \forall l \tag{13.21}$$

$$e_{tl_1} = e_{tl_2} \qquad \forall t \, \forall (l_1, l_2) \in \Omega \tag{13.22}$$

$$c_{tl_1} = c_{tl_2} \qquad \forall t \, \forall (l_1, l_2) \in \Omega \tag{13.23}$$

$$x_{stl} + q_s c_{tl} \geq v_{tl} \qquad \forall s \, \forall t \, \forall l \tag{13.24}$$

$$c_{tl} \geq 0 \qquad \forall t \, \forall l \tag{13.25}$$

$$x_{stl} \geq 0 \qquad \forall s \, \forall t \, \forall l \tag{13.26}$$

$$v_{tl} \geq 0 \qquad \forall t \, \forall l \tag{13.27}$$

$$a_{tfp} \in \{0, 1\} \qquad \forall t \, \forall f \, \forall p \in \rho_f \tag{13.28}$$

$$e_{tl} \in \{0, 1\} \qquad \forall t \, \forall l \tag{13.29}$$

Hasslinger and Schnitter (2004) present a heuristic for capacity expansion that takes into account the fact that traffic engineering can exploit the expanded capacity. They assume a traffic engineering strategy that minimises the maximum link utilisation and aim at maximising the average utilisation of the network. On the basis of these goals, their heuristic preferably upgrades links on a cut through the network. Their approach does not consider cost terms and the traffic engineering objectives are different from those in this section, TMCE: The TMCE strategy works with any of the path selection traffic engineering strategies discussed in the previous chapter and explicitly considers the trade-off between capacity costs and QoS. In Model 13.2, the path selection algorithm that minimises congestion costs was selected because it showed the best performance in the previous chapter. It explicitly showed better performance than strategies that minimise the maximum utilisation. In addition, TMCE is not a heuristic; it calculates the optimal capacity expansion plan and leads to the optimal solution in the absence of uncertain demands. It might be of higher computational complexity[4] but that should be relatively unimportant for a problem that only has to be solved once a month or once every three months. Because of the different goals and assumptions, it does not make sense to include that heuristic in this evaluation.

13.2.3 Performance Evaluation

13.2.3.1 Experiment Set-up

The same simulation environment and problem generation method as in Chapter 12 are used to evaluate the performance of the different capacity expansion strategies. Contrary to the single period evaluation of Chapter 12, 24 periods are considered here with one period representing one month. The size of the traffic flows r_{tf} is increased with a certain growth rate; the growth rate of the first period is drawn randomly from the interval [4%, 8%] and changed randomly by [–2%, 2%] points per period. The average growth rate of 6% leads to an average increase of roughly 100% per 12 periods; this expected increase is consistent with Odlyzko (2003) and Hasslinger and Schnitter (2004).

In a period t_0, the size of the traffic flows r_{tf} can be predicted with a maximal error $\pm 10\%$ for the following period; the maximal error increases by 3% per period $t > t_0 + 1$.

The expansion time Δe is set to $\Delta e = 3$ in the beginning, it will also be varied below. The decision that links to upgrade is made every $\Delta p = 3$ periods; that means we analyse a situation where the INSP is making the decision of when to expand its network every three periods.

As traffic engineering strategy, the CC strategy is used a maximum number of $n = 5$ paths between each node pair and maximal $\Delta l = 2$ additional hops. This strategy showed very good performance in the previous chapter.

The default congestion cost function from the previous chapter is used here (Function (1) from Figure 12.1). For evaluating the strategies, the absolute interest and congestion

---

[4] On a 2 GHz Pentium III with 512 MB RAM the TMCE strategy rarely needed more than one hour for the problems presented in this section.

costs are irrelevant as the results only depend on the relationship between those costs. The relationship between the interest for the network equipment and the congestion costs $c = \frac{p^i}{p^c(60\%)}$ is set to 1 in the beginning, it is varied later.

The threshold strategy T_{th}^{la} is evaluated with look-ahead la values of 0, 3 and 6 as well as various thresholds th that are depicted on the x axis of the graphs in this section. The absolute cost-minimal capacity expansion plan for the network can be calculated with the TMCE strategy if the uncertainty is switched off and Δp is set to 1; that is, if capacity expansion is planned every period based on the real future traffic. We call this the reference strategy REF[5].

13.2.3.2 Basic Results

The average congestion costs, the interest costs and the sum of both are shown in Figure 13.4. The 95% confidence intervals are also shown. Each experiment was repeated 20 times for different problem instances; all strategies and all different experiment set-ups solved the same 20 problems so the results are directly comparable. Because of the computational complexity, the experiment was restricted to the Telekom topology (see Appendix A). Selected experiments were repeated for the DFN topology and lead to very similar results. All costs are normalised relative to the costs of the REF strategy.

- The TMCE strategy that is executed only every $\Delta p = 3$ periods on the uncertain traffic predictions leads to only less than 1% higher total costs than when it is executed every period without uncertainty (REF). This strategy is obviously robust against the uncertainty and performs very well even if run only every third period.
- Comparing the CE with the TMCE strategy, there is a significant difference in costs. CE leads to more than 6.5% higher total costs than TMCE. With respect to the individual cost terms, CE leads to only slightly higher congestion costs than TMCE but to much higher interest costs. This results from CE not accounting for the fact that the traffic engineering algorithm can use the additional capacity of an expanded link to decrease the overall congestion in the subsequent periods. Therefore, CE overestimates the true congestion and invests too much in capacity leading to the relatively high interest costs and relative low congestion.
- Looking at the T strategies, one can first notice that all of these strategies reach the performance of the CE strategy if the threshold value th is set correctly. If it is set too high, the congestion costs explode and ruin the performance because capacity is expanded too late. This explosion becomes smaller for high look-ahead periods.
  If the threshold is set too low, too much capacity is bought and the interest costs increase. At the same time, the congestion costs decrease but that decrease becomes smaller and smaller because of the convex shape of the congestion cost function (see Figure 12.1). For decreasing values of th the congestion costs in Figure 13.4 approach a linear function with a small steepness corresponding to the lowest segment of the congestion cost function (1) in Figure 12.1.

---

[5] The optimal expansion plan can also be calculated by running TMCE once with ΔT encompassing all 24 periods. This, however, leads to a much higher overall computational complexity than solving TMCE with smaller ΔT every period.

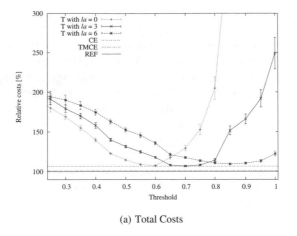

(a) Total Costs

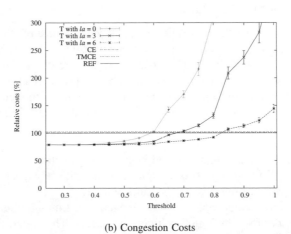

(b) Congestion Costs

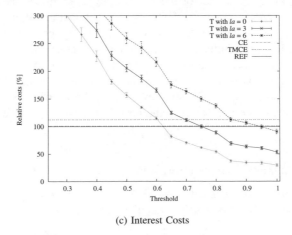

(c) Interest Costs

Figure 13.4   Costs of the Different Capacity Expansion Strategies for $c = 1$, $\Delta e = 3$, $\Delta p = 3$

- Comparing the different look-ahead values la for the T strategies, the lower the look-ahead value, the lower the optimal capacity expansion threshold. For $la = 0$ – that is, if the capacity expansion is based purely on measurements of the current period and no traffic predictions – the optimal threshold is around 60%, while it is close to 70% for $la = 3$ and 80–85% for $la = 6$. Obviously, as the traffic volume is generally increasing from period to period, higher look-ahead values lead to higher predicted utilisations and therefore higher optimal thresholds ceteris paribus.

  For a given threshold, the threshold strategy with the highest look-ahead time la leads to the lowest congestion costs and the highest interest costs because it triggers expansions significantly earlier because of the higher la value. As a result of that, this strategy leads to the highest total costs for low thresholds because the interest costs dominate in that region and to the lowest total costs for high thresholds because the congestion costs dominate in that region.

13.2.3.3 Variation of the Cost Ratio

Next, the effect of changing the cost ratio c is analysed. c measures the ratio between the interest costs for the equipment and the congestion costs. The interest costs are determined by the prices for the network hardware and the interest rate of the financial market. The congestion costs, however, are largely determined by the provider itself depending on how important QoS (low congestion) is for its network, its business model, and its customers. In Figure 13.4, the results for a cost ratio of $c = 1$ are depicted, Figure 13.5 shows the results for lower and higher cost ratios:

- The general shapes of the congestion and interest cost functions remain the same but as they are added in different ratios to the total cost function now, the total cost function is distorted compared to the original one in Figure 13.4.
- If c is set to 0.2, the congestion costs are judged five times higher than before. This resembles a provider for which QoS is highly important. The congestion costs dominate the overall performance and the total costs more closely resemble the congestion cost function. The optimal threshold for the T strategies is significantly lower now as can be expected. The TMCE strategy offers a 3% cost advantage compared to the best T strategies and the CE strategy; it leads to only 0.35% higher costs than the optimum.
- If c is increased to 5, the influence of the congestion costs is five times smaller than before. The general shape of the total cost function is now strongly influenced by the shape of the interest cost function. The optimal expansion threshold of the T strategies is higher than before. The TMCE strategy offers a 16% cost advantage compared to CE and a 12% advantage compared to the best T strategies. It comes as close as 2.2% to the optimum.

13.2.3.4 On the Capacity Expansion Process

Next, the parameters of the capacity planning process are changed. So far, for every $\Delta p = 3$ periods the capacity planning strategies were run and a single expansion took

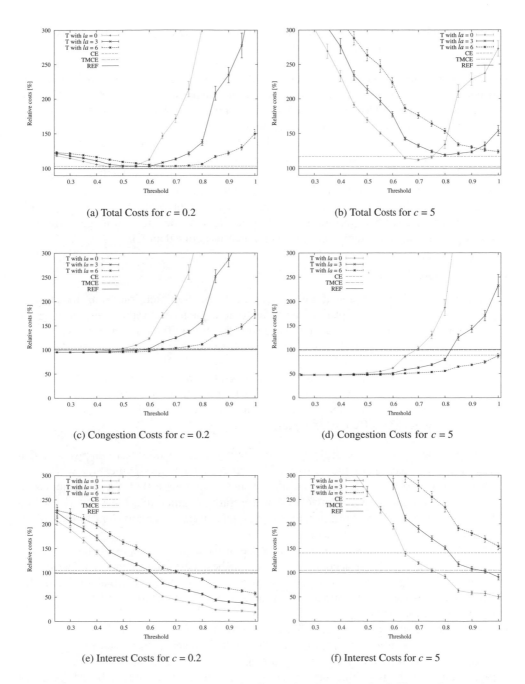

Figure 13.5 Costs of the Different Capacity Expansion Strategies for Different Cost Ratios c; $\Delta e = 3$, $\Delta p = 3$

$\Delta e = 3$ periods to take effect. Figure 13.6 shows the resulting total costs for different values of Δe and for different values of Δp.

- If the expansion time Δe increases, the thresholds when an expansion should be triggered obviously decrease as visible on the left-hand side of Figure 13.6. The performance of CE and TMCE is not influenced significantly. The same holds true for the respective optimal values of the T strategies.
- On the right-hand side of Figure 13.6, the effect of an increasing time between two planning periods Δp is visible. An increase in Δp leads to a higher planning uncertainty that should be countered by decreasing the expansion threshold of the T strategies. The overall performance of all strategies decreases with an increasing Δp. TMCE for $\Delta p = 1$ leads to optimal performance in almost all cases uninfluenced by the uncertainty, while for $\Delta p = 6$ it loses 5% performance. CE loses 13% while the T strategies lose 6%. For high Δp, the T strategies perform significantly better than CE.

13.2.4 Recommendations

In the face of traffic volumes that are growing in the medium and long run, the capacity expansion decision is not about *whether* to upgrade capacity but rather *when* to upgrade capacity. This decision is directly influenced by the trade-off between the costs of the network (therefore the network efficiency) and the QoS. This trade-off was modelled by the price ratio c.

We evaluated different capacity expansion strategies with respect to their total costs. The total costs are the interest costs for the networking equipment and the congestion costs, a fictive cost term describing the ill-effects of a congested network. We now summarise the conclusions for the different strategies:

- The CE strategy bases its decision on the solution of an optimisation problem that assumes fixed routing for the network. This strategy leads to significantly worse performance than the TMCE strategy and in some cases worse than the T strategies. It cannot be recommended for networks that use traffic engineering. For networks with a fixed routing (e.g. plain shortest-path routing), this strategy is equivalent to the TMCE strategy and can be recommended.
- The TMCE strategy takes the mutual influence of the capacity expansion and the traffic engineering strategy into account. It led to the best performance in all experiments. Depending on the settings, this comes as close as 0 to 5% to the optimal solution. This strategy can be clearly recommended. In the absence of uncertainty and for $\Delta p = 1$, it yields the optimal solution.
- The threshold strategies (T) are simple rules of thumb used by today's INSPs that expand a link once a certain utilisation threshold is reached in the current period or predicted to be reached in a certain future period. These strategies can lead to good performance if the threshold parameter is set to the correct value. The performance degrades rapidly if it is set too high, especially when using current and not predicted future link utilisations. These strategies can be recommended only if the threshold value is set correctly.

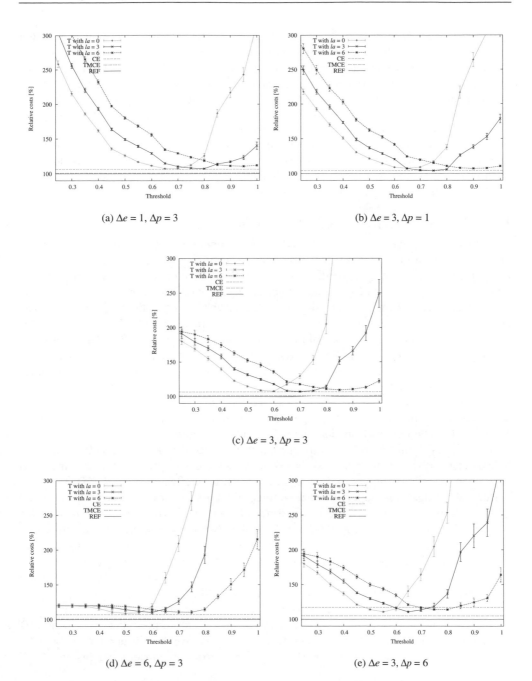

(a) $\Delta e = 1$, $\Delta p = 3$

(b) $\Delta e = 3$, $\Delta p = 1$

(c) $\Delta e = 3$, $\Delta p = 3$

(d) $\Delta e = 6$, $\Delta p = 3$

(e) $\Delta e = 3$, $\Delta p = 6$

Figure 13.6 Total Costs for Variation of the Parameters Δe and Δp of the Capacity Expansion Process; $c = 1$

If the T strategy is used with predicted demands, the overall performance does not increase significantly, therefore it is probably not worth the effort for predicting the future demands. However, if a provider is unsure about the correct setting of the threshold parameter, it is worth considering a higher look-ahead time because it can significantly reduce the ill-effects of a too high threshold value.

With respect to the overall capacity expansion process, the expansion time Δe for a link has no massive influence on the overall performance but the time interval between two planning periods Δp (when capacity expansions are considered) has. For the given parameter settings, a capacity expansion planning every three months yielded satisfactory results that were improved by only 1% if reduced to every month.

13.3 On the Influence of Elastic Traffic*

As argued before, traffic matrices are fundamental for network design and traffic engineering problems. Normally, the traffic matrix entry r_{ij} is expressed statically as a scalar – we call a traffic matrix with static predictions r_{ij} a *static traffic matrix*. However, Internet traffic is dominantly TCP traffic that adapts to changing network conditions like routing or the link capacity. This effect is systematically neglected when using static traffic matrices. The effect of capacity changes was probably negligible in times when the Internet was dominated by web traffic that consisted of huge numbers of short-lived TCP connections dominated by the slow start and not the elastic congestion avoidance phase. Traffic matrix entries at these times mainly increased if the customer base or browsing behaviour changed.

Nowadays, however, most of the traffic is generated by peer-to-peer (P2P) applications, see Chapter 5. As discussed in Section 5.2, these applications use mainly long-lived TCP connections for file transfers. This supports the assumption of this chapter that long-lived reactive TCP connections start dominating the Internet traffic.

Besides P2P traffic, future multimedia Internet traffic like streaming videos can also be expected to be TCP friendly and therefore show similar reactive effects as long-lived TCP connections that we are looking at in this section, see Handley *et al.* (2003).

Because of this, it is important to investigate the effect of the elasticity of long-lived TCP connections in their congestion avoidance phase on traffic matrices used as input for *network design* and *network engineering problems*. Normally, these problems are based on a static traffic matrix and ignore the effect that the new capacity (or capacity change) has on the amount of traffic matrix itself. We use the term *elastic traffic matrix* for a traffic matrix M with entries $r_{ij} = f(\ldots)$ that capture the elasticity of the TCP traffic and investigate the use of these elastic matrices in this section.

We developed three different network models to analyse this effect. They consist of a combination of the TCP formula and queueing theory. They are presented in Appendix D and form the analytical foundation for further analysis. We first generally analyse the elasticity of traffic matrices and then determine the impact on capacity expansion.

13.3.1 Elasticity of Traffic Matrices

The influence of the elasticity of a traffic matrix when the capacity of the network changes while all other conditions remain the same (ceteris paribus) is being analysed in this section. The effects described here are neglected when static traffic matrices are used.

---

* Reproduced by permission of VDE Verlag GMBH.

We base our analysis on the different network models derived and described in Appendix D.

13.3.1.1 Single Link Experiments

We start our analysis with an extensive series of experiments on a single link. Figure 13.7 shows the rate increase $(r_{ij}^{new} - r_{ij}^{old})/r_{ij}^{old}$ of the symmetrical macroflows over the single link topology for different queue lengths B (measured in packets) and different values for the external loss $\tilde{p}$ and delay $\tilde{q}$ when the link capacity μ_l is doubled $\mu_l^{new} = 2 \cdot \mu_l^{old}$. Figure 13.7(a) lists the results for the basic model of Section D.1, Figure 13.7(b) shows the results for the model with discrete service times of Section D.2. We used two different service time distributions. Distribution A consists of 50% packets with a size of 40 bytes and 50% packets with a size of 1500 bytes. Distribution B consists of packets of size 1000 bytes only. We assumed a line rate of 1 Mbps and had to use a rather low queue length of $B = 10$ packets because the loss probability formula gets too complicated for larger values of B to be handled analytically.

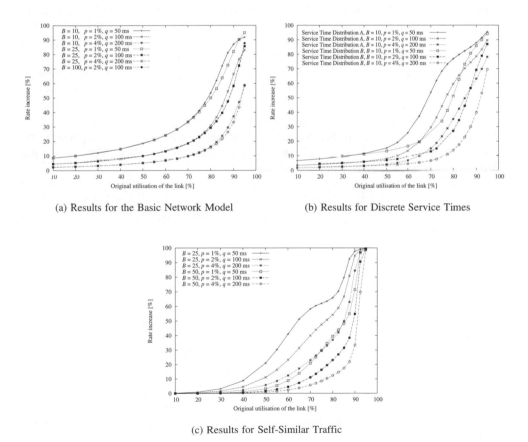

(a) Results for the Basic Network Model

(b) Results for Discrete Service Times

(c) Results for Self-Similar Traffic

Figure 13.7   Single Link Experiment Results. (Reproduced by permission of VDE Verlag GMBH)

Figure 13.7(c) shows the results obtained if we apply the model for self-similar traffic from Section D.3. A Hurst parameter of $H = 0.75$, a line rate of 1 Mbps, an average service packets size of 1000 bytes and the corresponding average service time were used.

Looking at the results, one notices that for all three different network models and most parameters, the general behaviour of the traffic is the same. Up to a certain utilisation threshold of the analysed link, the traffic is affected by the increase in capacity only slightly. Then, the traffic increases very quickly. If the initial utilisation of the link is high enough, the analysed link forms a strong bottleneck and all additional capacity is used up completely by a rate increase of 100%.

The step is steeper for the $M/M/1/B$ network model than for the other two models that can be deemed more realistic.

13.3.1.2 Different Topologies

We now analyse the elasticity in the form of the rate increase for more complex topologies than the single link topology of the previous experiments. Figure 13.8 summarises the results for three different topologies, the backbone of the Deutsche Telekom, a dumb-bell topology with a single bottleneck link and three nodes on each side of the bottleneck and a simple star-shaped topology with one internal and four external nodes. The value of t_{ij} is varied between 10^{-1} and 10^2. The network capacity for each t_{ij} is doubled and the rate increase recorded. As one can see, the different topologies lead to similar results. While most of the rate increases are very small (more than 50% of the times the rate increase was below 10%), there are a significant number of times where the rate increase was very high. Because of the different paths the different flows take through the topology, the rate increase can be higher than 100% if a series of links is doubled in capacity for a flow.

If a traffic matrix is used in the context of network design or capacity expansion, the elasticity of the traffic can be neglected up to a certain utilisation of a link. Once that threshold is passed, the error can be significant.

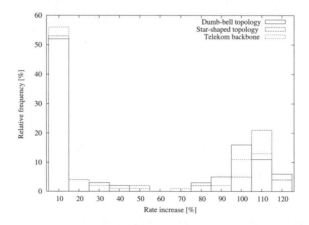

Figure 13.8 Rate Increase for Different Topologies. (Reproduced by permission of VDE Verlag GMBH)

13.3.2 Impact on Capacity Expansion

We now address the question of how the elasticity of the traffic matrix affects capacity expansion and how the capacity expansion strategies of Section 13.2 can be adapted.

If the network models of Appendix D are combined with the MIP model of the CE or TMCE strategy of Section 13.2, the resulting optimisation problem becomes nonlinear and can no longer be solved easily. For these strategies, an iterative approach could be used to take the elasticity of the traffic matrix into account. The threshold heuristic T of Section 13.2, however, can be combined directly with the network models of Appendix D. We do so exemplarily for the threshold heuristic T with a look-ahead value of $la = 0$. Using that heuristic, we can evaluate the impact of the elastic traffic matrices on capacity expansion: If the utilisation ρ_l exceeds a certain threshold th on a link l, the capacity expansion for that link is triggered. For this analysis, we assume that the link capacity is effectively doubled to the beginning of the next period after the one that triggered the expansion.

Traffic is given in form of the parameter t_{ij} of equation D.1. The actual traffic volume passed through the network is elastic and thus reacts to changes in capacity.

In 'classical' network design and capacity expansion algorithms, the elasticity of the traffic is ignored. The problem is that by increasing the capacity of a link, the traffic flows through that link will increase their rate and therefore also the utilisation of the other links they are flowing through. This can lead to the situation (a) that immediately after the expansion the threshold th on other links is exceeded and not predicted by the classical model with static traffic matrices. It will take an additional period until these links too can be expanded. Furthermore, if a link is an extreme bottleneck for some flows, it is possible that the utilisation will not significantly decrease if the link is doubled. This effect (b) can also not be predicted with static matrices. This effect was, for example, observed when the UK ISP Rednet quadrupled their DSL access link capacity, as reported to the author.

Using the models of Appendix D, we can predict the traffic increase and utilisation change of a planned network expansion and avoid the effects (a) and (b). We use the following simulation as a proof of concept:

Using the backbone topology of the Deutsche Telekom again, we generate a traffic matrix with random entries r_{ij} between 1.0 and 5.0. We use this for the initial parameters t_{ij}. A starting line rate of 1 Mbps is used for all links; it is doubled for each link before the actual simulation until all link utilisations are below 70%. We then simulate 10 periods; at the beginning of each period each traffic matrix entry is increased randomly between 5 and 20%. The basic model of Appendix D is used to calculate the link utilisations – we assume that the result of these initial calculations represents the SNMP (simple network management protocol) data collected by the provider. An external loss of 2% and delay of 100 ms is assumed; this results in a not too aggressive behaviour of TCP. In the experiment, the expansion of a link l is triggered if it has a utilisation of $\rho_l \geq th = 0.75$.

In order to capture the elasticity of the traffic matrix, we can again use our basic model to predict the effect of the triggered capacity expansions in order to avoid the effects (a) and (b) described above. We do so and measure how often these effects occur.

Effect (b) was not observed. Because we increase the rates only in moderate steps and allow the capacity to increase in each period effect, (b) does not occur in our simulations and can therefore not be avoided by the model. Effect (a), however, occurred 12 times (in 23% of all expansions) in the experiment and can be avoided by using our prediction of the elastic traffic. This example demonstrates that our concept works and helps in capacity expansion decisions.

13.4 Summary and Conclusions

Capacity expansion is an important and frequent task in today's IP networks because the traffic volume is increasing steadily. In this chapter, the influence of capacity expansion on the performance of the different QoS architectures of Chapter 8 was analysed first. If capacity is abundant, the differences between the QoS architectures vanish. However, if capacity is scarce, the systems with a strict admission control manage to maintain QoS while the other systems suffer to different extents.

Different capacity expansion algorithms were presented and evaluated. One of the introduced algorithms considers the effect of traffic engineering and capacity expansion at the same time. It leads to the best performance and is very robust against uncertain demand predictions. The simple heuristics that are often used by actual INSPs also show good performance – but only if their parameters are set correctly. The effects of several parameters on these parameters were also studied in this chapter.

Finally, the effects of elastic TCP traffic on traffic matrices and capacity expansion were discussed with some analytical models. It influences the capacity expansion measures if the network is highly utilised before the expansion. It was shown how this effect can be predicted and reacted upon accordingly.

Part V
Appendices

A

Topologies Used in the Experiments

Throughout the various experiments in this book, different network topologies have been used. As the properties of the topology influence the outcome of the experiments, we base most of the experiments on real-world topologies of Internet network service provider (INSP) networks.

These topologies are depicted in Figures A.1 to A.3. The properties of the underlying graphs are more important than the shape of these topologies. Their basic graph properties are listed in Table A.1.

- The number of nodes and links are listed.
- *Dia.* denotes the diameter of the graph; the diameter is the length (number of hops) of the longest path of the set of all the shortest paths between all nodes.
- The hop-plot is the proportion of all nodes that can be reached within a certain number of hops. For example, for the Deutsches Forschungsnetz (DFN) topology from a random node, on average 97% of all other nodes can be reached within four or less hops. The diameter and the hop-plot are a measure of the connectivity properties of the graph and the lengths of the shortest paths between node pairs.
- The outgoing node-degree of a node is the number of outgoing links connected to that node. The average outgoing node-degree and the standard deviation of the outgoing node-degree distribution are also listed in the table.

We tried to include many real-world topologies; as providers are reluctant to reveal their true network topology in every detail, some of the topologies had to be altered slightly compared to the true topology. However, this mostly affects the node placement and not the connectivity properties of the topologies needed for the experiments. Also, some artificially created topologies are included (Artificial-1 to 3). They were created using Tiers V1.2 with the parameters listed in Table A.2. The setting of these parameters was based on the findings of Heckmann *et al.* (2003). Finally, star and cross topologies were added for the Quality of Service (QoS) experiments because they allow creating a lot of cross traffic at well-defined points.

The Competitive Internet Service Provider: Network Architecture, Interconnection, Traffic Engineering and Network Design
Oliver Heckmann © 2006 John Wiley & Sons, Ltd

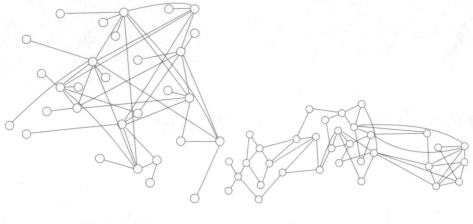

(a) DFN-like Topology   (b) Cable Wireless–like Topology

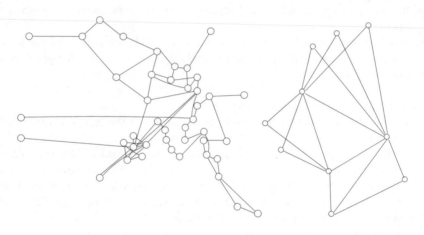

(c) Colt Telecom–like Topology   (d) Deutsche Telekom
                                                   Backbone-like Topology

Figure A.1   Topologies (1)

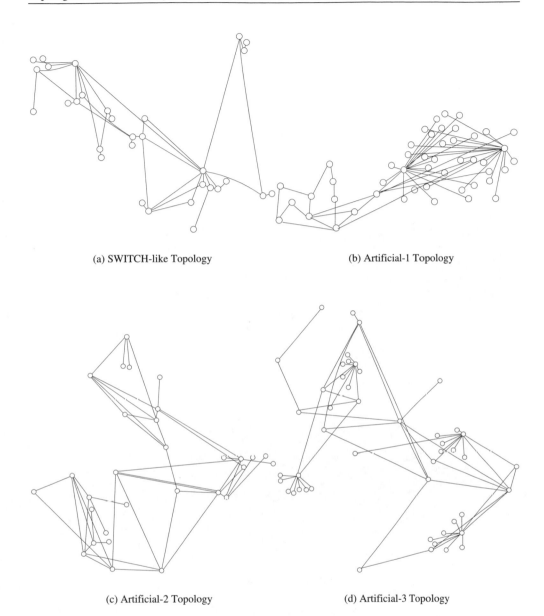

(a) SWITCH-like Topology

(b) Artificial-1 Topology

(c) Artificial-2 Topology

(d) Artificial-3 Topology

Figure A.2 Topologies (2)

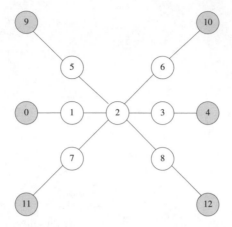

(a) Star Topology

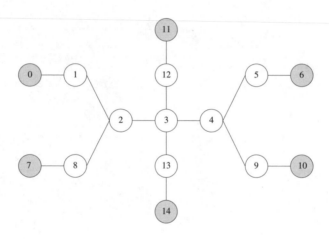

(b) Cross Topology

Figure A.3   Topologies (3)

Table A.1 Properties of the Topology Graphs

Topology	# Nodes/ Edge Nodes	# Links	Dia.	Hop-plot$_{1;2;3;4;5}$	Degree$_{av}$	Degree$_{stddev}$
DFN	30/30	97	5	0.11; 0.41; 0.79; 0.97; 1.0	3.13	2.84
C & W	33/33	107	12	0.11; 0.25; 0.39; 0.51; 0.62	6.48	2.64
Colt	43/43	107	17	0.07; 0.16; 0.26; 0.36; 0.47	4.98	2.34
Telekom	10/10	34	2	0.38; 1.0; 1.0; 1.0; 1.0	6.80	4.31
SWITCH	31/31	80	9	0.09; 0.28; 0.57; 0.81; 0.92	5.16	6.89
Artificial-1	47/47	106	7	0.05; 0.33; 0.50; 0.85; 0.95	4.51	6.88
Artificial-2	30/30	150	10	0.09; 0.26; 0.4; 0.55; 0.65	10.0	9.50
Artificial-3	47/47	180	10	0.05; 0.23; 0.35; 0.56; 0.66	7.66	10.06
Star	13/6	12	4	0.15; 0.42; 0.81; 1.0; 1.0	1.85	1.3
Cross	15/6	14	6	0.133; 0.3; 0.55; 0.81; 0.96	1.86	0.88

Cable & Wireless (C&W), Colt Telekom Europe (Colt), Deutsches Forschungsnetz (DFN), Deutsche Telekom Backbone (Telekom), Swiss Education and Research Network (SWITCH)

Table A.2 Tiers Parameters Used for the Generation of the Artificial Topologies

Abbreviation	Function	Artificial-1	Artificial-2	Artificial-3
NW	Number of WANs$_*$	1	1	1
NM	Number of MANs per WAN	1	2	2
NL	Number of LANs per MAN	2	3	2
SW	Number of nodes per WAN	9	6	7
SM	Number of nodes per MAN	4	4	4
SL	Number of nodes per LAN	17	4	8
RW	Redundancy of the links within a WAN	3	6	5
RM	Redundancy of the links within a MAN	2	3	3
RL	Redundancy of the links within a LAN$_*$	1	1	1
RMW	Redundancy of the links between MANs and WANs	4	9	7
RLM	Redundancy of the links between LANs and MANs	1	6	6

* In Tiers V1.2 only $NW = 1$ and $RL = 1$ are supported

B

Experimental Comparison of Quality of service Systems

Figures B.1 to B.22 depict the results obtained for the various experiments of Chapter 8. The average and the 95% confidence interval are marked in the figures. *s_TCP* stands for short-lived TCP flows and *l_TCP*, for long-lived TCP connections respectively. *CBR* stands for constant bit-rate and *VBR* for variable bit-rate traffic. The abbreviations for the different Quality of service (QoS) systems are listed in Table B.1.

Table B.1 Abbreviations for the Different Quality of service Systems

QoS System	Abbreviation	Parameters
Intserv	$IS - \alpha_{GS}$	α_{GS} = Maximum proportion of the link resources available for the guaranteed service class
Standard Diffserv	$sDS - bb - p$	bb = Bandwidth broker type (c = central, d = decentral, n = none) p = Bandwidth broker parameters for the central BB: p = Overbooking factor ob for the decentral BB: p = Overbooking factor times scaling factor ($ob \cdot \gamma$)
Olympic Diffserv	$oDS - bb - p$	bb = Bandwidth broker type (c = central, d = decentral, n = none) p = Bandwidth broker parameters, same as above
Best-Effort	$BE - OF$	OF = Overprovisioning factor

The Competitive Internet Service Provider: Network Architecture, Interconnection, Traffic Engineering and Network Design
Oliver Heckmann © 2006 John Wiley & Sons, Ltd

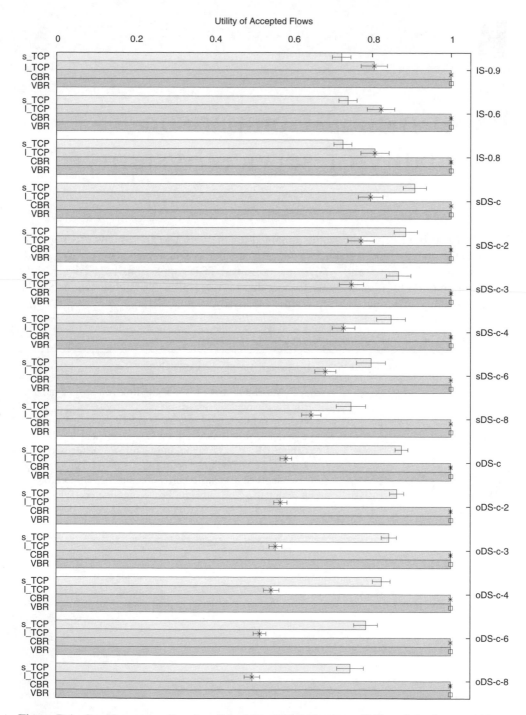

Figure B.1 Per-flow versus Per-class Scheduling, DFN Topology, Utility of the Accepted Flows

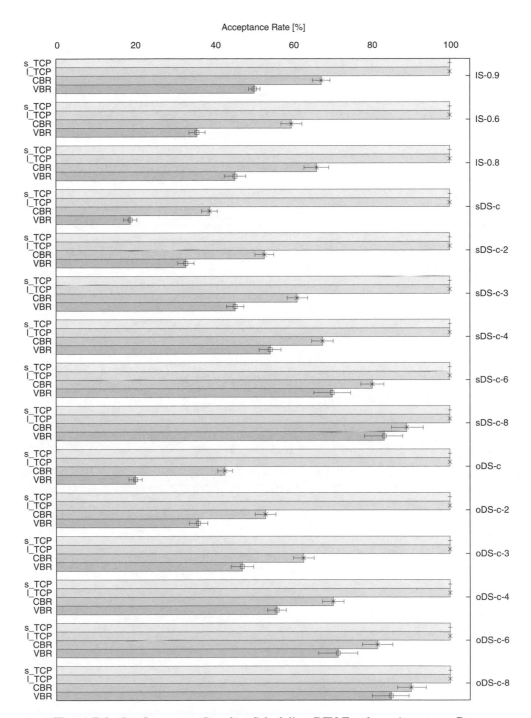

Figure B.2 Per-flow versus Per-class Scheduling, DFN Topology, Acceptance Rate

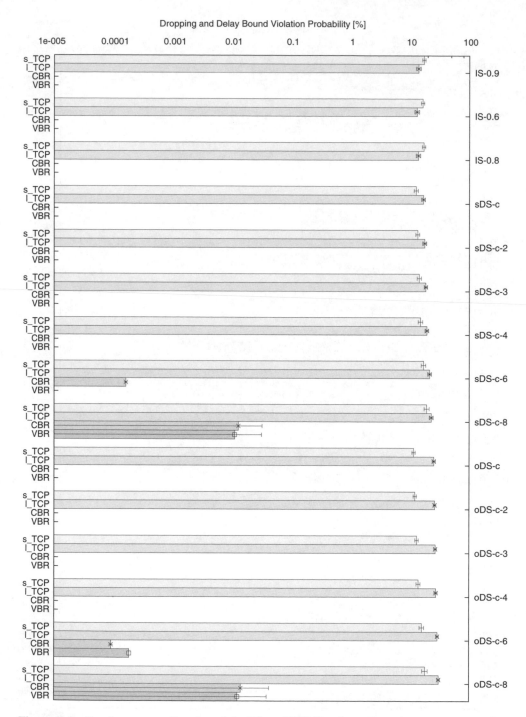

Figure B.3 Per-flow versus Per-class Scheduling, DFN Topology, Dropping and Delay Bound Violation Probability

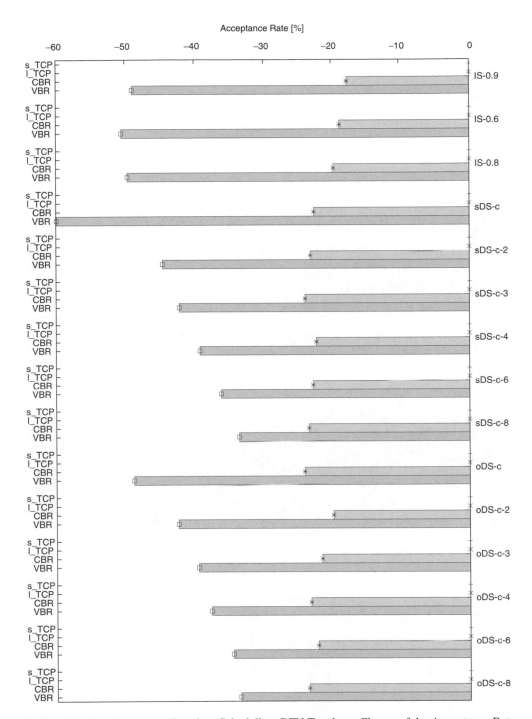

Figure B.4   Per-flow versus Per-class Scheduling, DFN Topology, Change of the Acceptance Rate when Decreasing the Delay Bound to 10 ms/hop

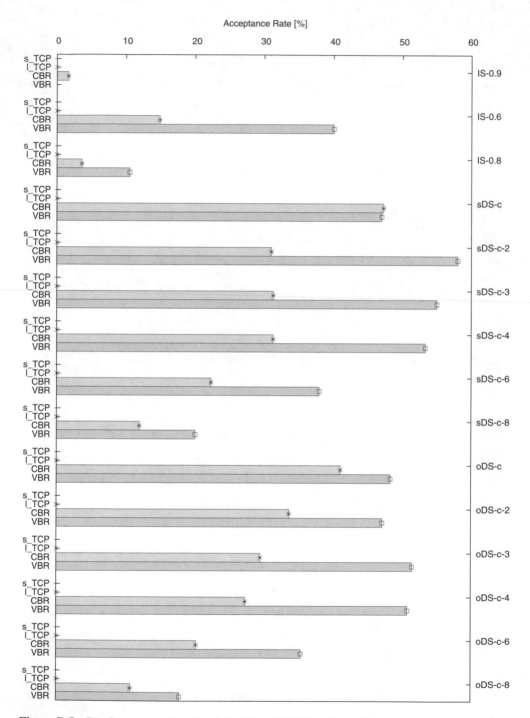

Figure B.5 Per-flow versus Per-class Scheduling, DFN Topology, Change of the Acceptance Rate when Increasing the Delay Bound to 40 ms/hop

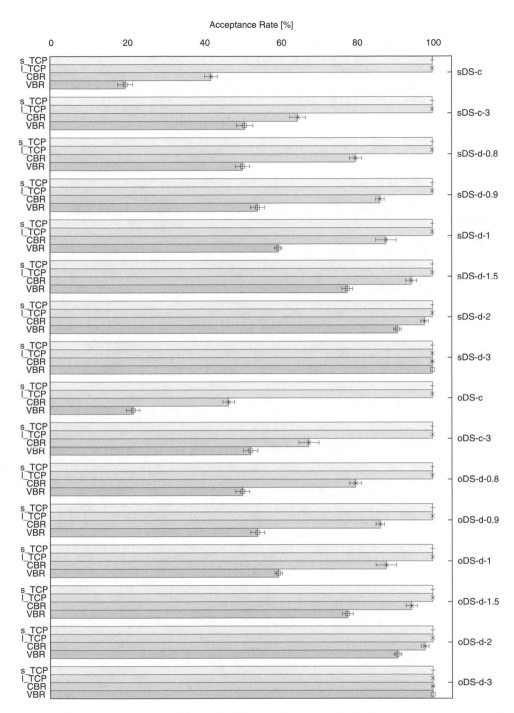

Figure B.6 Central versus Decentral Admission Control, DFN Topology, Acceptance Rate in Situation A (Contingents Match Flow Distribution)

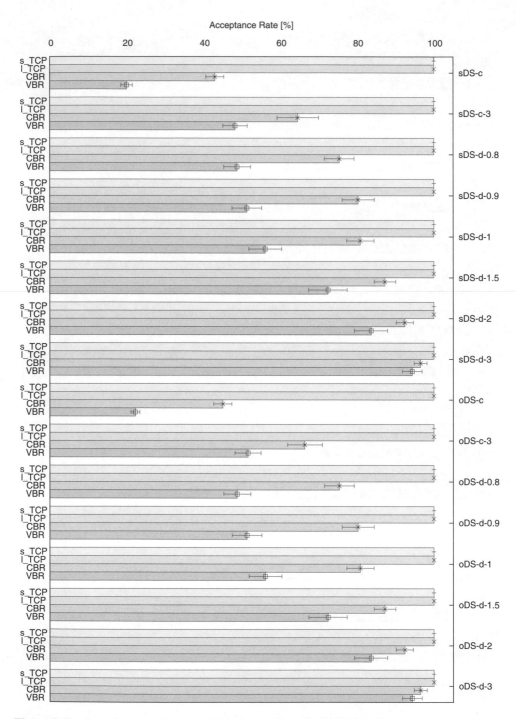

Figure B.7 Central versus Decentral Admission Control, DFN Topology, Acceptance Rate in Situation B (Contingents do not Match Flow Distribution)

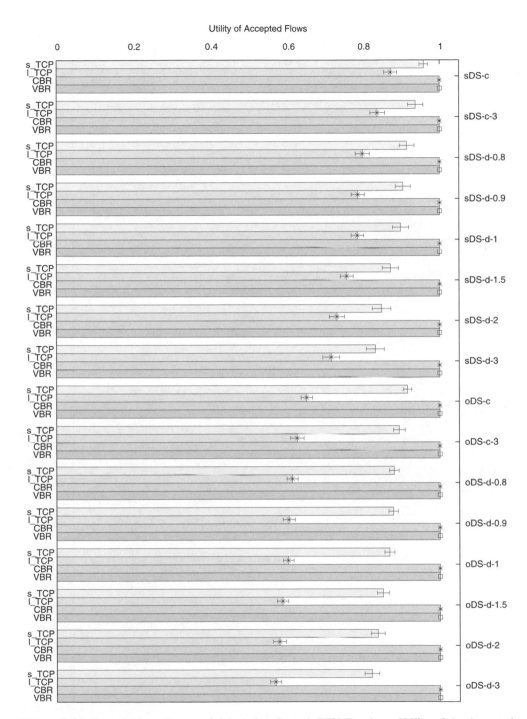

Figure B.8 Central versus Decentral Admission Control, DFN Topology, Utility of the Accepted Flows in Situation A (Contingents Match Flow Distribution)

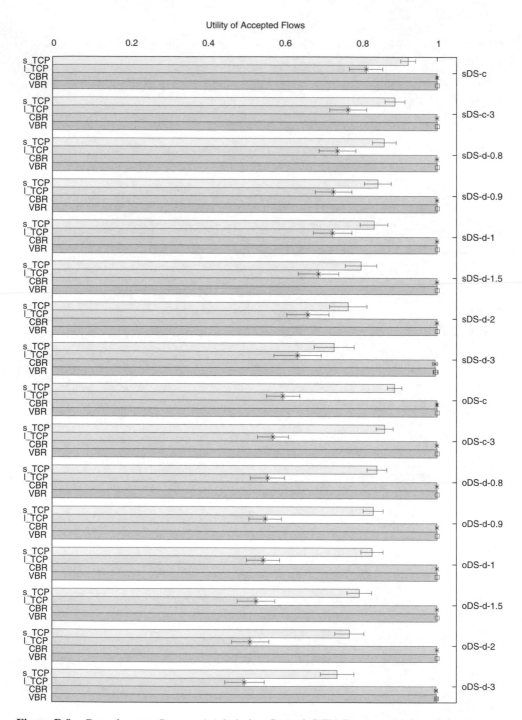

Figure B.9 Central versus Decentral Admission Control, DFN Topology, Utility of the Accepted Flows in Situation B (Contingents do not Match Flow Distribution)

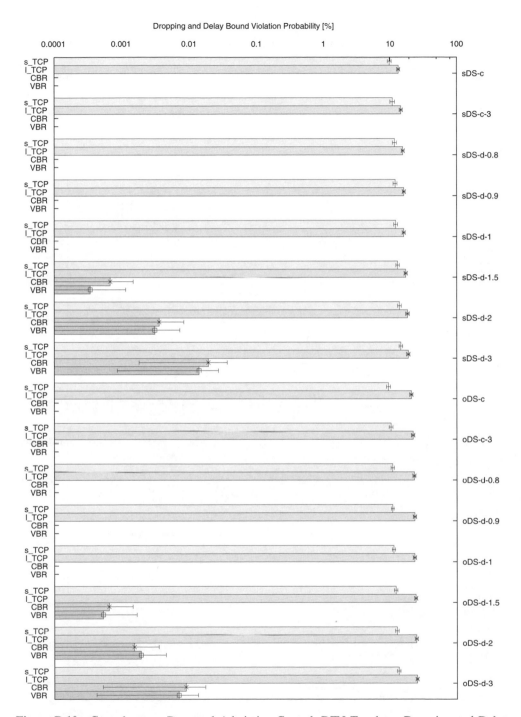

Figure B.10   Central versus Decentral Admission Control, DFN Topology, Dropping and Delay Bound Violation Probability in Situation A (Contingents Match Flow Distribution)

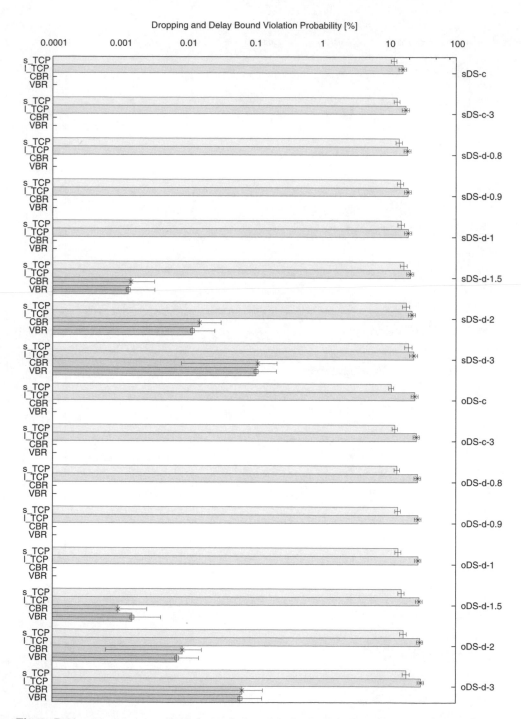

Figure B.11 Central versus Decentral Admission Control, DFN Topology, Dropping and Delay Bound Violation Probability in Situation B (Contingents do not Match Flow Distribution)

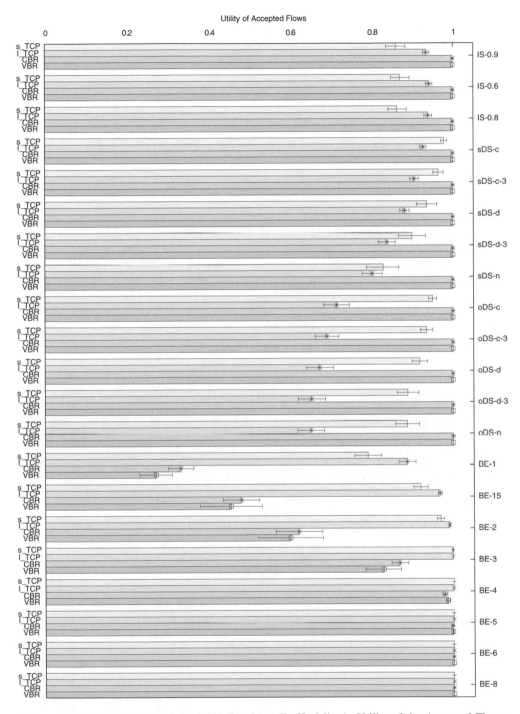

Figure B.12   Direct Comparison, DFN Topology, Traffic Mix A, Utility of the Accepted Flows

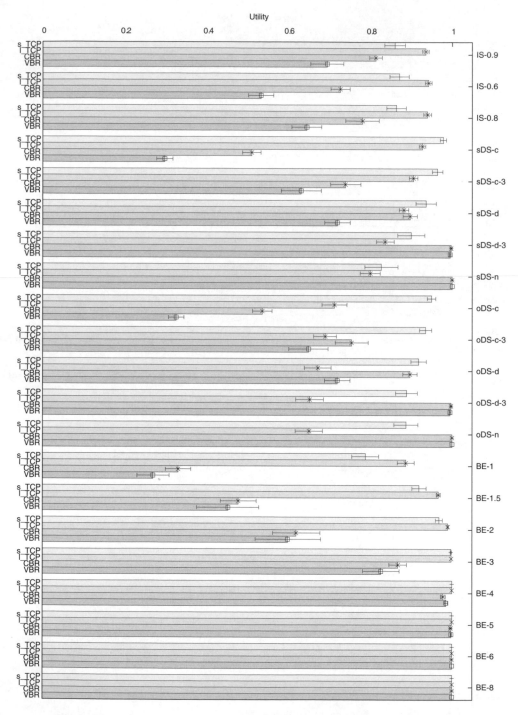

Figure B.13 Direct Comparison, DFN Topology, Traffic Mix A, Overall Utility

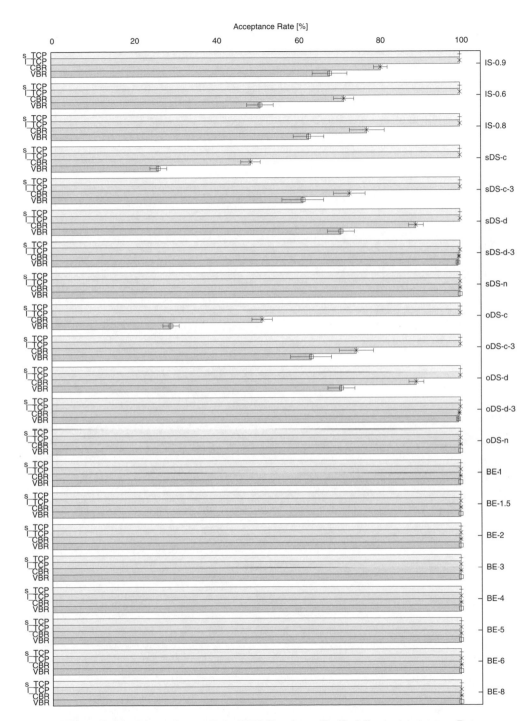

Figure B.14 Direct Comparison, DFN Topology, Traffic Mix A, Acceptance Rate

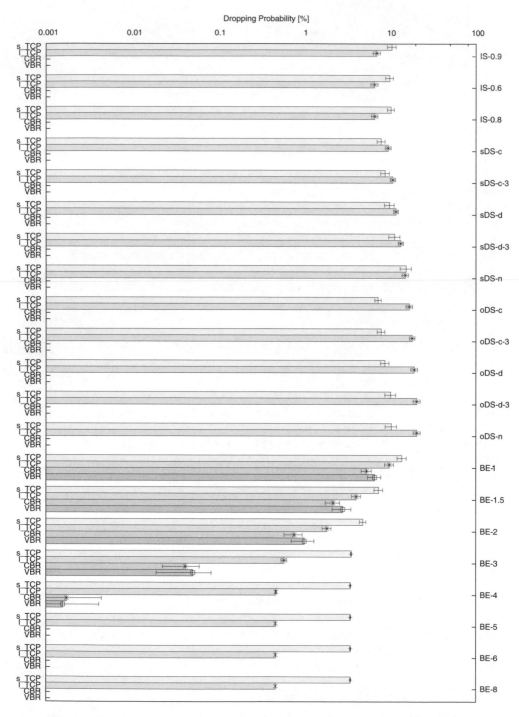

Figure B.15   Direct Comparison, DFN Topology, Traffic Mix A, Dropped Packets

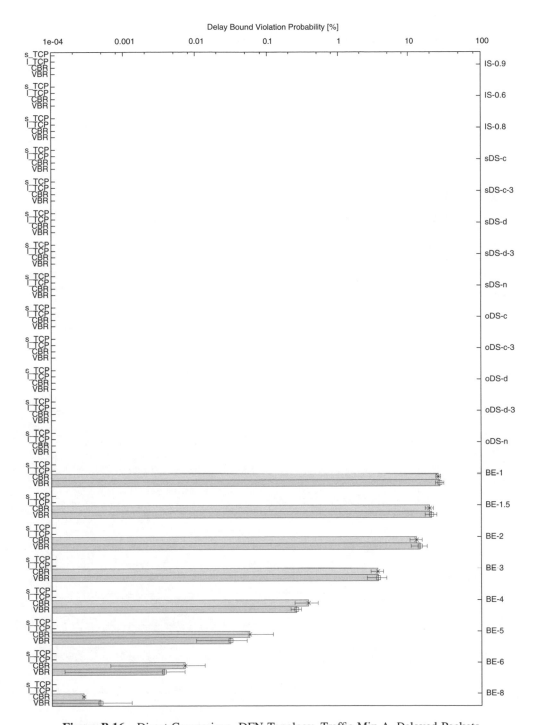

Figure B.16 Direct Comparison, DFN Topology, Traffic Mix A, Delayed Packets

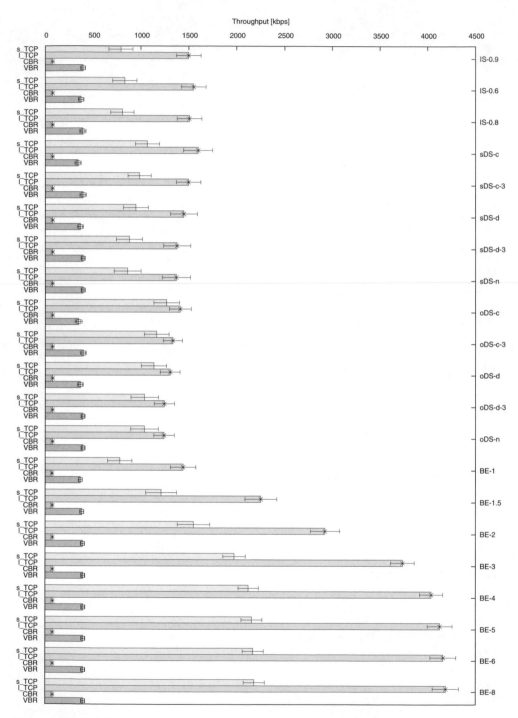

Figure B.17   Direct Comparison, DFN Topology, Traffic Mix A, Throughput

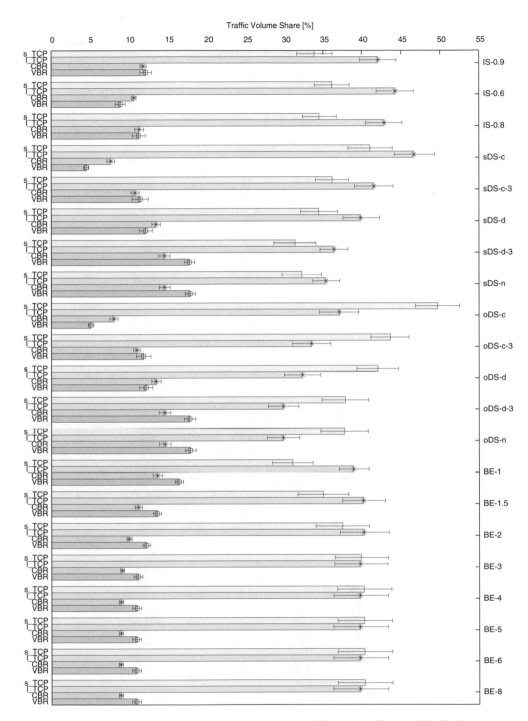

Figure B.18 Direct Comparison, DFN Topology, Traffic Mix A, Share of Traffic Volume

Table B.2 Direct Comparison, DFN Topology, Traffic Mix B and C Utility of the Accepted Flows and Overall Utility

| | | Utility of Accepted Flows | | | | | | | |
| | | Traffic Mix B | | | | Traffic Mix C | | | |
System		s_TCP	l_TCP	CBR	VBR	s_TCP	l_TCP	CBR	VBR
IS	0.8	0.87	0.95	1.00	1.00	0.93	0.97	1.00	1.00
sDS	c	0.98	0.94	1.00	1.00	1.00	0.95	1.00	1.00
	c-3	0.97	0.91	1.00	1.00	0.99	0.94	1.00	1.00
	n	0.67	0.61	0.94	0.94	0.99	0.94	1.00	1.00
oDS	c	0.96	0.71	1.00	1.00	0.99	0.81	1.00	1.00
	c-3	0.94	0.76	1.00	1.00	0.99	0.78	1.00	1.00
	n	0.73	0.51	0.95	0.95	0.99	0.78	1.00	1.00
BE	1	0.66	0.75	0.31	0.26	0.92	0.97	0.35	0.30
	1.5	0.86	0.95	0.47	0.42	0.98	0.99	0.58	0.51
	2	0.95	0.98	0.61	0.57	1.00	1.00	0.67	0.70
	3	1.00	1.00	0.86	0.84	1.00	1.00	0.78	0.97
	4	1.00	1.00	0.97	0.97	1.00	1.00	1.00	1.00
	5	1.00	1.00	1.00	1.00	1.00	1.00	1.00	1.00
	6	1.00	1.00	1.00	1.00	1.00	1.00	1.00	1.00
	8	1.00	1.00	1.00	1.00	1.00	1.00	1.00	1.00

| | | Overall Utility | | | | | | | |
| | | Traffic Mix B | | | | Traffic Mix C | | | |
System		s_TCP	l_TCP	CBR	VBR	s_TCP	l_TCP	CBR	VBR
IS	0.8	0.87	0.95	0.59	0.45	0.93	0.97	0.99	0.93
sDS	c	0.98	0.94	0.36	0.22	1.00	0.95	0.76	0.56
	c-3	0.97	0.91	0.55	0.42	0.99	0.94	0.98	0.98
	n	0.66	0.61	0.94	0.94	0.99	0.94	1.00	1.00
oDS	c	0.96	0.71	0.38	0.23	0.99	0.81	0.77	0.61
	c-3	0.92	0.67	0.56	0.44	0.99	0.78	0.98	0.98
	n	0.73	0.51	0.95	0.95	0.99	0.78	1.00	1.00
BE	1	0.66	0.75	0.31	0.26	0.92	0.97	0.35	0.30
	1.5	0.86	0.95	0.47	0.42	0.98	0.99	0.58	0.51
	2	0.95	0.98	0.61	0.57	1.00	1.00	0.78	0.70
	3	1.00	1.00	0.86	0.84	1.00	1.00	0.97	0.97
	4	1.00	1.00	0.97	0.97	1.00	1.00	1.00	1.00
	5	1.00	1.00	1.00	1.00	1.00	1.00	1.00	1.00
	6	1.00	1.00	1.00	1.00	1.00	1.00	1.00	1.00
	8	1.00	1.00	1.00	1.00	1.00	1.00	1.00	1.00

Table B.3   Direct Comparison, DFN Topology, Traffic Mix B and C Acceptance Rate and Dropping respectively Delay Bound Violation Probability

		Acceptance Rate [%]							
		Traffic Mix B				Traffic Mix C			
System		s_TCP	l_TCP	CBR	VBR	s_TCP	l_TCP	CBR	VBR
IS	0.8	100	100	57	42	100	100	99	93
sDS	c	100	100	32	18	100	100	75	54
	c-3	100	100	53	39	100	100	98	98
	n	100	100	100	100	100	100	100	100
oDS	c	100	100	35	19	100	100	76	58
	c-3	100	100	53	41	100	100	98	98
	n	100	100	100	100	100	100	100	100
BE	1	100	100	100	100	100	100	100	100
	1.5	100	100	100	100	100	100	100	100
	2	100	100	100	100	100	100	100	100
	3	100	100	100	100	100	100	100	100
	4	100	100	100	100	100	100	100	100
	5	100	100	100	100	100	100	100	100
	6	100	100	100	100	100	100	100	100
	8	100	100	100	100	100	100	100	100

		Dropping and Delay Bound Violation Probability [%]							
		Traffic Mix B				Traffic Mix C			
System		s_TCP	l_TCP	CBR	VBR	s_TCP	l_TCP	CBR	VBR
IS	0.8	9.99	6.89	0.00	0.00	6.76	4.29	0.00	0.00
sDS	c	7.52	9.05	0.00	0.00	5.06	7.87	0.00	0.00
	c-3	8.55	10.85	0.00	0.00	5.23	8.42	0.00	0.00
	n	22.96	23.59	0.87	0.89	5.65	8.44	0.00	0.00
oDS	c	6.71	16.84	0.00	0.00	4.31	12.46	0.00	0.00
	c-3	7.83	18.68	0.00	0.00	4.39	13.32	0.00	0.00
	n	18.74	28.11	0.76	0.78	4.41	13.26	0.00	0.00
BE	1	21.56	17.06	29.88	32.69	7.16	4.59	30.09	32.97
	1.5	10.19	6.34	21.42	22.77	4.18	1.62	16.21	17.25
	2	5.88	2.65	13.87	14.52	3.58	0.76	7.05	6.72
	3	3.66	0.67	3.76	3.75	3.46	0.47	0.63	0.47
	4	3.48	0.47	0.55	0.56	3.46	0.46	0.06	0.05
	5	3.48	0.46	0.04	0.04	3.46	0.46	0.01	0.01
	6	3.48	0.46	0.01	0.01	3.46	0.46	0.00	0.00
	8	3.47	0.46	0.00	0.00	3.46	0.46	0.00	0.00

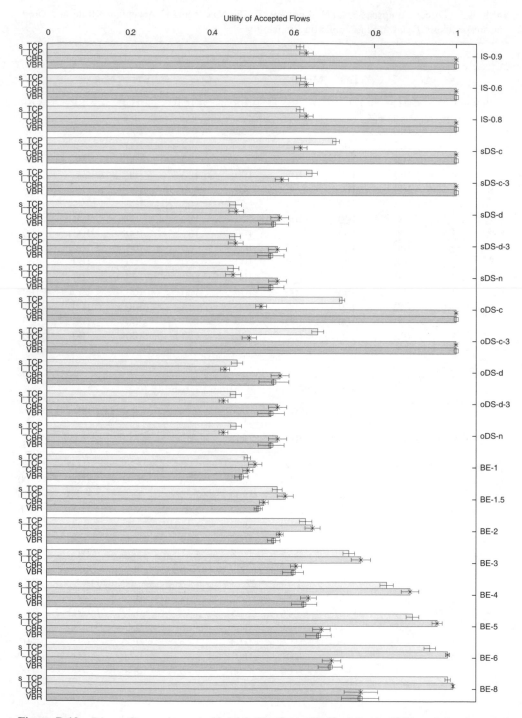

Figure B.19   Direct Comparison, Artificial-3 Topology, Traffic Mix A, Utility of the Accepted Flows

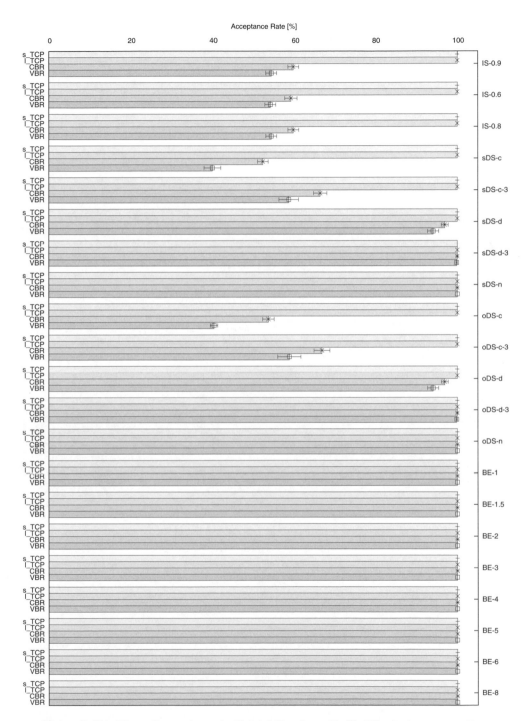

Figure B.20   Direct Comparison, Artificial-3 Topology, Traffic Mix A, Acceptance Rate

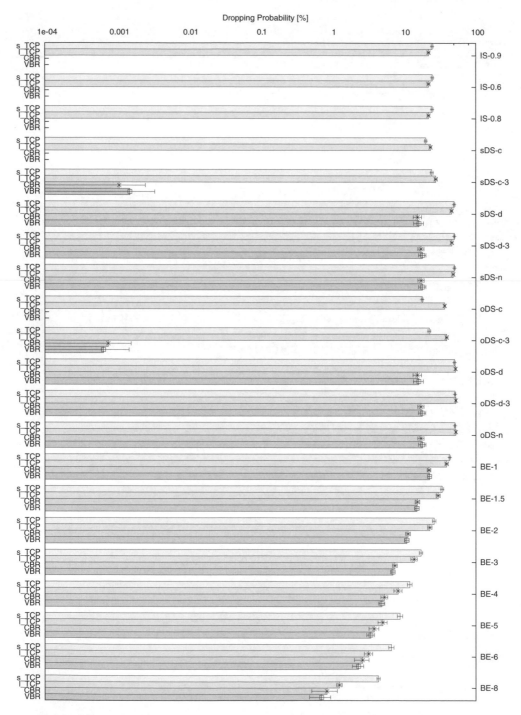

Figure B.21 Direct Comparison, Artificial-3 Topology, Traffic Mix A, Dropped Packets

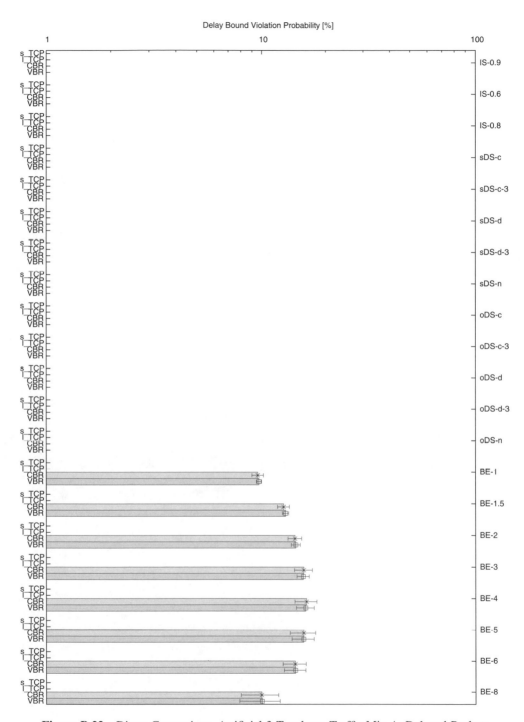

Figure B.22 Direct Comparison, Artificial-3 Topology, Traffic Mix A, Delayed Packets

C

Analytical Comparison
of Interconnection Methods

In this chapter, we shed some light on the costs of the different interconnection methods. As elaborated in Chapter 9 (Section 9.2), there are two basic interconnection methods:

- A *direct line* to connect two interconnection partners directly.
- An *Internet Exchange Point (IXP)* that a larger number of providers are connected to. A large number of interconnections can be realized via a single IXP. Three theoretical types of IXPs can be distinguished by whether they are based on
  - an exchange router.
  - an exchange Local Area Network (LAN) (switch).
  - an exchange Metropolitan Area Network (MAN).

We use some very simple analytical models to investigate the cost structure of the different IXP types (Section C.1), to investigate when the use of an IXP is cost efficient (Section C.2) and which type of IXP is more cost efficient, depending on the number of connected parties (Section C.3).

C.1 Internet Exchange Point Cost Models

In this section, simple cost models for the different IXP structures are elaborated. Table C.1 lists the variables and parameters used in these models.

C.1.1 Exchange Router

If an IXP uses an exchange router, each Internet Network Service Provider (INSP) has to spend the full costs for the lease of the connection line (c_L) to the IXP and part of the costs for the central exchange router (c_{ER}) at the IXP location. The cost function for the exchange router model shown in Figure C.1 is

$$c_{INSP}^{ExchRouter} = c_L + \frac{1}{N} \cdot c_{ER} \tag{C.1}$$

The Competitive Internet Service Provider: Network Architecture, Interconnection, Traffic Engineering and Network Design
Oliver Heckmann   © 2006 John Wiley & Sons, Ltd

Table C.1   Variables and Parameters of the Cost Models

N	Number of INSPs
c_{INSP}	Total cost of one INSP within an existing set of N INSPs
c_{ER}	Cost for one exchange router
c_L	Cost for a connection line
c_{EN}	Costs of the exchange network
c_{SW}	Costs for a switch in the exchange network

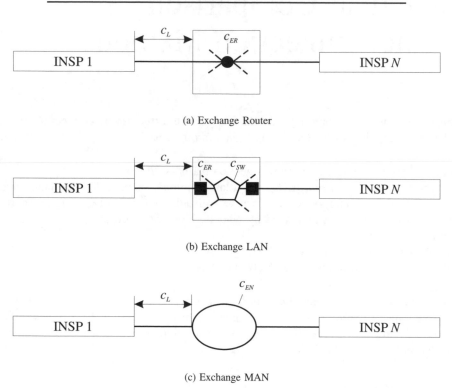

(a) Exchange Router

(b) Exchange LAN

(c) Exchange MAN

Figure C.1   Internet Exchange Point Costs Models

The exchange router model is the most cost efficient structure for IXP interconnection, but is vulnerable to congestion and has some structural drawbacks additionally. It has insufficient support for Quality of Service (QoS) as well as individual peering and routing policies. For example, the IXP managing the exchange router selects a single route to one destination that then has to be used by all connected providers (as seen in Figure C.2). This is a huge drawback for INSPs and therefore the exchange router is practically not used nowadays.

C.1.2 Exchange LAN

For the exchange LAN structure (see Figure C.3), N lines are needed in total to connect the N INSPs to the IXP LAN. Additionally, one edge router per INSP is necessary. The

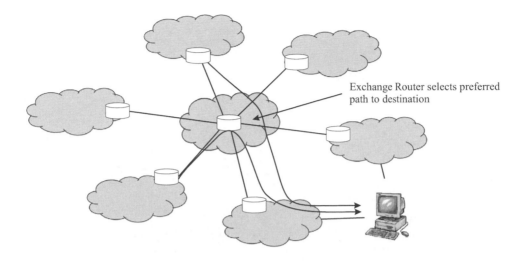

Figure C.2   Exchange Router Structure

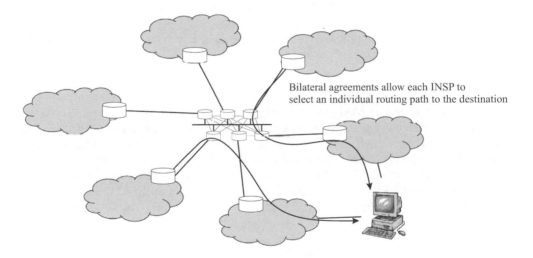

Figure C.3   The Exchange LAN Structure

edge router is owned by each INSP. It enables the INSP to choose its own routing and QoS policies and to decide which INSP to cooperate with.

The IXP has to operate one central network switch (c_{SW}). This results in the following cost function, see also Figure C.1:

$$c_{INSP}^{LAN} = c_L + c_{ER} + \frac{1}{N} \cdot c_{SW} \tag{C.2}$$

This model of exchange colocation enables connection with diverse access media, as the provider's colocated router undertakes the media translation between access link protocol

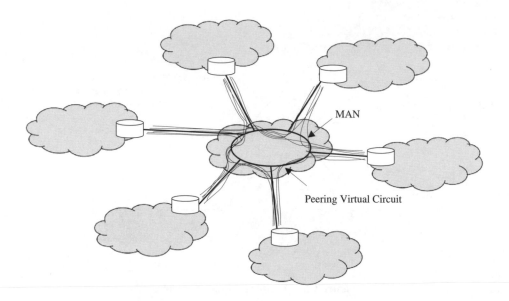

Figure C.4   The Exchange MAN Structure

and the common exchange protocol (usually Border Gateway Protocol (BGP), see Rekhter and Li (1995)). A drawback of this model is that of imposed traffic[1].

C.1.3 Exchange MAN

The costs of an exchange MAN IXP (see Figure C.4) consist of the line costs to connect the IXP to the next entry point of the MAN. These line costs c_L are typically smaller than those in the exchange LAN model because the geographical distance to the next access point of the distributed MAN will typically be smaller than that to the central LAN. This is a cost shift from the INSP to the IXP which results in lower line costs for the Internet Service Provider (ISP) but higher access costs for connecting to the IXP network.

The resulting cost function for the exchange MAN model is as follows, see also Figure C.1:

$$c_{INSP}^{MAN} = c_L + \frac{1}{N} \cdot c_{EN} \tag{C.3}$$

Exchange MAN structures enforce the use of a uniform access technology, see Huston (1999a).

C.2 Cost Efficiency of an Internet Exchange Point

It is quite intuitive that for a larger number of INSPs a fully meshed interconnection structure where every INSP is directly connected with all others (see Figure 9.2 (b)) is

---

[1] In the absence of a defensive mechanism a router accepts all traffic forwarded to it, even if there are no interconnection agreements between the two parties. Therefore, exchange routers require careful configuration management to ensure that the traffic matches the interconnection agreements, see Huston (1999a).

not as cost effective as a structure where all INSPs are connected with each other indirectly via an IXP. With a simple analytical model, we show now that an IXP is already cost effective for a very small number of INSPs.

We compare the costs of a fully meshed structure without an IXP (C.4) with those of a structure using an IXP. The IXP is modelled as exchange LAN in (C.5).

$$c_{INSP}^{FM} = c_L^{FM} \cdot \frac{N-1}{2} \qquad (C.4)$$

$$c_{INSP}^{LAN} = c_L^{LAN} + c_{ER} + \frac{1}{N} \cdot c_{SW} \qquad (C.5)$$

The terms c_L express the average line costs. We assume that they are proportional to the Euclidean distance d between the connecting parties:

$$c_L = p_c \cdot d \qquad (C.6)$$

p_c is the price per distance; it is assumed to be identical for the fully meshed and the IXP LAN models. The distance d will be different for the two models. We elaborate the distance assuming that the INSPs are uniformly distributed over a quadratic, circular area, see Figure C.5:

1. Let the positions of the INSPs be distributed uniformly in a *quadratic area* with the dimension $2R$, as illustrated in Figure C.5. It is assumed that the IXP is located in the middle of the distribution.

   (a) The expected Euclidean distance *between two INSPs i and j* is defined as

$$d_q^{FM} = \sqrt{(x_i - x_j)^2 + (y_i - y_j)^2} = \sqrt{(\Delta x)^2 + (\Delta y)^2} \qquad (C.7)$$

The expected distance between two uniformly distributed independent random variables x, y on interval [0,1] is

$$\int_0^1 \left(x \cdot \frac{x}{2} + (1 - x) \cdot \frac{1 - x}{2} \right) dx = 1/3 \qquad (C.8)$$

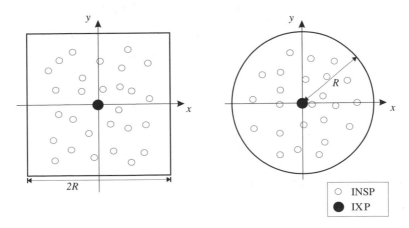

Figure C.5   Quadratic and Circular Distribution

therefore, $\Delta x = \Delta y = \frac{2}{3}R$ and the average distance d_q^{FM} between two INSPs in the quadratic model is

$$d_q^{FM} = \frac{\sqrt{2} \cdot 2R}{3} \tag{C.9}$$

(b) The expected Euclidean distance d_q^{LAN} between one INSP i and the IXP is defined accordingly as

$$d_q^{LAN} = \sqrt{(x_i - x_{IXP})^2 + (y_i - y_{IXP})^2} \tag{C.10}$$

$$= \sqrt{(\Delta x')^2 + (\Delta y')^2} \tag{C.11}$$

The expected distance between a uniformly distributed random variable and the origin on interval [-1, 1] is

$$\int_{-1}^{0} -\frac{x}{2}\,dx + \int_{0}^{1} \frac{x}{2}\,dx = 1/2 \tag{C.12}$$

therefore, $\Delta x' = \Delta y' = \frac{1}{2} \cdot R$ and the average distance d_q^{LAN} between an INSP and the IXP in the quadratic model is

$$d_q^{LAN} = \frac{R}{\sqrt{2}} \tag{C.13}$$

2. Let the positions of the INSPs be distributed uniformly in a *circular area* with diameter $2R$, as illustrated in Figure C.5. Again, it is assumed that the IXP is located in the middle of the distribution.

(a) The expected Euclidean distance d_c^{FM} between INSPs i and j is defined as

$$d_c^{FM} = \sqrt{(x_i - x_j)^2 + (y_i - y_j)^2} \tag{C.14}$$

$$\text{with } p_i = (x_i, y_i) \text{ and } p_j = (x_j, y_j)$$

Let $C = \{p = (x, y) \mid x^2 + y^2 \le R\}$ denote the set of points in a circle with radius R. The expected distance between two INSPs in the circular model is (see Santaló (2004))

$$d_c^{FM} = \left(\frac{1}{\pi}\right)^2 \int_C \int_C d_{ij}(p_i, p_j)\,dp_i\,dp_j$$

$$= \frac{128}{45\pi}R = 0.9054 \cdot R \tag{C.15}$$

(b) The expected Euclidean distance d_c^{LAN} between an INSP and the IXP is

$$d_c^{LAN} = \frac{\int_0^R 2\pi r \cdot r\,dr}{\int_0^R 2\pi r\,dr} = \frac{2}{3}R \tag{C.16}$$

Now the equations (C.4) and (C.5) can be compared with each other to calculate the value of N at which the exchange LAN structure is more cost effective than the fully meshed structure:

$$c_{INSP}^{FM} \geq c_{INSP}^{LAN} \tag{C.17}$$

$$(N-1) \cdot \frac{c_L^{FM}}{2} \geq c_L^{LAN} + c_{ER} + \frac{1}{N} \cdot c_{SW} \tag{C.18}$$

$$0 \leq N \cdot (N-1) \cdot \frac{c_L^{FM}}{2} + N(-c_L^{LAN} - c_{ER}) - c_{SW} \tag{C.19}$$

$$0 \leq N^2 \cdot \frac{c_L^{FM}}{2} + N(-\frac{c_L^{FM}}{2} - c_L^{LAN} - c_{ER}) - c_{SW} \tag{C.20}$$

with $\frac{c_L^{FM}}{2} \geq 0$ (C.20) is an open parable $f(N) = aN^2 + bN + c$ with the minimal $N = -\frac{b}{2a}$

$$N \geq -\frac{-\frac{c_L^{FM}}{2} - c_L^{LAN} - c_{ER}}{2 \cdot \frac{c_L^{FM}}{2}} \tag{C.21}$$

$$N \geq \frac{1}{2} + \frac{c_L^{LAN}}{c_L^{FM}} + \frac{c_{ER}}{c_L^{FM}} \tag{C.22}$$

With $c_L^{type} = p_c \cdot d^{type}$

$$N \geq \frac{1}{2} + \frac{d^{LAN}}{d^{FM}} + \frac{c_{ER}}{p_c d^{FM}} \tag{C.23}$$

For the quadratic distribution

$$N_q \geq \frac{5}{4} + \frac{c_{ER}}{\frac{\sqrt{2} \cdot 2R}{3} \cdot p_c} \tag{C.24}$$

For the circular distribution

$$N_c \geq \frac{1}{2} + \frac{15}{64} \cdot \pi + \frac{c_{ER}}{\frac{128}{45\pi} R \cdot p_c} \tag{C.25}$$

Assuming that the exchange router is a Cisco Catalyst 7206 with an approximate value of $c_{ER} = 20,000$ EUR and the fibre price per kilometre and year of approximate $p_c = 1,000$ EUR/km. Assuming further that the connecting INSPs are within the boundaries of a city the size of Frankfurt/Main, the value of R is approximately 7 km. These assumptions lead to the value of N at which the exchange LAN structure is more cost efficient.

$$N_q \geq 4.28 \tag{C.26}$$

$$N_c \geq 4.39 \tag{C.27}$$

With at least five connecting INSPs, the exchange LAN structure is already more cost efficient than the fully meshed structure. While this result is based on a lot of assumptions and varies depending on the chosen R and the assumed costs, it still points out well that using an IXP is cost efficient for a very small number of providers within a city's boundary. With an increasing R the number of providers N becomes even smaller.

C.3 LAN versus MAN IXP Structure

Next, we compare the exchange LAN and MAN structures for a single IXP. For simplification purposes, it is assumed that the exchange MAN forms a circle with radius R' through the circular area of the previous section so that the expected distance d_c^{MAN} between an INSP and the MAN is $R/4$.

$$c_{INSP}^{LAN} = p_c \cdot \frac{2R}{3} + c_{ER} + \frac{1}{N} \cdot c_{SW} \tag{C.28}$$

$$c_{INSP}^{MAN} = p_c \cdot \frac{R}{4} + \frac{1}{N} \cdot c_{EN} \tag{C.29}$$

Combining equations (C.28) and (C.29) leads to the value of N at which the exchange MAN structure is more cost effective than the exchange LAN:

$$c_{INSP}^{LAN} \geq c_{INSP}^{MAN} \tag{C.30}$$

$$p_c \cdot \frac{2R}{3} + c_{ER} + \frac{1}{N} \cdot c_{SW} \geq \frac{1}{N} \cdot c_{EN} + p_c \cdot \frac{R}{4} \tag{C.31}$$

$$\text{with } c_{EN} \gg c_{SW}$$

$$N \geq \frac{c_{EN}}{\frac{5}{12} \cdot p_c \cdot R + c_{ER}} \tag{C.32}$$

Let us assume that the exchange MAN has costs roughly similar to the DE-CIX IXP in Frankfurt. DE-CIX has three main locations with redundant switches whose approximate value is 300,000 EUR and about 20 km fibre lines connect the locations with each other. For a city the size of Frankfurt, the value of R is 7 km. These assumptions lead to the value of N at which the exchange MAN structure is more cost efficient.

$$N \geq \frac{3 \cdot 2 \cdot 300,000 + 1,000 \cdot 20}{\frac{5}{12} \cdot 1,000 \cdot 7 + 20,000}$$

$$\geq 79.42$$

$$\geq 80$$

When 80 or more INSPs use the IXP, the exchange MAN structure is more cost efficient than the exchange LAN structure. As an example, consider the real DE-CIX which is mostly a MAN and has currently 128 connected customers (see German Internet Exchange DE-CIX (2004)), which is more than enough for this simple model to make exchange MAN cost efficient.

Next, we assume that the exchange MAN has costs similar to the London Internet Exchange (LINX) IXP in London. LINX has four main locations with redundant switches and five smaller locations with small redundant switches with approximately 100 km fibre lines connecting the locations with each other. The costs of small switches are approximately 100,000 EUR and the costs of bigger switches are approximately 300,000 EUR. For a city the size of London, the value of R is 25 km. These assumptions lead to the value of N at which the exchange MAN structure is more cost efficient.

$$N_q \geq \frac{4 \cdot 2 \cdot 300,000 + 5 \cdot 2 \cdot 100,000 + 1,000 \cdot 100}{\frac{5}{12} \cdot 1,000 \cdot 25 + 20,000}$$

$$\geq 115.07$$

$$\geq 116$$

The difference is that DE-CIX involves greater network costs which are partly offset by the greater area covered with the exchange MAN structure. The LINX is mostly a MAN and at the time of writing had 143 connected customers (see LINX (2003)) and is cost efficient within the limitations of this simple model.

D

Elasticity of Traffic Matrices – Network Models*

Here, we present the analytical foundation for the analysis of elastic traffic matrices described in Section 13.3. Several network models of increasing complexity that describe the behaviour of the traffic flows through a network with respect to the capacity of the links and nodes of that network are described.

D.1 Basic Model

We model a subnetwork Λ of the Internet consisting of N nodes and L directed links. The traffic through the network consists of long-lived greedy Transmission Control Protocol (TCP) connections and is represented by TCP *macroflows*. A TCP macroflow represents a number of TCP connections that have the same ingress node i and egress node j of Λ. We assume that the connections of a macroflow experience on average the same loss $\tilde{p}$ and delay $\tilde{q}$ when traversing the other networks that are not modelled in detail with this model. The macroflows are assumed to be small compared to the other flows flowing through the external networks; therefore, the external loss $\tilde{p}$ and delay $\tilde{q}$ are independent of the rate of the macroflows. The macroflows are elastic traffic; their rate is described by a TCP formula and adapts to the network conditions of Λ. There are a number of works about predicting the average TCP throughput depending on the loss and delay properties of a flow, see Section 4.1.3. As we are not interested in details such as the duration of the connection establishment etc., we use the rather simple square-root formula (4.2) here.

An output queue is attached to each link. In the basic model, we describe the queues as *M/M/1/B* queues. This is not the most realistic approach: First, because Internet traffic is not described very well by a Poisson arrival process, see Paxson and Floyd (1995). Second, since packet sizes are not exponentially distributed, an exponential service rate is also not realistic, see AIX – NASA Ames Internet Exchange (2000); Claffy *et al.* (1998). However, the *M/M/1/B* model is one of the simplest queueing models and is used in related works like Garetto *et al.* (2001); Gibbens *et al.* (2000). We will investigate more realistic queueing models later in this section.

* Reproduced by permission of VDE Verlag GMBH

The basic network model with elastic traffic is described by the non-linear equation system in Model D.1.

Model D.1 Basic Network Model for Elastic Traffic Matrices

Indices:

$i, j = 1, ..., N$ Node i or j

$l = 1, ..., L$ Link or output queue l

Parameters:

ψ_{ij} Path from node i to node j and back (set of links)

t_{ij} Size of macroflow between node pair i, j [pkts]

μ_l Service rate of link or queue l [pkts/s]

B Buffer size [pkts]

$\tilde{q}$ Average external queueing and total propagation delay [s]

$\tilde{p}$ Average external loss probability

Variables:

r_{ij} Rate of macroflow between node pair i, j [pkts/s]

ρ_l Utilisation of link or queue l

p_l Loss probability of link or queue l

q_l Queueing delay of link or queue l [s]

$$r_{ij} = \frac{t_{ij}}{[(\sum_{l \in \psi_{ij}} q_l) + \tilde{q}] \cdot \sqrt{\frac{2}{3}} \cdot \sqrt{1 - [\prod_{l \in \psi_{ij}} (1 - p_l)] \cdot (1 - \tilde{p})}} \qquad \forall i, j \, | i \neq j \quad \text{(D.1)}$$

$$\rho_l = (\sum_{(i,j) \, | l \in \psi_{ij}} r_{ij}) \cdot \frac{1}{1 - p_l} \cdot \frac{1}{\mu_l} \qquad \forall l \qquad \text{(D.2)}$$

$$p_l = (1 - \rho_l) \cdot \frac{\rho_l^B}{1 - \rho_l^{B+1}} \qquad \forall l \qquad \text{(D.3)}$$

$$q_l = \frac{1 + B\rho^{B+1} - (B+1)\rho^B}{\mu_l (1 - \rho_l)(1 - \rho_l^B)} \qquad \forall l \qquad \text{(D.4)}$$

The total loss probability of a macroflow ij can be approximated by

$$p_{ij} = \tilde{p} + \sum_{l \in \psi_{ij}} p_l \qquad \qquad \text{(D.5)}$$

Table D.1   Assessment of the Approximations

Approximation	Maximal Error [%]
For p_{ij}	0.0004795
For ρ_l	0.0009097

for small loss probabilities. Similarly, for small loss probabilities at a link l the utilisation (D.2) can be approximated as

$$\rho_l = (\sum_{(i,j)\,|\,l\in\psi_{ij}} r_{ij}) \cdot \frac{1}{\mu_l} \tag{D.6}$$

These simplifications can reduce the computational effort to solve the resulting non-linear equation system by up to 25%. In order to assess the systematic error of these approximations we ran a number of experiments on the Deutsche Telekom backbone topology (see Appendix A) with different parameters of t_{ij}, B and μ_l. We solve the non-linear equation system from Model D.1 using Maplesoft (2004) and compare the difference in ρ_l. The maximum errors of 25 different settings are listed in Table D.1. They are extremely small and can be neglected.

Next we discuss the possible extensions of the basic model.

D.2 Discrete Service Times

We first investigate how the basic model from Section D.1 can be extended to account for more realistic service times. IP packets can differ drastically in their size (40 to 1500 bytes), see AIX – NASA Ames Internet Exchange (2000); Claffy *et al.* (1998). We assume a service time proportional to the packet size and use a discrete distribution with $c = 1, \ldots, C$ classes of differently sized packets to model the service time; si_c is the packet size of class c and h_c the relative frequency of class c with $\sum_c h_c = 1$. Using sp_l as the line speed of link l, the probability density function of the service time distribution is given as

$$pdf(x) = \sum_c h_c \cdot \delta(x - \frac{si_c}{sp_l}) \tag{D.7}$$

where $\delta(x)$ is the Dirac impulse $\delta(x) = 1$ for $x = 0$ and 0 otherwise. The probability distribution function is

$$PDF(x) = \sum_c h_c \cdot u(x - \frac{si_c}{sp_l}) \tag{D.8}$$

where $u(x)$ is the unit function $u(x) = 1$ for $x \geq 0$ and 0 otherwise. In order to model the *queueing delay*, we use the Pollaczek–Khinchin formula for the queueing delay of an *M/G/1* queue

$$q_l = E(x) \cdot (1 + \frac{1+C_v^2}{2} \frac{\rho_l}{1-\rho_l}) \tag{D.9}$$

with the expected service time[1]

$$E(x) = \frac{1}{\mu} = \int_{-\infty}^{\infty} x \cdot pdf(x) \, dx = \sum_{c} h_c \cdot \frac{si_c}{sp_l} \tag{D.10}$$

and the square of the coefficient of variation

$$C_v^2 = \frac{Var(x)}{E(x)^2} = \frac{\int_{-\infty}^{\infty} (x - E(x))^2 \cdot pdf(x) \, dx}{E(x)^2} \tag{D.11}$$

For the *loss probability* p_l we turn to the *M/G/1/B* queue. There is no general closed form for the loss probability of the *M/G/1/B* or the queue length distribution of the *M/G/1* queue. We can derive the loss probability of the *M/G/1/B* queue exactly if we know the state probabilities $\pi_{lk}^{(\infty)}$ for queue length k of the corresponding *M/G/1* queue l.

Cooper (1981); Virtamo (2003) list an iterative algorithm based on Markov chains that can be used to numerically derive $\pi_{lk}^{(\infty)}$. We do not want to use this Markov chain algorithm; first, because it does not give us a closed form for the loss probability that we need for our equation system and, second, because for that approach we would have to solve several complex integrals numerically, while we are interested in an analytical form. Therefore, we use a different method to derive the state probabilities $\pi_{lk}^{(\infty)}$ of the *M/G/1* queue: The Laplace transform of the service time distribution $pdf(x)$ is

$$b_l^*(s) = \sum_{c} h_c \cdot e^{-s\frac{si_c}{sp_l}} \tag{D.12}$$

Kleinrock (1975); Virtamo (2003) show that the transformed state probabilities follow the Pollaczek–Khinchin transform formula for the queue length

$$Q_l(z) = (1 - \rho_l)\frac{b_l^*(\lambda - \lambda z)}{b_l^*(\lambda - \lambda z) - z}(1 - z) \tag{D.13}$$

With the inverse Z-transformation on $Q_l(z)$, we can derive the state probabilities $\pi_{lk}^{(\infty)}$ analytically. We can use the Taylor series expansion to analytically transform the somewhat complex term $Q_l(z)$ back:

$$\pi_{lk}^{(\infty)} = \frac{1}{k!}\frac{d^k}{dz^n}Q_l(z)\,|_{z=0} \tag{D.14}$$

The loss probability of the related *M/G/1/B* queue is now given as

$$p_l = 1 - \frac{1}{\rho_l + \pi_{l0}^{(B)}} \tag{D.15}$$

using the state probability $\pi_{l0}^{(B)}$ of the finite queue as in Virtamo (2003)

$$\pi_{l0}^{(B)} = \frac{\pi_{l0}^{(\infty)}}{\sum_{j=0}^{B-1} \pi_{lj}^{(\infty)}} \tag{D.16}$$

This leaves us with closed-form non-linear equations for loss and delay of the *M/G/1/B* queue with a discrete service time distribution.

---

[1] We continue using μ for the inverse of the expected service time as we did with the *M/M/1/B* queue.

D.3 Self-similar Traffic

Internet traffic measurements show self-similar, heavy-tailed and long-range dependent properties as discussed in Section 5.1. The burstiness of Internet traffic on larger timescales can significantly influence the loss probability. To take this effect into account we use the Gaussian approximation of aggregate traffic and the following loss formula based on Addie *et al.* (2002); Mannersalo and Norros (2002):

$$
p_l = \frac{C}{\frac{B-\lambda\hat{t}}{\sigma_{\hat{t}}^2}\sqrt{2\pi\sigma_{\hat{t}}^2}} \cdot e^{-\inf_{t\in\Re^+}\frac{(B+(\mu-\lambda)\cdot t)^2}{2\cdot\sigma_t^2}}
\tag{D.17}
$$

$\hat{t}$ is the optimiser from the infimum condition, t is the timescale, B is the buffer size, λ and μ are the arrival and service rates. For a given Hurst parameter, σ_t^2 is given as $\sigma_t^2 = \sigma^2 \cdot t^{2H}$.

D.4 Related Work

Some works use network models similar to our models. The performance models of Gibbens *et al.* (1999, 2000); May *et al.* (1999) are used to analyse quality of service (QoS) in Diffserv IP networks with two service classes. They assume a Poisson arrival process and exponential service times (*M/M/1/B*). The fixed point model of Gibbens *et al.* (2000) combine the Diffserv resource models with the TCP formula. We use a similar approach but we also investigate non-exponential service times and non-Poisson arrivals. Also, we investigate the performance in the context of network design and capacity expansion and do not use different service classes.

Garetto *et al.* (2001) present an analytical TCP model for multiple flows and verifies it against NS2 simulations. Similar to our model, they combine a TCP and a network model and calculate the fixed point of the two models. Their TCP model, however, is more fine grained and complex than our TCP formula–based TCP model. This, however, comes at the cost of losing a closed-form formulation of the whole model. The authors investigate different network models and find that the simple *M/M/1/B* gives sufficiently accurate results.

Schwefel (2001) introduces a queueing model that is based on multiple ON/OFF arrival processes; this allows accounting for long-range dependency. It is extended to be reactive to congestion by slowing down the rate similar to the way TCP reacts and can thus be used for the performance analysis of TCP-generated bursty traffic. Contrary to this approach, we combine the TCP formula with the standard queueing theory.

Bibliography

Abreu E 2001 Down on the Server Farm. *The Industry Standard* **4**(7), 4.

Access eCommerce 2003 Recent Developments – Consolidation and Dominance. Source: Webmergers.com "Internet Companies Three Years After the Height of the Bubble" http://www.access-ecom. info/article.cfm?id=24&xid=pa.

Addie R, Mannersalo P and Norros I 2002 Most Probable Paths and Performance Formulae for Buffers with Gaussian Input Traffic. *European Transactions on Telecommunications* **13**(3), 183–196.

Ahuja RK, Magnanti TL and Orlin JB 1993 *Network Flows: Theory, Algorithms, and Applications.* Prentice Hall. ISBN: 013617549.

Aiello W, Chung F and Lu L 2001 A Random Graph Model for Power Law Graphs. *Experimental Mathematics* **10**(1), 53–66.

AIX – NASA Ames Internet Exchange 2000 Packet Length Distributions. http://www.caida.org/analysis/AIX/ plen_hist/.

Akamai Technologies Inc. 2002 Applications for Akamai: EdgeScape (White Paper). http://www.akamai. com/en/resources/pdf/edgescape_applications_whitepaper.pdf.

Akamai Technologies Inc. 2006 Homepage. http://www.akamai.com/.

America Online 2006 Homepage. http://www.aol.com/.

AMS-IX Membership 2004 Amsterdam Internet Exchange, Membership Requirements – 5, Requirements for Becoming a Member of AMS-IX. http://www.ams-ix.net/joining/steps/step1.html.

Andersson L, Doolan P, Feldman N, Fredette A and Thomas B 2001 LDP Specification, RFC 3036.

ANSI T1.105 1995 Synchronous Optical Network (SONET) – Basic Description including Multiplex Structure, Rates and Formats.

ANSI T1.119 1995 SONET – Operations, Administration, Maintenance, and Provisioning.

Antila J and Luoma M 2003 Scheduling and Quality Differentiation in Differentiated Services. *Proceedings of the International Workshop on Multimedia Interactive Protocols and Systems (MIPS 2003)*, pp. 119–130.

Antila J and Luoma M 2004 Adaptive Scheduling for Improved Quality Differentiation. *Proceedings of the International Workshop on Multimedia Interactive Protocols and Systems (MIPS 2004)*.

Armitage G 2000 *Quality of Service in IP Networks.* MacMillan Technical Publishing. ISBN: 1578701899.

Armitage G 2001 Sensitivity of Quake 3 Players to Network Latency Poster. *ACM SIGCOMM Internet Measurement Workshop*.

Armitage G, Carpenter B, Casati A, Crowcroft J, Halpern J, Kumar B and Schnizlein J 2002 A Delay Bound Alternative Revision of RFC 2598, RFC 3248.

ASPnews 2004 January ASPnews Top 50: Top 25 Providers. http://www.aspnews.com/top50/article.php/ 11307_3300671_2.

Athuraliya S, Li VH, Low SH and Yin Q 2001 REM: Active Queue Management. *IEEE Network* **15**(3), 48–53.

Awduche D, Agogbua J and McManus J 1998 An Approach to Optimal Peering Between Autonomous Systems in the Internet. *Proceedings of the International Conference on Computer Communications and Networks (IC3N 1998)*, pp. 346–351.

Awduche D, Berger L, Gan D, Li T, Srinivasan V and Swallow G 2001 RSVP-TE: Extensions to RSVP for LSP Tunnels, RFC 3209.

Azzouna NB and Guillemin F 2003 Analysis of ADSL Traffic on an IP Backbone Link. *Proceedings of the IEEE Global Communications Conference (Globecom 2003)*, pp. 3742–3746.

Baake P and Wichmann T 1998 On the Economics of Internet Peering. Technical report, Berlecon Research Papers. http://www.berlecon.de/tw/peering.pdf.

Baccelli F, Cohen G, Olsder GJ and Quadrat JP 1992 *Synchronization and Linearity: An Algebra for Discrete Event Systems*. John Wiley & Sons. ISBN: 047193609X.

Badasyan N and Chakrabarti S 2003 Private Peering Among Internet Backbone Providers. Industrial Organization / Economics Working Paper Archive at WUSTL, http://econwpa.wustl.edu/eps/mic/papers/0301/0301003.pdf.

Bailey JP 1997 Economics and Internet Interconnection Agreements, In *Internet Economics*, L McKnight and JP Bailey (Editors), MIT Press, pp. 155–168. ISBN: 0262133369.

Bailey RW 1989 *Human Performance Engineering – Using Human Factors/Ergonomics to Achieve Computer System Usability*, 2nd edn. Prentice Hall, Englewood Cliffs, New York. ISBN: 0134451805.

Baiocchi A, Melazzi N, Listani M, Roveri A and Winkler R 1991 Loss Performance Analysis of an ATM Multiplexer Loaded with High-Speed On-Off Sources. *IEEE Journal on Selected Areas in Communications* **9**(3), 388–393.

Baker F 1995 Requirements for IP Version 4 Routers, RFC 1812.

Baker F, Iturralde C, Faucheur FL and Davie B 2001 Aggregation of RSVP for IP4 and IP6 Reservations, RFC 3175.

Banerjee A, Drake J, Lang J, Turner B, Awduche D, Berger L, Kompella K and Rekhter Y 2001 Generalized Multiprotocol Label Switching: An Overview of Signaling Enhancements and Recovery Techniques. *IEEE Communications Magazine* **39**(7), 144–151.

Barabasi AL and Albert R 1999 Emergence of Scaling in Random Networks. *Science* **286**, 509–512.

Barford P, Bestavros A, Bradley A and Crovella M 1999 Changes in Web Client Access Patterns: Characteristics and Caching Implications. *World Wide Web Journal* **2**(1-2), 15–28.

Barford P and Crovella M 1998 Generating Representative Web Workloads for Network and Server Performance Evaluation. *Proceedings of the ACM Joint International Conference on Measurement and Modeling of Computer Systems (SIGMETRICS 1998)*, pp. 151–160.

Beigbeder T, Coughlan R, Lusher C, Plunkett J, Agu E and Claypool M 2004 The Effects of Loss and Latency on User Performance in Unreal Tournament 2003. *Proceedings of ACM SIGCOMM 2004 NetGames Workshop*, pp. 144–151.

Benameur N and Roberts JW 2002 Traffic Matrix Inference in IP Networks. *Proceedings of Networks 2002*, pp. 151–156.

Benameur N, Fredj S, Oueslati-Boulahia S and Roberts J 2002 Quality of Service and Flow Level Admission Control in the Internet. *Computer Networks* **40**(1), 57–71.

Bennett JCR and Zhang H 1996 WF2Q: Worst-case Fair Weighted Fair Queuing. *Proceedings of the IEEE Conference on Computer Communications and Networking (INFOCOM 1996)*, pp. 120–127.

Bennett JCR and Zhang H 1997 Hierarchical Packet Fair Queuing Algorithms. *IEEE/ACM Transactions on Networking* **5**(5), 675–689.

Beran J 1994 *Statistics for Long-Memory Processes*. Chapman & Hall/CRC. ISBN: 0412049015.

Berger L 2003 Generalized Multi-Protocol Label Switching (GMPLS) Signaling Functional Description, RFC 3471.

Berger L, Gan D, Swallow G, Pan P, Tommasi F and Molendini S 2001 RSVP Refresh Overhead Reduction Extensions, RFC 2961.

Berry LTM, Murthagh BA, McMahon GB, Sugden SJ and Welling LD 1998 Genetic Algorithms in the Design of Complex Distribution Networks. *International Journal on Physical Distribution and Logistics Management* **28**(5), 377–381.

Berry LTM, Murtagh BA, McMahon GB, Sugden SJ and Welling LD 1999 An Integrated GA-LP Approach to Communication Network Design. *Telecommunication Systems Journal* **12**(2-3), 265–280.

Bessler S 2002 Label Switched Paths Re-configuration under Time-Varying Traffic Conditions. *Proceedings of the 15th ITC Specialist Seminar*.

Bessler S and Reichl P 2003 Auction-Based Optimal Dimensioning of MPLS Tunnels. *Proceedings of KiVS (Kommunikation in Verteilten Systemen) 2003*.

Bhatnagar S and Nath B 2003 Distributed Admission Control to Support Guaranteed Services in Core-Stateless Networks. *Proceedings of the IEEE Conference on Computer Communications and Networking (INFOCOM 2003)*, pp. 1659–1669.

Bhatnagar S and Vickers B 2001 Providing Quality of Service Guarantees using Only Edge Routers. *Proceedings of the IEEE Global Communications Conference (Globecom 2001)*.

Bhattacharyya S, Diot C, Jetcheva J and Taft N 2001 Pop-Level and Access-Link-Level Traffic Dynamics in a Tier-1 POP. *Proceedings of the First ACM SIGCOMM Workshop on Internet Measurement*, pp. 39–53.

Bhatti SN and Crowcroft J 2000 QoS-Sensitive Flows: Issues in IP Packet Handling. *IEEE Internet Computing* **4**(4), 48–57.

Bhatti N, Bouch A and Kuchinsky A 2000 Integrating User-Perceived Quality into Web Server Design. *Proceedings of the Ninth International World Wide Web Conference (WWW 9)* .

Bianchi G, Capone A and Petrioli C 2000 Packet Management Techniques for Measurement Based End-to-End Admission Control in IP Networks. *Journal of Communications and Networks* **2**, 147–156.

Bianchi G, Borgonovo F, Capone A and Petrioli C 2002 Endpoint Admission Control with Delay Variation Measurements for QoS in IP Networks. *ACM Computer Communication Review* **32**, 61–69.

Bitorika A, Robin M, Huggard M and Mc Goldrick C 2004 A Comparative Study of Active Queue Management Schemes. *under submission*.

Black D, Blake S, Carlson M, Davies E, Wang Z and Weiss W 1998 An Architecture for Differentiated Services, RFC 2475.

Bless R, Nichols K and Wehrle K 2003 A Lower Effort Per-Domain Behavior (PDB) for Differentiated Services, RFC 3662.

Bley A, Koch T and Wessäly R 2004 Large-scale Hierarchical Networks: How to Compute an Optimal Architecture. *Proceedings of Networks 2004*, pp. 429–434.

Boardwatch 2003
(i) B2 Innovates to Deliver VOIP
Url:(http://www.boardwatch.com) with Document ID = 43436
(ii) Bell Thinks Outside the Box
Url:(http://www.boardwatch.com) with Document ID = 41651
(iii) Ovum: Don't Slash DSL Prices
Url:(http://www.boardwatch.com) with Document ID = 36749
(iv) Content Delivery Picks Up in '03
Url:(http://www.boardwatch.com) with Document ID = 32687
(v) W2Forum: Caveat Content Provider
Url:(http://www.boardwatch.com) with Document ID = 36680
(vi) Aurora Unifies Billing for Vodat
Url:(http://www.boardwatch.com) with Document ID = 32243
(vii) VOIP Stocks: How High?
Url:(http://www.boardwatch.com) with Document ID = 44448
(viii) Linx Hits 30 Gbit/S Record
Url:(http://www.boardwatch.com) with Document ID = 43327.

Boardwatch 2004
(i) Report: Big ISPs Dominate
Url:(http://www.boardwatch.com) with Document ID = 45880
(ii) POTS Down But Not Out
Url:(http://www.boardwatch.com) with Document ID = 45917
(iii) Broadband Ventures Into Value
Url:(http://www.boardwatch.com) with Document ID = 50845.

Bodamer S and Charzinski J 2000 Evaluation of Effective Bandwidth Schemes for Self-Similar Traffic. *Proceeding of the 13th International Teletraffic Congress Specialist Seminar*.

Bolch G, Greiner S, DeMeer H and Trivedi K 1998 *Queueing Networks and Markov Chains: Modeling and Performance Evaluation With Computer Science Applications*. Wiley-Interscience. ISBN: 0471193666.

Boorstyn R, Burchard A, Liebeherr J and Oottamakorn C 2000 Effective Envelopes: Statistical Bounds on Multiplexed Traffic in Packet Networks. *Proceedings of the IEEE Conference on Computer Communications and Networking (INFOCOM 2000)*, pp. 1223–1232, Tel Aviv, Israel.

Bouras C, Campanella M and Sevasti A 2002 SLA Definition for the Provision of an EF-based Service. *Proceedings of the 16th International Workshop on Communications Quality & Reliability (CQR 2002)*, pp. 17–21.

Boyer J, Guillemin F, Robert P and Zwart B 2003 Heavy Tailed M/G/1-PS Queues with Impatience and Admission Control in Packet Networks. *Proceedings of the IEEE Conference on Computer Communications and Networking (INFOCOM 2003)*.

Braden R, Clark D and Shenker S 1994 Integrated Services in the Internet Architecture: an Overview, RFC 1633.

Braden B, Zhang L, Berson S, Herzog S and Jamin S 1997 Resource ReSerVation Protocol (RSVP) – Version 1 Functional Specification, RFC 2205.

Brakmo L and Peterson L 1995 TCP Vegas: End to End Congestion Avoidance on a Global Internet. *IEEE Journal on Selected Areas in Communication* **13**(8), 1465–1480.

Breslau L and Shenker S 1998 Best-Effort versus Reservations: A Simple Comparative Analysis. *Proceedings of the ACM Special Interest Group on Data Communication Conference (SIGCOMM 1998)*, pp. 3–16.

Breslau L, Jamin S and Shenker S 2000a Comments on the Performance of Measurement-Based Admission Control Algorithms. *Proceedings of the IEEE Conference on Computer Communications and Networking (INFOCOM 2000)*, pp. 1233–1242.

Breslau L, Knightly E, Shenker S, Stoica I and Zhang H 2000b Endpoint Admission Control: Architectural Issues and Performance. *ACM Computer Communications Review* **30**(4), 57–69.

Briscoe B, Darlagiannis V, Heckmann O, Oliver H, Siris V, Songhurst D and Stiller B 2003 A Market Managed Multi-Service Internet (M3I). *Computer Communications* **26**(4), 404–414.

BRITE 2004 Boston University Representative Internet Topology Generator. http://www.cs.bu.edu/brite/.

Brittain P and Farrel A 2000 MPLS Traffic Engineering: A Choice of Signaling Protocols. White paper, Dataconnection. http://www.dataconnection.com.

Bu T and Towsley D 2002 On Distinguishing between Internet Power Law Topology Generators. *Proceedings of the IEEE Conference on Computer Communications and Networking (INFOCOM 2002)*, pp. 638–647.

Bu T, Gao L and Towsley D 2002 On Characterizing Routing Table Growth. *Proceedings of the IEEE Global Communications Conference (Globecom 2002)*.

Burchard A, Liebeherr J and Patek S 2002 A Calculus for End-to-end Statistical Service Guarantees. Technical Report CS-2001-19, University of Virginia, Department of Computer Science.

CAIDA – Cooperative Association for Internet Data Analysis 2003 The AS Internet Graph: A Macroscopic Visualisation of the Internet. http://www.caida.org/analysis/topology/as_core_network/.

CAIDA – Cooperative Association for Internet Data Analysis 2006. http://www.caida.org/.

Cain B, Deering S, Kouvelas I, Fenner B and Thyagarajan A 2002 Internet Group Management Protocol, Version 3, RFC 3376.

Callon R 1990 Use of OSI IS-IS for Routing in TCP/IP and Dual Environments, RFC 1195.

Cao J, Davis D, Wiel SV and Yu B 2000 Time-Varying Network Tomography. *Journal of the American Statistical Association* **95**, 1063–1075.

Cardwell N, Savage S and Anderson T 2000 Modeling TCP Latency. *Proceedings of the IEEE Conference on Computer Communications and Networking (INFOCOM 2000)*, pp. 1742–1751.

Carlberg K, Gevros P and Crowcroft J 2001 Lower than Best Effort: A Design and Implementation. *ACM SIGCOMM Computer Communication Review* **31**(2), 244–265.

Carlson J, Langner P, Hernandez-Valencia E and Manchester J 2000 PPP over Simple Data Link (SDL) using SONET/SDH with ATM-like Framing, RFC 2823.

Carpenter B, IETF Network Working Group 1996 Architectural Principles of the Internet, RFC 1958.

Carpenter B and Nichols K 2001 A Bulk Handling Per-Domain Behavior for Differentiated Services, draft-ietf-diffserv-pdb-bh-02.txt.

Cavendish D 2000 Evolution of Optical Transport Technologies: From SONET/SDH to WDM. *IEEE Communications Magazine* **38**(6), 164–172.

Cetinkaya C and Knightly E 2000 Egress Admission Control. *Proceedings of the IEEE Conference on Computer Communications and Networking (INFOCOM 2000)*, pp. 1471–1480.

Chang CS 1994 Stability, Queue Length and Delay of Deterministic and Stochastic Queueing Networks. *IEEE Transaction on Automatic Control* **39**(5), 913–931.

Chang CS 2000 *Performance Guarantees in Communication Networks*. Springer-Verlag. ISBN: 1852332263.

Chang SG and Gavish B 1993 Telecommunications Network Topological Design and Capacity Expansion: Formulations and Algorithms. *Telecommunication Systems* **1**, 99–131.

Chang SG and Gavish B 1995 Lower Bounding Procedures for Multiperiod Telecommunication Network Expansion Problems. *Operations Research* **43**(1), 43–57.

Charny A and Le Boudec JY 2000 Delay Bounds in a Network with Aggregate Scheduling. *Proceedings of the International Workshop on Quality of Future Internet Services (QoFIS 2000)*, pp. 1–13.

Charny A, Bennet JCR, Benson K, Le Boudec JY, Chiu A, Courtney W, Davari S, Firoiu V, Kalmanek C and Ramakrishnan K 2002 Supplemental Information for the New Definition of the EF PHB (Expedited Forwarding Per-Hop Behavior), RFC 3247.

Chen KT, Huang P, Huang CY and Lei CL 2005 Game Traffic Analysis: An MMORPG Perspective. *Proceedings of the International Workshop on Network and Operating Systems Support for Digital Audio and Video (NOSSDAV)*, pp. 19–24.

Chen Q, Chang H, Govindan R, Jamin S, Shenker S and Willinger W 2002 The Origin of Power Laws in Internet Topologies Revisited. *Proceedings of the IEEE Conference on Computer Communications and Networking (INFOCOM 2002)*.

Choe J and Shroff N 1998 A Central Limit Theorem Based Approach to Analyze Queue Behavior in ATM Networks. *IEEE/ACM Transactions on Networking* **6**(5), 659–671.

Choi HK and Limb JO 1999 A Behavioral Model of Web Traffic. *Proceedings of the Seventh Annual International Conference on Network Protocols (ICNP 1999)*, pp. 327–342.

Choi BK, Xuan D, Li C, Bettati R and Zhao W 2000 Scalable QoS Guaranteed Communication Services for Real-Time Applications. *Proceedings of the 20th IEEE International Conference on Distributed Computing Systems*, pp. 180–187.

Choudhury G, Lucantoni D and Whitt W 1996 Squeezing the Most Out of ATM. *IEEE Transactions on Communications* **44**(2), 203–217.

Christin N, Liebeherr J and Abdelzaher T 2002 A Quantitative Assured Forwarding Service. *Proceedings of the IEEE Conference on Computer Communications and Networking (Infocom 2002)*.

Cisco 2003 Enhanced Interior Gateway Routing Protocol Cisco White Paper, Document ID: 16406. http://www.cisco.com/warp/public/103/eigrp-toc.html.

CiscoWorks 2004 CiscoWorks2000 Service Management Solution. http://www.cisco.com/warp/public/cc/pd/wr2k/svmnso/index.shtml.

Ciucu F, Burchard A and Liebeherr J 2005 A Network Service Curve Approach for the Stochastic Analysis of Network. *Proceedings of ACM SIGMETRICS*, pp. 279–290.

Claffy K, Miller G and Thompson K 1998 The Nature of the Beast: Recent Traffic Measurements from an Internet Backbone. *Proceedings of the Internet Society Internet Global Summit (Inet 1998)*. http://www.caida.org/Papers/Inet98/index.html.

Clark D 1988 The Design Philosophy of the DARPA Internet Protocols. *ACM Computer Communication Review* **18**(4), 106–114. Proceedings of the SIGCOMM 1988 Symposium.

Clark D and Fang W 1998 Explicit Allocation of Best-Effort Packet Delivery Service. *IEEE/ACM Transactions on Networking* **6**(4), 362–373.

Claypool M, LaPoint D and Winslow J 2003 Network Analysis of Counter-strike and Starcraft. *Proceedings of the IEEE International Performance Computing and Communications Conference (IPCCC)* .

Constantiou I and Altmann J 2003 Towards a Profitable ISP Business in a Competitive Environment, In *Social and Economic Transformation in the Digital Era* GI Doukidis, N Mylonopoulos and N Pouloudi (Editors), Idea Group Publishing , pp. 182–200. ISBN: 1591402670.

Cooper RB 1981 *Introduction to Queueing Theory*, 2nd edn. Elsevier, North-Holland. ISBN: 0444003797.

Courcoubetis C and Weber R 1995 Effective Bandwidths for Stationary Sources. *Probability in Engineering and Informational Sciences* **9**(2), 285–294.

Courcoubetis C and Weber R 2003 *Pricing Communication Networks: Economics, Technology and Modelling*. John Wiley & Sons. ISBN: 0470851309.

Courcoubetis C, Dramitinos MP and Stamoulis GD 2001 An Auction Mechanism for Bandwidth Allocation Over Paths. *Proceedings of the 17th International Teletraffic Congress (ITC 2001)*, pp. 1163–1174.

Courcoubetis C, Siris V and Stamoulis G 1998 Application and Evaluation of Large Deviation Techniques for Traffic Engineering in Broadband Networks. *ACM SIGMETRICS 1998*, pp. 212–221.

Crovella ME and Bestavros A 1997 Self-similarity in World Wide Web Traffic: Evidence and Possible Causes. *IEEE/ACM Transactions on Networking* **5**(6), 835–846.

Crowcroft J and Oechslin P 1998 Differentiated End-to-End Internet Services using a Weighted Proportional Fair Sharing TCP. *ACM SIGCOMM Computer Communication Review* **28**(3), 53–69.

Crowcroft J, Handley M and Wakeman I 1999 *Internetworking Multimedia*. Morgan Kaufmann. ISBN: 1558605843.

Cruz R 1991 A Calculus for Network Delay. *IEEE Transactions on Information Theory* **37**(1), 114–141.

Cruz RL 1998 SCED+: Efficient Management of Quality of Service Guarantees. *Proceedings of the IEEE Conference on Computer Communications and Networking (INFOCOM 1998)*, pp. 625–634.

d' Halluin Y, Forsyth PA and Vetzal KR 2002 Managing Capacity for Telecommunications Networks under Uncertainty. *IEEE/ACM Transactions on Networking* **10**(4), 579–587.

Dantzig GB 1951 Maximization of a Linear Function of Variables Subject to Linear Inequalities, In *Activity Analysis of Production and Allocation* TC Koopmans (Editor), John Wiley & Sons , pp. 359–373.

Danzig PB and Jamin S 1991 tcplib: A Library of TCP/IP Traffic Characteristics. Technical Report TR CS-SYS-91-01, USC Networking and Distributed Systems Laboratory.

Davie B, Charny A, Bennet JCR, Benson K, Le Boudec JY, Courtney W, Davari S, Firoiu V and Stiliadis D 2002 An Expedited Forwarding PHB (Per-Hop Behaviour), RFC 3246.

Davin J and Heybey A 1994 A Self-clocked Fair Queueing Scheme for High Speed Applications. *Proceedings of the IEEE Conference on Computer Communications and Networking (INFOCOM 1994)*, pp. 636–646.

Deering S and Hinden R 1998 Internet Protocol, Version 6 (IPv6) Specification, RFC 2460.

Degermark M, Brodnik A and Pink S 1997 Small Forwarding Table for Fast Routing Lookups. *ACM SIGCOMM Computer Communication Review* **27**(4), 3–14.

Delgrossi L and Berger L 1995 Internet Stream Protocol Version 2 (ST2) Protocol Specification – Version ST2+, RFC 1819.

Delgrossi L, Herrtwich R, Vogt C and Wolf L 1993 Reservation Protocols for Internetworks: A Comparison of ST-II and RSVP. *Proceedings of Network and Operating System Support for Digital Audio and Video (NOSSDAV)*, pp. 195–203.

Demers A, Keshav S and Shenker S 1989 Analysis and Simulation of a Fair Queueing Algorithm. *Proceedings of the ACM Special Interest Group on Data Communication Conference (SIGCOMM 1989)*.

Dewan R, Freimer M and Gundepudi P 1999 Evolution of Internet Infrastructure in the 21st Century: The Role of Private Interconnection Agreements. *Proceedings of the International Conference on Information Systems (ICIS 1999)*, pp. 144–154.

Dewan R, Freimer M and Gundepudi P 2000 Interconnection Agreements between Competing Internet Service Providers. *Proceedings of the 33rd Hawaii International Conference of System Sciences (HICSS 2000)*, p. 6014.

Dharmapurikar S, Krishnamurthy P and Taylor DE 2003 Longest Prefix Matching using Bloom Filters. *Proceedings of the 2003 Conference on Applications, Technologies, Architectures, and Protocols for Computer Communications (SIGCOMM 2003)*, pp. 201–212.

Dijkstra EW 1959 A Note on Two Problems in Connection with Graphs. *Numerische Mathematik* **1**(1), 269–271.

Dominion Telecom 2004 Homepage. http://www.dominiontel.com/.

Dovrolis C and Ramanathan P 1999 A Case for Relative Differentiated Services and the Proportional Differentiation Model. *IEEE Network* **13**(5), 26–34.

Duffield NG and O'Connell N 1995 Large Deviations and Overflow Probabilities for the General Single Server Queue, with Application. *Cambridge Philosophical Society* **118**, 363–374.

Durham D, Boyle J, Cohen R, Herzog S, Rajan R and Sastry A 2000 The COPS (Common Open Policy Service) Protocol, RFC 2748.

Durresi A, Jain R, Chandhok N, Jagannathan R, Seetharaman S and Vinodkrishnan K 2001 IP over All-Optical Networks – Issues. *Proceedings of the IEEE Global Communications Conference (Globecom 2001)*, Vol. 4, pp. 2144–2149.

Dutta A and Lim JI 1992 A Multiperiod Capacity Planning Model for Backbone Computer Communication Networks. *Operations Research* **40**(4), 689–705.

Eatherton W, Dittia Z and Varghese G 2002 Tree Bitmap: Hardware/Software IP Lookups with Incremental Updates. Technical Report. http://www.eathertons.com/sigcomm-withnames.PDF.

ECIN 2003 Electronic Commerce Info Net. eMail-Services im Vergleichstest (Comparing Email-Services) Url:(http://www.ecin.de/news/2003/08/29/06157/).

Edell R and Varaiya P 1999 Providing Internet Access: What We Learn From INDEX. *IEEE Network* **13**(5), 18–25.

Elwalid A and Mitra D 1993 Effective Bandwidth of General Markovian Traffic Sources and Admission Control of High Speed Networks. *IEEE/ACM Transactions on Networking* **1**(3), 329–343.

Elwalid A, Heyman D, Lakshman T, Mitra D and Weiss A 1995a Fundamental Bounds and Approximations for ATM Multiplexers with Applications to Video Teleconferencing. *IEEE Journal on Selected Areas in Communications* **13**(6), 1004–1016.

Elwalid A, Mitra D and Wentworth R 1995b A New Approach for Allocating Buffers and Bandwidth to Heterogeneous, Regulated Traffic in an ATM Node. *IEEE Journal on Selected Areas in Communications* **13**(6), 1115–1127.

ESA 2005 Computer and Video Game Software Sales Reach Record 7.3 Billion Dollars in 2004. http://www.theesa.com/archives/2005/02/computer_and_vi.php.

Eschenauer H, Koski J and Osyczka A 1990 *Multicriteria Design Optimization: Procedures and Applications.* Springer-Verlag. ISBN: 0387506047.

EURO-IX 2006 European Internet Exchange Homepage. http://www.euro-ix.net/.

Faloutsos M, Faloutsos P and Faloutsos C 1999 On Power-law Relationships of the Internet Topology. *Proceedings of the ACM Special Interest Group on Data Communication Conference (SIGCOMM 1999), Computer Communications Review 25*, pp. 251–262.

Fang W and Peterson LL 1999 Inter-AS Traffic Patterns and Their Implications. *Proceedings of the 4th Global Internet Symposium (GIS 1999)*.

Fang W, Seddigh N and Nandy B 2000 A Time Sliding Window Three Colour Marker, RFC 2859.

Fankhauser G 2000 *A Network Architecture Based on Market Principles.* PhD thesis, Swiss Federal Institute of Technology Zürich (ETH Zürich).

Federal Networking Council (FNC) Resolution 1995 Definition of "Internet". http://www.itrd.gov/fnc/Internet_res.html.

Fehér G, Németh K and Czslényi I 2002 Performance Evaluation Framework for IP Resource Reservation Signalling. *Performance Evaluation* **48**(1-4), 131–156.

Fehér G, Németh K, Maliosz M, Czslényi I, Bergkvist J, Ahlard D and Engborg T 1999 Boomerang – A Simple Protocol for Resource Reservation in IP Networks. *IEEE Workshop on QoS Support for Real-Time Internet Applications (RTAS 99)*.

Feldmann A, Greenberg AG, Lund C, Reingold N, Rexford J and True F 2000 Deriving Traffic Demands for Operational IP Networks: Methodology and Experience. *Proceedings of the ACM Special Interest Group on Data Communication Conference (SIGCOMM 2000)*, pp. 257–270.

Feng W, Chang F, Feng W and Walpole J 2005 A Traffic Characterization of Popular On-line Games. *IEEE/ACM Transactions on Networking* **13**(March), 488–500.

Feng W, Kandlur D, Saha D and Shin K 1999 Adaptive Packet Marking for Maintaining End-to-End Throughput in a Differentiated-Services Internet. *IEEE/ACM Transactions on Networking* **7**(5), 685–697.

Feng W, Kandlur D, Saha D and Shin KG 2002 The BLUE Active Queue Management Algorithms. *IEEE/ACM Transactions on Networking* **10**(4), 513–528.

Fenner W 1997 Internet Group Management Protocol, Version 2, RFC 2236.

Ferrari D and Verma D 1990 A Scheme for Real-Time Channel Establishment in Wide-Area Networks. *IEEE Journal on Selected Areas in Communications* **8**(3), 368–379.

Fidler M 2003 On the Impacts of Traffic Shaping on End-to-End Delay Bounds in Aggregate Scheduling Networks. *Proceedings of the Workshop on Quality of Future Internet Services (QoFIS 2003)*, pp. 1–10.

Fidler M and Recker S 2005 Transformation-Based Network Calculus Applying Convex/Concave Conjugates. *Proceedings of KiVS 2005*, pp. 181–192.

Fidler M, Heckmann O and Steinmetz R 2005 Preserving the Independence of Flows in General Topologies Using Turn-Prohibition. *Proceedings of the International Workshop on Quality of Service (IWQoS 2005)*, pp. 193–205.

Floyd S 1991 Connections with Multiple Congested Gateways in Packet-Switched Networks, Part 1: One-way Traffic. *Computer Communications Review* **21**(5), 30–47.

Floyd S 2003 HighSpeed TCP for Large Congestion Windows, RFC 3649.

Floyd S and Jacobson V 1993 Random Early Detection Gateways for Congestion Avoidance. *IEEE/ACM Transactions on Networking* **1**(4), 397–413.

Floyd S and Jacobson V 1995 Link-sharing and Resource Management Models for Packet Networks. *IEEE/ACM Transactions on Networking* **3**(4), 365–386.

Floyd S, Gummadi R and Shenker S 2001 Adaptive RED: an Algorithm for Increasing the Robustness of RED's Active Queue Management. Technical Report, The ICSI Center for Internet Research. http://www.icir.org/floyd/papers/adaptiveRed.pdf.

Ford LR and Fulkerson DR 1956 Maximal Flow Through a Network. *Canadian Journal of Mathematics* **8**, 399–404.

Fortz B and Thorup M 2002 Optimizing OSPF/IS-IS Weights in a Changing World. *IEEE Journal on Selected Areas in Communications* **20**(4), 756–767.

Fraleigh C, Moon S, Lyles B, Cotton C, Khan M, Moll D, Rockell R and Seely T 2003 Packet-Level Traffic Measurements from the Sprint IP Backbone. *IEEE Networks* **17**(6), 6–16.

Frankel S 2001 *Demystifying the Ipsec Puzzle*. Artech House Publishers. ISBN: 1580530796.

Fredkin E 1960 Trie Memory. *Communications of the ACM* **3**, 490–499.

Freeman RL 2004 *Telecommunication System Engineering*. John Wiley & Sons. ISBN: 0471451339.

Friedmann J and Mills-Scofield D 1997 Internet Settlements Pricing Model and Implications. *Proceedings of the Conference on the Impact of the Internet on Communications Policy (IIComPol 1997)*.

Fritsch T, Ritter H and Schiller J 2005 The Effect of Latency and Network Limitations on MMORPGs (A Field Study of Everquest2). *Proceedings of Workshop on Network & System Support for Games (NetGames 2005)*.

Fuller V, Li T, Yu J and Varadhan K 1993 Classless Inter-Domain Routing (CIDR): An Address Assignment and Aggregation Strategy, RFC 1519.

Garetto M, Cigno RL, Meo M and Marsan MA 2001 A Detailed and Accurate Closed Queueing Network Model of Many Interacting TCP Flows. *Proceedings of the IEEE Conference on Computer Communications and Networking (INFOCOM 2001)*, pp. 1706–1715.

Gavish B 1992 Topological Design of Communication Networks – The Overall Design Problem. *European Journal of Operational Research* **58**, 149–172.

Gelenbe E 1993 G-Networks with Instantaneous Customer Movement. *Journal of Applied Probability* **30**(3), 742–748.

Georgia Tech Internetwork Topology Models 2004 GT-ITM Topology Generator. http://www.cc.gatech.edu/projects/gtitm/.

German Internet Exchange DE-CIX 2004 Information about: Connections, Pricing, Services Included and Prerequisites. http://www.de-cix.net/info/DE-CIX_at_a_glance.html#services.

Gibbens R and Kelly F 1997 Measurement-based Connection Admission Control. *Proceedings of the 15th International Teletraffic Congress*, pp. 879–888.

Gibbens R and Kelly F 1999 Resource Pricing and the Evolution of Congestion Control. *Automatica* **35**, 1969–1985.

Gibbens R and Teh Y 1999 Critical Time and Space for Statistical Multiplexing in Multiservice Networks. *Proceeding of the 16th International Teletraffic Congress (ITC 16)*, pp. 87–96.

Gibbens R, Kelly F and Key P 1995 A Decision-Theoretic Approach to Call Admission Control in ATM Networks. *IEEE Journal on Selected Areas in Communications* **13**(6), 1101–1114.

Gibbens R, Sargood S, Kelly F, Azmoodeh H, Macfadyen R and Macfadyen N 1999 An Approach to Service Level Agreements for IP Networks with Differentiated Services. *Royal Society Discussion Meeting on Network Modelling in the 21st Century*.

Gibbens R, Sargood S, Eijl CV, Kelly F, Azmoodeh H, Macfadyen R and Macfadyen N 2000 Fixed-Point Models for the End-to-End Performance Analysis of IP Networks. *Proceedings of the 13th ITC Specialist Seminar on IP Traffic Management, Modeling and Management*.

Giovannetti E 2002 Interconnection, Differentiation and Bottlenecks in the Internet. *Information Economics and Policy* **14**(3), 385–404.

Giovannetti E, Neuhoff K and Spagnolo G 2003 Agglomeration in the Internet: Does Space Still Matter? The MIX-IXP Case, *Online Proceedings of the Cambridge Economics for the Future Conference 2003*. Oxford University Press. http://www.econ.cam.ac.uk/cjeconf/delegates/giovannetti.pdf.

Glover F and Laguna M 1998 *Tabu Search*. Kluwer Academic Publishers. 0792381874.

Goldschmidt O 2000 ISP Backbone Traffic Inference Methods to Support Traffic Engineering. *Proceedings of the Internet Statistics and Metrics Analysis (ISMA) Workshop*.

Gondran M and Minoux M 1984 *Graphs and Algorithms*. John Wiley & Sons. ISBN: 0471103748.

Goralski WJ 2002 *Sonet/SDH*, 3rd edn. Osborne/McGraw-Hill. ISBN: 0072225246.

Goyal P, Vin HM and Cheng H 1997 Start-time Fair Queuing: A Scheduling Algorithm for Integrated Services Packet Switching Networks. *IEEE/ACM Transactions on Networking* **5**(5), 690–704.

Greenstein S 1999 Understanding the Evolution Structure of Commercial Internet Markets, *Understanding the Digital Economy: Data, Tools and Research*. MIT Press, Washington DC.

Gupta P 2000 *Algorithms for Routing Lookups and Packet Classification*. PhD thesis, Stanford University.

Guérin R, Ahmadi H and Naghshineh M 1991 Equivalent Capacity and its Application to Bandwidth Allocation in High-Speed Networks. *IEEE Journal on Selected Areas in Communications* **9**(7), 968–981.

Guérin R, Orda A and Williams D 1997 QoS Routing Mechanisms and OSPF Extensions. *Proceedings of the 2nd Global Internet Miniconference (joint with Globecom 1997)*.

Gwehenberger G 1968 Anwendung einer binären Verweiskettenmethode beim Aufbau von Listen (Use of a Binary Tree Structure for Processing Files). *Elektronische Rechenanlagen* **10**, 223–226.

Haberman B 2003 Source Address Selection for the Multicast Listener Discovery (MLD) Protocol, RFC 3590.

Han J, McMahon G and Sugden S 2000 A Model for Nonlinear Cost Problems in Optimal Design of Communication Networks. *IEEE International Conference on Computer Communications and Networks (ICCCN 2000)*.

Handley M, Floyd S, Padhye J and Widmer J 2003 TCP Friendly Rate Control (TFRC): Protocol Specification, RFC 3448, Proposed Standard.

Hansen B and Naevdal E 2000 Competition in the Internet and Dynamic Pricing by ECN Marks. Internet Charging Workshop (in Association with QoFIS 2000 – Quality of Future Internet Services Workshop). http://www.m3i.org/workshop/workshop/pdf/HansenBerlin.pdf.

Hashmani M, Yoshida M, Ikenaga T and Oie Y 2001 Management and Realization of SLA for Providing Network QoS. *Proceedings of the IEEE International Conference on Networking (ICN 2001)*, pp. 398–408.

Hasslinger G 2005 ISP Platforms Under a Heavy Peer-to-Peer Workload, *Peer-to-Peer Systems and Applications*, Lecture Notes in Computer Science 3485. Springer-Verlag, pp. 369–381. ISBN: 354029192X.

Hasslinger G and Schnitter S 2002a How Service Providers Can Profit From Traffic Engineering. *Eurescom Summit 2002 "Powerful Networks for Profitable Services"*.

Hasslinger G and Schnitter S 2002b Optimized Traffic Load Distribution in MPLS Networks. *Proceedings of the Sixth Telecommunications Conference of the Institute for Operations Research and the Management Sciences (INFORMS 2002)*.

Hasslinger G and Schnitter S 2004 IP Network Expansion for Growing Traffic Demand with Shortest Path Routing Compared to Traffic Engineering. *Proceedings of Networks 2004*, pp. 81–86.

Heckmann O, Schmitt J and Steinmetz R 2001 Multi-Period Resource Allocation at System Edges – Capacity Management in a Multi-Provider Multi-Service Internet. *Proceedings of the IEEE Conference on Local Computer Networks (LCN 2001)*.

Heckmann O, Schmitt J and Steinmetz R 2002 Multi-Period Resource Allocation at System Edges. *Proceedings of the 10th International Conference on Telecommunication Systems Modelling and Analysis (ICT SM10)*, pp. 1–25, Monterey, USA.

Heckmann O, Bock A, Mauthe A and Steinmetz R 2004 The eDonkey File-Sharing Network. *Proceedings of the Workshop on Algorithms and Protocols for Efficient Peer-to-Peer Applications, GI Jahrestagung*, pp. 224–228.

Heckmann O, Piringer M, Schmitt J and Steinmetz R 2003 On Realistic Network Topologies for Simulation. *Proceedings of the ACM SIGCOMM Workshop on Models, Methods and Tools for Reproducible Network Research (SIGCOMM MoMeTools)*, pp. 28–32.

Hedrick C 1988 Routing Information Protocol, RFC 1058.

Heegaard P 2000 GenSyn – a Java Based Generator of Synthetic Internet Traffic Linking User Behaviour Models to Real Network Protocols. *Proceedings of the ITC Specialist Seminar on IP Traffic Measurement, Modeling and Management*.

Heinanen J and Guérin R 1999a A Single Rate Three Color Marker, RFC 2697.

Heinanen J and Guérin R 1999b A Two Rate Three Color Marker, RFC 2698.

Heinanen J, Baker F, Weiss W and Wroclawski J 1999 Assured Forwarding PHB Group, RFC 2597.

Henderson T 2001 Latency and User Behaviour on a Multiplayer Game Server. *Proceedings of the 3rd International Workshop on Networked Group Communications (NGC)*.

Henderson T and Bhatti S 2003 Networked Games – a QoS-sensitive Application for QoS-insensitive Users? *Proceeding of the ACM SIGCOMM 2003 RIPQoS Workshop*.

Henderson T, Crowcroft J and Bhatti S 2001 Congestion Pricing: Paying Your Way in Communication Networks. *IEEE Internet Computing* **5**(5), 85–89.

Herzog S 2000 COPS Usage for RSVP, RFC 2749.

Heying Z, Baohong L and Wenhua D 2003 Design of a Robust Active Queue Management Algorithm Based on Feedback Compensation. *Proceedings of the ACM Special Interest Group on Data Communication Conference (SIGCOMM 2003)*, pp. 277–285.

Hillier FS and Lieberman GJ 2001 *Introduction to Operations Research*. McGraw-Hill. ISBN: 0072321695.

Hinden R and Deering S 1998 IP Version 6 Addressing Architecture, RFC 2373.

Hollot CV, Misra V, Gong WB and Towsley D 2001 On Designing Improved Controllers for AQM Routers Supporting TCP Flows. *Proceedings of the IEEE Conference on Computer Communications and Networking (INFOCOM 2001)*, pp. 1726–1734.

Hubermann BA, Pirolli PL, Pitkow JE and Lukose RM 1998 Strong Regularities in World Wide Web Surfing. *Science* **280**(5360), 95–97.

Hurley P 2001 *The Provision of a Low Delay Service within the Best-Effort Internet*. PhD thesis, EPFL Lausanne, Switzerland.

Hurley P, Kara M, Le Boudec JY and Thiran P 2001 ABE: Providing a Low-Delay Service within Best Effort. *IEEE Network Magazine* **15**(3), 60–69.

Hurley P, Iannaccone G, Kara M, Le Boudec JY, Thiran P and Diot C 2000 The ABE Service, draft-hurley-alternative-best-effort-01.txt.

Huston G 1998 *ISP Survival Guide: Strategies for Running a Competitive ISP*, 1st edn. John Wiley & Sons. ISBN: 0471314994.

Huston G 1999a Interconnection Peering and Settlement – Part I. *Internet Protocol Journal (Cisco Publications)* **2**(1), 2–16.

Huston G 1999b Interconnection Peering and Settlement – Part II. *Internet Protocol Journal (Cisco Publications)* **2**(2), 2–23.

Hwang J and Weiss MBH 2000 Market-Based QoS Interconnection Economy in the Next-Generation Internet. *Journal of Information Economics and Policy*.

IANA (Internet Assigned Numbers Authority) 2004 Autonomous System Numbers. http://www.iana.org/assignments/as-numbers.

IEEE 802.3 High Speed Study Group 2002 802.3ae: 5 Criteria. http://grouper.ieee.org/groups/802/3/ae/criteria.pdf.

ILOG CPLEX 2004 Mathematical Programming Optimizer. http://www.ilog.com/products/cplex/.

INET 2004 Inet Topology Generator V3.0. University of Michigan, http://topology.eecs.umich.edu/inet/.

Institute of Electrical and Electronic Engineers 1996 *1278.2-1995, IEEE Standard for Distributed Interactive Simulation – Communication Services and Profiles*. IEEE, New York, USA.

International Telecommunication Union 2000 *ITU-T Recommendation G.114: International Telephone Connections and Circuits – General Recommendations on the Transmission Quality for an Entire International Telephone Connection – One-Way Transmission Time*. International Telecommunication Union, Geneva, Switzerland.

ISO DP 10589 1990 Intermediate System to Intermediate System Intra-Domain Routing Exchange Protocol for Use in Conjunction with the Protocol for Providing the Connectionless-mode Network Service (ISO 8473).

ITU Recommendation G.707 1996 Network Node Interface For The Synchronous Digital Hierarchy.

Jacobson V 1988 Congestion Avoidance and Control. *Proceedings of ACM SIGCOMM 1998*, pp. 314–329.

Jacobson V, Nichols K and Poduri K 1999 An Expedited Forwarding PHB, RFC 2598.

Jacobson V, Nichols K and Poduri K 2000 The Virtual Wire Per-Domain Behavior, draft-ietf-diffserv-pdb-vw-00.txt.

Jain R, Mullen T and Hausman R 2001 Analysis of Paris Metro Pricing Strategy for QoS with a Single Service Provider. *Proceedings of the International Workshop on Quality of Service (IWQoS 2001)*, pp. 44–58.

Jamin S, Shenker S and Danzig P 1997b Comparison of Measurement-based Admission Control Algorithms for Controlled-Load Service. *Proceedings of the Sixteenth Annual Joint Conference of the IEEE Computer and Communications Societies (INFOCOM 1997)*, pp. 973–981.

Jamin S, Danzig PB, Shenker S and Zhang L 1997a A Measurement-Based Admission Control Algorithm for Integrated Services Packet Networks. *IEEE/ACM Transactions on Networking* **5**(1), 56–70.

Jamoussi B, Andersson L, Callon R, Dantu R, Wu L, Doolan P, Worster T, Feldman N, A.Fredette, Girish M, Gray E, Heinanen J, Kilty T and Malis A 2002 Constraint-Based LSP Setup Using LDP, RFC 3212.

Jiang Y and Emstad PJ 2005a Analysis of Stochastic Service Guarantees in Communication Networks: A Server Model. *Thirteenth International Workshop on Quality of Service (IWQoS 2005)*, pp. 233–245.

Jiang Y and Emstad PJ 2005b Analysis of Stochastic Service Guarantees in Communication Networks: A Traffic Model. *International Teletraffic Congress (ITC19)*.

Jin C, Wei DX, Low SH, Buhrmaster G, Bunn J, Choe DH, Cottrell RLA, Doyle JC, Feng W, Martin O, Newman H, Paganini F, Ravot S and Singh S 2005 FAST TCP: From Theory to Experiments. *IEEE Network* **19**(1), 4–11.

Johari R and Tan D 2001 End-to-End Congestion Control for the Internet: Delays and Stability. *IEEE/ACM Transactions on Networking* **9**(2), 818–832.

Karagiannis T and Faloutsos M 2002 SELFIS: A Tool for Self-Similarity and Long-Range Dependence Analysis. *Proceedings of the 1st Workshop on Fractals and Self-Similarity in Data Mining*.

Karagiannis T, Papagiannaki K and Faloutsos M 2005 BLINC: Multilevel Traffic Classification in the Dark. *Proceedings of the Conference on Applications, Technologies, Architectures, and Protocols for Computer Communications (SIGCOMM 2005)*, pp. 229–240.

Karagiannis T, Broido A, Faloutsos M and Claffy K 2004 Transport Layer Identification of P2P Traffic. *Proceedings of the Internet Measurement Conference 2004*.

Karlsson G 1998 Providing Quality for Internet Video Services. *CNIT/ITWoDC 98*, pp. 133–146.

Karmarkar N 1984 A New Polynomial-Time Algorithm for Linear Programming. *Combinatorica* **4**, 373–395.

Karn P and Partridge C 1991 Improving Round-Trip Time Estimates in Reliable Transport Protocols. *ACM Transactions on Computer Systems* **9**(4), 364–373.

Karsten M 2000 *QoS Signalling and Charging in a Multi-service Internet using RSVP*. PhD thesis, Technische Universität Darmstadt.

Karsten M 2004 KOM RSVP Engine. http://www.kom.tu-darmstadt.de/rsvp/.

Karsten M and Schmitt J 2002 Admission Control Based on Packet Marking and Feedback Signalling. Mechanisms, Implementation and Experiments. Technical Report TR-KOM-2002-03, Technische Universität Darmstadt. http://www.kom.e-technik.tu-darmstadt.de/publications/abstracts/KS02-5.html.

Karsten M, Schmitt J and Steinmetz R 2001 Implementation and Evaluation of the KOM RSVP Engine. *Proceedings of the IEEE Conference on Computer Communications and Networking (INFOCOM 2001)*, pp. 1290–1299. ISBN: 0780370163.

Karsten M, Berier N, Wolf LC and Steinmetz R 1999 A Policy-Based Service Specification for Resource Reservation in Advance. *Proceedings of the International Conference on Computer Communications (ICCC 1999)*, pp. 82–88.

Kaufman C 2004 The Internet Key Exchange (IKEv2) Protocol, draft-ietf- ipsec-ikev2-14.txt.

Kaur J and Vin HM 2001 Core-stateless Guaranteed Rate Scheduling Algorithms. *Proceedings of the IEEE Conference on Computer Communications and Networking (INFOCOM 2001)*, pp. 1484–1492.

Kaur J and Vin HM 2003 Core-stateless Guaranteed Throughput Networks. *Proceedings of the IEEE Conference on Computer Communications and Networking (INFOCOM 2003)*, pp. 2155–2165.

Kelly F 1996 Notes on Effective Bandwidths. *Stochastic Networks: Theory and Applications*, Oxford University Press, pp. 141–168.

Kelly F 2000 Models for a Self-Managed Internet. *Philosophical Transactions of the Royal Society* **A358**, pp. 2335–2348.

Kelly F 2001a Mathematical Modelling of the Internet. *Mathematics Unlimited – 2001 and Beyond*, Springer-Verlag, pp. 685–702.

Kelly T 2001b An ECN Probe-Based Connection Acceptance Control. *ACM Computer Communication Review* **31**, 14–25.

Kelly F, Key P and Zachary S 2000 Distributed Admission Control. *IEEE Journal on Selected Areas in Communications* **18**(2), 2617–2628.

Kelly F, Maulloo A and Tan D 1998 Rate Control in Communication Networks: Shadow Prices, Proportional Fairness and Stability. *Journal of the Operational Research Society* **49**(3), 237–252.

Kencl L 1998 IP Routing Lookups Algorithms Evaluation Project Report. http://icapeople.epfl. ch/lkencl/lookups.ps.

Kende M 2000 The Digital Handshake: Connecting Internet Backbones. Federal Communications Commission Office of Plans and Policy, Working Paper No. 32, http://www.fcc.gov/Bureaus/OPP/working_papers/oppwp32.pdf.

Kent S 1998a IP Authentication Header, RFC 2402.

Kent S 1998b IP Encapsulating Security Payload (ESP), RFC 2406.

Kent S and Seo K 1998 Security Architecture for the Internet Protocol, RFC 2401.

Kershenbaum A 1993 *Telecommunications Network Design Algorithms*. McGraw-Hill. ASIN 0070342288.

Kesidis G, Walrand J and Chang C 1993 Effective Bandwidths for Multiclass Markov Fluids and Other ATM Sources. *IEEE/ACM Transactions on Networking* **1**(4), 424–428.

Key P 1999 Service Differentiation: Congestion Pricing, Brokers and Bandwidth Futures. *Proceedings of the International Workshop on Network and Operating System Support for Digital Audio and Video (NOSSDAV 99)*.

Khachiyan LG 1979 A Polynomial Algorithm in Linear Programming [in Russian]. *Doklady Akademii Nauk SSSR* **244**, 1093–1096. English translation: Soviet Mathematics Doklady 20, pages 191–194, 1979.

Khalil I 2003 *Dynamic Service Provisioning in IP Networks*. PhD thesis, University Bern.

Khalil I and Braun T 2000 Implementation of a Bandwidth Broker for Dynamic End-to-End Resource Reservation in Outsourced Virtual Private Networks. *Proceedings of the IEEE Conference on Local Computer Networks (LCN)*, pp. 511–519.

Kim H and Shroff N 2001 Loss Probability Calculations in a Finite Buffer Multiplexer. *IEEE/ACM Transactions on Networking* **9**(6), 765–768.

Kim J, Choi J, Chang D, Kwon T, Choi Y and Yuk E 2005 Traffic Characteristics of a Massively Multiplayer Online Role Playing Game and Its Implications. *Proceedings of Workshop on Network & System Support for Games (NetGames)*.

Kleinrock L 1975 *Queueing Systems – Theory*, Vol. 1. Wiley-Interscience, New York. ISBN: 0471491101.

Kleinrock L 1976 *Queueing Systems – Computer Applications*, Vol. 2. Wiley-Interscience, New York. ISBN: 047149111X.

Knightly E 1997 Second Moment Resource Allocation in Multi-Service Networks. *ACM SIGMETRICS 1997*, pp. 181–191.

Knightly E and Shroff N 1999 Admission Control for Statistical QoS: Theory and Practice. *IEEE Network Magazine* **13**(2), 20–29.

Knightly EW, Wrege DE, Liebeherr J and Zhang H 1995 Fundamental Limits and Tradeoffs of Providing Deterministic Guarantees to VBR Video Traffic. *Proceedings of the ACM Conference on Measurement and Modeling of Computer Systems (SIGMETRICS 1995)*, pp. 98–107.

Kodialam M and Lakshman TV 2000 Minimum Interference Routing with Applications to MPLS Traffic Engineering. *Proceedings of the IEEE Conference on Computer Communications and Networking (INFOCOM 2000)*.

Kohler E, Handley M and Floyd S 2005 Datagram Congestion Control Protocol (DCCP), draft-ietf-dccp-spec-11.txt Internet Draft.

Korte B and Vygen J 2002 *Combinatorial Optimization: Theory and Algorithms*, 2nd edn. Springer. ISBN: 3540431543.

Kouvatsos DD and Denazis SG 1994 A Universal Building Block for the Approximate Analysis of a Shared Buffer ATM Switch Architecture. *Annals of Operations Research, Special Issue on Methodologies for Performance Analysis of High Speed Networks* **49**, 241–278.

Kramer G 2004 UC Davis Generator of Self-Similar Traffic. http://wwwcsif.cs.ucdavis.edu/%7Ekramer/code/trf_gen3.html.

Kunniyur S and Srikant R 2004 An Adaptive Virtual Queue (AVQ) Algorithm for Active Queue Management. *IEEE Transactions on Networking* **12**(2), 286–299.

Lakelin P and Wood R 2000 The Western European Internet Market: Key Developments and Prospects (Analysys 2000 Market Study). http://www.isp-planet.com/research/isps_western_europe2a.html.

Lakelin P, Martin D and Sherwood K 1999 Internet Service Providers in Western Europe: The Dynamics of an Evolving Market. http://www.isp-planet.com/business/lakelin-exec.html.

Lakshman TV and Madhow U 1997 The Performance of TCP/IP for Networks with High Bandwidth-Delay Products and Random Loss. *IEEE/ACM Transactions on Networking* **5**(3), 336–350.

Le Boudec JY and Thiran P 2001 *Network Calculus: A Theory of Deterministic Queuing Systems for the Internet*, Lecture Notes in Computer Science 2050. Springer-Verlag. ISBN: 354042184X.

Le L, Aikat J, Jeffay K and Smith FD 2003 The Effects of Active Queue Management on Web Performance. *Proceedings of the ACM Special Interest Group on Data Communication Conference (SIGCOMM 2003)*, pp. 265–276.

Lee T, Lai K and Duann S 1996 Design of a Real-Time Call Admission Controller for ATM Networks. *IEEE/ACM Transactions on Networking* **4**(5), 758–765.

Leibowitz N, Ripeanu M and Wierzbicki A 2003 Deconstructing the Kazaa Network. *Proceedings of the Third IEEE Workshop on Internet Applications (WIAPP 2003)*.

Leighton T, Makedon F, Plotkin S, Stein C, Tardos E and Tragoudas S 1995 Fast Approximation Algorithms for Multicommodity Flow Problems. *Journal of Computer and System Sciences* **50**, 228–243.

Leiner B, Cole R, Postel J and Mills D 1985 The DARPA Internet Protocol Suite. *IEEE Communications Magazine* **23**(3), 29–34.

Leland WE, Taqqu MS, Willinger W and Wilson DV 1994 On the Self-similar Nature of Ethernet Traffic. *IEEE/ACM Transactions on Networking* **2**(1), 1–15.

Li L, Alderson D, Willinger W and Doyle J 2004 A First-Principles Approach to Understanding the Internet's Router-Level Topology. *Proceedings of the ACM Special Interest Group on Data Communication Conference (SIGCOMM 2004)*.

Liao R and Campbell A 2001 Dynamic Core Provisioning for Quantitative Differentiated Service. *Proceedings of the International Workshop on Quality of Service (IWQoS 2001)*, pp. 9–26.

Liebau N, Heckmann O, Hubbertz I and Steinmetz R 2005 A Peer-to-Peer Webcam Network. *Proceedings of Peer-to-Peer Systems and Applications Workshop associated with KiVS'05*, pp. 151–154.

Lin D and Morris R 1997 Dynamics of Random Early Detection. *Proceedings of the ACM Special Interest Group on Data Communication Conference (SIGCOMM 1997)*, pp. 127–137.

Lin F and Wang J 1993 Minimax Open Shortest Path First Routing Algorithms in Networks Supporting the SMDS Service. *Proceedings of the International Conference on Communications*, pp. 666–670.

LINX 2003 The London Internet Exchange – Memorandum of Understanding – Joining Requirements. http://www.linx.net/joining/mou.html.

Little JD 1961 A Proof of the Queueing Formula $L = \lambda W$. *Operations Research* **9**(3), 383–387.

Liu C and Layland J 1973 Scheduling Algorithms for Multiprogramming in a Hard-Real-Time Environment. *Journal of ACM* **20**(1), 46–61.

Liu Z, Gu Y and Medhi D 1998 On Optimal Location of Switches/Routers and Interconnection. Technical Report 20, Computer Science Telecommunications, University of Missouri-Kansas City. http://www.sice.umkc.edu/%7Edmedhi/abs/lgm_tr_20_98_abs.html.

Liu Z, Niclausse N and Jalpa-Villanueva C 2001 Traffic Model and Performance Evaluation of Web Servers. *Performance Evaluation* **46**(2-3), 77–100.

Loshin P 2004 *IPv6: Theory, Protocol, and Practice*, 2nd edn. Morgan Kaufmann. ISBN: 1558608109.

Lucent 2004 Lucent Technologies, VitalSuite Network Performance Management Software. http://www.lucent.com/.

Machiraju S, Seshadri M and Stoica I 2002 A Scalable and Robust Solution for Bandwidth Allocation. *Proceedings of the 10th International Workshop on Quality of Service*, pp. 148–157.

MacKenzie IS and Ware C 1993 Lag as a Determinant of Human Performance in Interactive Systems. *Proceedings of the Conference on Human Factors in Computing Systems (CHI) 1993*, pp. 488–493.

MacKie-Mason JK and Varian H 1995 Pricing Congestible Network Resources. *IEEE Journal of Selected Areas in Communications* **13**(7), 1141–1149.

Mah B 1997 An Empirical Model of HTTP Network Traffic. *Proceedings of the IEEE Conference on Computer Communications and Networking (INFOCOM 1997)*, pp. 592–600, Kobe, Japan.

Mahajan R, Floyd S and Wetherall D 2001 Controlling High-Bandwidth Flows at the Congested Router. *Proceedings of the 9th International Conference on Network Protocols (ICNP 2001)*, pp. 192–202.

Malis A and Simpson W 1999 PPP over SONET/SDH, RFC 2615.

Malkin G 1996 Internet Users' Glossary, RFC 1983.

Mallory T and Kullberg A 1990 Incremental Updating of the Internet Checksum, RFC 1141.

Mannersalo P and Norros I 2002 A Most Probable Path Approach to Queueing Systems with General Gaussian Input. *Computer Networks* **40**(3), 399–412.

Maplesoft 2006 MAPLE: Advanced Mathematical Problem Solving and Programming Environment. http://www.maplesoft.com/.

Massoulie L 2000 Stability of Distributed Congestion Control with Heterogeneous Feedback Delays. Technical Report Microsoft Research 2000-11, Microsoft Research.

Mathis M, Semke J, Mahdavi J and Ott T 1997 The Macroscopic Behavior of the TCP Congestion Avoidance Algorithm. *Computer Communication Review*, Vol. 27. Issue 3.

Mathworks 2004 MATLAB: Software Package for Mathematical and Technical Computing and Visualization. http://www.mathworks.com/products/matlab/.

May M, Bolot JC, Jean-Marie A and Diot C 1999 Simple Performance Models of Differentiated Services Schemes for the Internet. *Proceedings of the IEEE Conference on Computer Communications and Networking (INFOCOM 1999)* pp. 1385–1394.

McAuley AJ and Francis P 1993 Fast Routing Table Lookup Using CAMs. *Proceedings of the IEEE Conference on Computer Communications (INFOCOM 1993)*, pp. 1382–1391.

McBride RD 1998 Progress Made in Solving the Multicommodity Flow Problem. *SIAM Journal on Optimization* **8**(4), 947–955.

McGarty TP 2002 Peering, Transit, Interconnection: Internet Access in Central Europe. The Merton Group, White Paper. http://mertongroup.com/PeeringCentralEurope.pdf.

MCI 2004 MCI Internet Service Level Agreement. http://global.mci.com/terms/sla/t_sla_terms.xml.

Medina A, Matta I and Byers J 2000 On the Origin of Power Laws in Internet Topologies. *ACM SIGCOMM Computer Communication Review* **30**(2), 18–28.

Medina A, Taft N, Salamatian K, Bhattacharyya S and Diot C 2002 Traffic Matrix Estimation: Existing Techniques and New Directions. *Proceedings of the ACM Special Interest Group on Data Communication Conference (SIGCOMM 2002)*, pp. 161–174.

Menth M 2004 *Admission Control and Resilience for Next Generation Networks*. PhD thesis, Bayerische Julius-Maximilians-Universität Würzburg.

Menth M, Kopf S and Charzinski J 2003 Impact of Network Topology on the Performance of Budget Based Network Admission Control Methods. *Proceeding of MIPS 2003*, pp. 195–206.

Merriam-Webster 2006 Merriam-Webster Dictionary. http://www.m-w.com.

Mitra D and Ramakrishnan KG 2001 Techniques for Traffic Engineering of Multiservice, Multipriority Networks. *Bell Labs Technical Journal* **6**(1), 139–151.

Mitzel D, Estrin D, Shenker S and Zhang L 1994 An Architectural Comparison of ST-II and RSVP. *Proceedings of the IEEE Conference on Computer Communications and Networking (INFOCOM 1994)*, pp. 716–725.

Moore A and Zuev D 2005 Internet Traffic Classification Using Bayesian Analysis Techniques. *ACM SIGMETRICS Performance Evaluation Review* **33**(1), 50–60.

Morrison DR 1968 PATRICIA – Practical Algorithm to Retrieve Information Coded in Alphanumeric. *Journal of the ACM (JACM)* **15**(4), 514–534.

Moy J 1998 OSPF Version 2, RFC 2328.

Más I and Karlsson G 2001 PBAC: Probe-Based Admission Control. *Proceedings of the Workshop on Quality of Future Internet Services (QoFIS 2001)*, pp. 99–109.

Más I, Fodor V and Karlsson G 2002 Probe-based Admission Control for Multicast. *Proceedings of the International Workshop on Quality of Service (IWQoS 2002)*, pp. 99–105.

Más I, Fodor V and Karlsson G 2003 The Performance of Endpoint Admission Control Based on Packet Loss, *Proceedings of the Workshop on Quality of Future Internet Services (QoFIS 2003)*, Lecture Notes in Computer Science 2811. Springer, pp. 93–101.

Nandy B, Seddigh N and Pieda P 1999 Diffserv's Assured Forwarding PHB: What Assurance does the Customer Have?. *Proceedings of the 9th International Workshop on Network and Operating Systems support for Digital Audio and Video (NOSSDAV 1999)*.

Narayan H, Govindan R and Varghese G 2003 The Impact of Address Allocation and Routing on the Structure and Implementation of Routing Tables. *Proceedings of the 2003 Conference on Applications, Technologies, Architectures, and Protocols for Computer Communications (SIGCOMM 2003)*, pp. 125–136.

NetFlow 2004 Cisco NetFlow Software, Homepage. http://www.cisco.com/warp/public/732/Tech/nmp/netflow/.

Nichols J and Claypool M 2003 The Effects of Latency on Online Madden NFL Football. *Proceedings of ACM International Workshop on Network and Operating Systems Support for Digital Audio and Video (NOSSDAV)*.

Nichols K, Jacobson V and Zhang L 1999 A Two-Bit Differentiated Services Architecture for the Internet, RFC 2638.

Nichols K, Blake S, Baker F and Black D 1998 Definition of the Differentiated Services Field (DS Field) in the IPv4 and IPv6 Headers, RFC 2474.

Nilsson S and Karlsson G 1999 IP-Address Lookup Using LC-Tries. *IEEE Journal on Selected Areas in Communications* **17**(6), 1083–1092.

Norton WB 2002 A Business Case for ISP Peering. Technical Report, Equinix White Paper, Equinix. http://www.equinix.com.

Norton WB 2004 Equinix, The Peering Simulation Game. http://www.equinix.com.

NS2 2004 Network Simulator 2, Homepage. http://www.isi.edu/nsnam/ns/.

Odlyzko AM 1999 Paris Metro Pricing for the Internet. *ACM Conference on Electronic Commerce*, pp. 140–147.

Odlyzko AM 2003 Internet Traffic Growth: Sources and Implications. *Proceedings of the SPIE Conference on Optical Transmission Systems and Equipment for WDM Networking II*, pp. 1–15.

Ossipov E and Karlsson G 2003 The Effect of Per-input Shapers on the Delay Bound in Networks with Aggregate Scheduling. *Proceedings of the International Workshop on Multimedia Interactive Protocols and Systems (MIPS 2003)*, pp. 107–118.

Oram A (eds) 2001 *Peer-to-Peer – Harnessing the Power of Disruptive Technologies*. O'Reilly. ISBN: 059600110X.

Ott T, Lakshman T and Wong L 1999 SRED: Stabilized RED. *Proceedings of the IEEE Conference on Computer Communications and Networking (INFOCOM 1999)*, pp. 1346–1355.

Padhye J, Firoiu V, Towsley D and Kurose J 1998 Modeling TCP Throughput: A Simple Model and its Empirical Validation. *Proceedings of the ACM Special Interest Group on Data Communication Conference (SIGCOMM 1998)*, pp. 303–314.

Palmer CC and Kershenbaum A 1995 An Approach to a Problem in Network Design using Genetic Algorithms. *Networks* **26**(3), 151–163.

Palmer CR and Steffan JG 2000 Generating Network Topologies that Obey Power Laws. *Proceedings of the IEEE Global Communications Conference (Globecom 2000)*.

Pan R, Prabhakar B and Psounis K 2000 A Stateless Active Queue Management Scheme For Approximating Fair Bandwidth Allocation. *Proceedings of the IEEE Conference on Computer Communications and Networking (INFOCOM 2000)*, pp. 942–951.

Pang P and Schulzrinne H 1999 YESSIR: A Simple Reservation Mechanism for the Internet. *Computer Communication Review* **29**(2), 89–101.

Pang P and Schulzrinne H 2000 Lightweight Resource Reservation Signaling: Design, Performance, and Implementation. Technical Memorandum 10009669-03, Bell Labs. http://www1.cs.columbia.edu/%7Epingpan/papers/BLtm_reservation.pdf.

Pantel L and Wolf L 2002 On the Impact of Delay on Real-time Multiplayer Games. *Proceedings of the 12th International Workshop on Network and Operating System Support for Digital Audio and Video (NOSSDAV)*, pp. 23–29.

Papagiannaki K, Thiran P, Crowcroft J and Diot C 2001 Preferential Treatment of Acknowledgement Packets in a Differential Services Network. *Proceedings of the International Workshop on Quality of Service (IWQoS 2001)*, pp. 187–202.

Parekh AK 1992 *A Generalized Processor Sharing Approach to Flow Control in Integrated Services Networks*. PhD thesis, Department of Electrical Engineering and Computer Science, MIT.

Park K and Willinger W 2000 *Self-Similar Network Traffic and Performance Evaluation*. John Wiley & Sons. ISBN: 0471319740.

Paxson V 1994 Empirically Derived Analytic Models of Wide-area TCP Connections. *IEEE/ACM Transactions on Networking* **2**(4), 316–336.

Paxson V and Floyd S 1995 Wide Area Traffic: the Failure of Poisson Modeling. *IEEE/ACM Transactions on Networking* **3**(3), 226–244.

Pop O, Mahr T, Dreilinger T and Szabo R 2001 Vendor-Independent Bandwidth Broker Architecture for DiffServ Networks. *Proceedings of the IEEE International Conference on Telecommunications (ICT 2001)*.

Poppe F, den Bosch SV, de La Vallée-Poussin P, Hove HV, Neve HD and Petit GH 2000 Choosing the Objectives for Traffic Engineering in IP Backbone Networks Based on Quality-of-Service Requirements, In *Proceedings of Quality of Future Internet Services/First COST 263 International Workshop (QoFIS 2000)* J Crowcroft, J Roberts and MI Smirnov (Editors), Lecture Notes in Computer Science. Springer, pp. 129–140.

Porter ME 1980 *Competitive Strategy: Techniques for Analyzing Industries and Competitors*. Free Press, New York. ISBN: 0684841487.

Postel J 1980 User Datagram Protocol, RFC 768.

Press WH, Flannery BP, Teukolsky SA and Vetterling WT 1992 *Numerical Recipes in C: The Art of Scientific Computing*, 2nd edn. Cambridge University Press. ISBN: 0521431085.

Qiu J and Knightly E 2001 Measurement-Based Admission Control with Aggregate Traffic Envelopes. *IEEE/ACM Transactions on Networking* **9**(2), 199–210.

Rajagopal S, Reisslein M and Ross K 1998 Packet Multiplexers with Adversarial Regulated Traffic. *Proceedings of the IEEE Conference on Computer Communications and Networking (INFOCOM 1998)*, pp. 347–355.

Ramakrishnan K, Floyd S and Black D 2001 The Addition of Explicit Congestion Notification (ECN) to IP, RFC 3168.

Randall M, McMahon GB and Sugden SJ 2000 Using Simulated Annealing to Solve Telecommunication Network Design Problems. *Proceedings of the Fifth Conference of the Association of Asian-Pacific Operations Research Societies (APORS 2000)*.

Rayward-Smith VJ, Osman IH, Reeves CR and Smith GD (eds) 1996 *Modern Heuristic Search Methods*. John Wiley & Sons. ISBN: 0471962805.

Rekhter Y and Li T 1993 An Architecture for IP Address Allocation with CIDR, RFC 1518.

Rekhter Y and Li T 1995 A Border Gateway Protocol 4 (BGP-4), RFC 1771.

Rijsinghani A 1994 Computation of the Internet Checksum via Incremental Update, RFC 1624.

Robertazzi TG 2000 *Computer Networks and Systems: Queueing Theory and Performance Evaluation*, 3rd edn. Springer Verlag. ISBN: 0387950370.

Roberts J 2001 Traffic Theory and the Internet. *IEEE Communications Magazine* (1), 94–99.

Robertson R 1997 *Who's MAE and Why's She So Slow?*. FAQ. http://info.ipinc.net/support/faqs/mae.html.

Rodellar D 2003 Eurescom TWIN (Testing WDM IP Networks) Project – Deliverable 2 – Experience of Testing IP over WDM Networks using Packet over SDH/SONET and Gigabit Ethernet http://www.eurescom.de/%7Epub-deliverables/P1000-series/P1014/D2/p1014-d2.pdf.

Rosen E, Viswanathan A and Callon R 2001 Multiprotocol Label Switching Architecture, RFC 3031.

Rosolen V, Bonaventure O and Leduc G 1999 A RED Discard Strategy for ATM Networks and its Performance Evaluation with TCP/IP Traffic. *Computer Communication Review* **29**(3), 23–43.

Roughan M, Sen S and Duffield OSN 2004 Class-of-service Mapping for QoS: A Statistical Signature-based Approach to IP Traffic Classification. *Proceedings of the 4th ACM SIGCOMM Conference on Internet Measurement*, pp. 135–148.

Roughan M, Thorup M and Zhang Y 2003 Traffic Engineering with Estimated Traffic Matrices. *Proceedings of the 2003 ACM SIGCOMM Conference on Internet Measurement*, pp. 248–258.

Ruiz-Sanchez M, Biersack E and Dabbous W 2001 Survey and Taxonomy of IP Lookup Algorithms. *IEEE Network* **15**(2), 8–23.

Rutgers CLH 1991 An Introduction to IGRP. Technical Report, The State University of New Jersey, Center for Computers and Information Services, Laboratory for Computer Science Research, Document ID: 26825. http://www.cisco.com/warp/public/103/5.html.

Sahu S, Nain P, Towsley D, Diot C and Firoiu V 2000 On Achievable Service Differentiation with Token Bucket Marking for TCP.

Saltzer JH, Reed DP and Clark DD 1984 End-To-End Arguments in System Design. *ACM Transactions on Computer Systems* **2**(4), 277–288.

Sandvine Incorporated 2003 Regional Characteristics of P2P – File Sharing as a Multi-Application, Multi-National Phenomenon. *White Paper*. http://www.sandvine.com/solutions/pdfs/Euro_Filesharing_DiffUnique.pdf.

Santaló LA 2004 *Integral Geometry and Geometric Probability*. Addison-Wesley Publishing Company. ISBN: 0521523443.

Schaefer C, Enderes T, Ritter H and Zitterbart M 2002 Subjective Quality Assessment for Multiplayer Real-time Games. *Proceedings of the 1st Workshop on Network and System Support for Games (NetGames)*, pp. 75–78.

Schelen O 1998 *Quality of Service Agents in the Internet*. PhD thesis, Lulea University of Technology.

Schelen O, Nilsson A, Norrgard J and Pink S 1999 Performance of QoS Agents for Provisioning Network Resources. *Proceedings of the Seventh IEEE/IFIP International Workshop on Quality of Service (IWQoS 1999)*, pp. 17–27.

Schmitt J 2001 *Heterogeneous Network Quality of Service Systems*. Kluwer Academic Publishers. ISBN: 07937410X.

Schmitt J and Roedig U 2005 Sensor Network Calculus – A Framework for Worst Case Analysis. *Proceedings of the Conference on Distributed Computing in Sensor Systems (DCOSS 2005)*, pp. 141–154.

Schmitt J, Heckmann O, Karsten M and Steinmetz R 2002 Decoupling Different Time Scales of Network QoS Systems. *Computer Communications* **25**(11-12), 1047–1057.

Schneier B 1995 *Applied Cryptography: Protocols, Algorithms, and Source Code in C*, 2nd edn. Wiley & Sons. ISBN: 0471117099.

Schnitter S and Horneffer M 2004 Traffic Matrices for MPLS Networks with LDP Traffic Statistics. *Proceedings of the 11th International Telecommunications Network Strategy and Planning Symposium (Networks 2004)*, pp. 231–236.

Schrijver A 2003 *Combinatorial Optimization*, 1st edn. Springer-Verlag. ISBN: 3540443894.

Schulzrinne H, Casner S, Frederick R and Jacobson V 1996 RTP – A Transport Protocol for Real-Time Applications.

Schwefel HP 2001 Behavior of TCP-like Elastic Traffic at a Buffered Bottleneck Router. *Proceedings of the IEEE Conference on Computer Communications and Networking (INFOCOM 2001)*, pp. 1698–1705.

Scoglio C, de Oliveira JC, Akyildiz IF and Uhl G 2001 A New Threshold-based Policy for Label Switching Path (LSP) Setup in MPLS Networks. *Proceedings of the 17th International Teletraffic Congress*.

Seddigh N, Nandy B and Heinanen J 2000 An Assured Rate Per-Domain Behaviour for Differentiated Services, draft-seddigh-pdb-ar-00.txt.

Sen S, Spatscheck O and Wang D 2004 Accurate, Scalable In-network Identification of P2P Traffic Using Application Signatures. *Proceedings of the 13th International Conference on World Wide Web*, pp. 512 – 521.

Shah D and Gupta P 2001 Fast Updating Algorithms for TCAMS. *IEEE Micro* **21**(1), 36–47.

Sheldon N, Girard E, Borg S, Claypool M and Agu E 2003 The Effect of Latency on User Performance in Warcraft III. *Proceedings of 2nd Workshop on Network and System Support for Games*.

Shenker S 1995 Fundamental Design Issues for the Future Internet. *IEEE Journal on Selected Areas in Communications* **13**(7), 1176–1188.

Shenker S and Breslau L 1995 Two Issues in Reservation Establishment *Proceedings of the ACM Special Interest Group on Data Communication Conference (SIGCOMM 1995), ACM Computer Communications Review*, pp. 14–26.

Shenker S, Partridge C and Guérin R 1997 Specification of Guaranteed Quality of Service, RFC 2212.

Shreedhar M and Varghese G 1996 Efficient Fair Queueing using Deficit Round-Robin. *IEEE Transactions on Networking* **4**(3), 375–385.

Shroff N and Schwartz M 1998 Improved Loss Calculations at an ATM Multiplexer. *IEEE/ACM Transactions on Networking* **6**(4), 411–422.

Siekkinen M, Urvoy-Keller G, Biersack E and En-Najjary T 2005 Root Cause Analysis for Long-Lived TCP Connections. *Proceedings of the ACM Conference on Emerging Network Experiment and Technology (CoNEXT 2005)*, pp. 200–210.

Singh S, Baboescu F, Varghese G and Wang J 2003 Packet Classification using Multidimensional Cutting. *Proceedings of the ACM Special Interest Group on Data Communication Conference (SIGCOMM 2003)*, pp. 213–224.

Sklower K 1993 A Tree-Based Packet Routing Table for Berkeley Unix. Technical Report, University of California, Berkeley. http://www.cs.berkeley.edu/~sklower/routing.ps.

Songhurst D 2001 M3I Deliverable 11 - Global Interaction Models. http://www.m3i.org/results/m3idel11.pdf.

Spring N, Mahajan R and Wetherall D 2002 Measuring ISP Topologies with Rocketfuel. *Proceedings of the ACM Special Interest Group on Data Communication Conference (SIGCOMM 2002)*, pp. 133–146.

Srinivasan V and Varghese G 1999a A Survey of Recent IP Lookup Schemes. *Proceedings of the IFIP Sixth International Workshop on Protocols for High Speed Networks (PfHSN 1999)*, pp. 9–24.

Srinivasan V and Varghese G 1999b Fast Address Lookups using Controlled Prefix Expansion. *Transactions on Computer Systems* **17**(1), 1–40.

Starobinski D, Karpovsky M and Zakrevski L 2002 Application of Network Calculus to General Topologies using Turn-Prohibition. *Proceedings of the IEEE Conference on Computer Communications and Networking (INFOCOM 2002)*, pp. 1151–1159.

Statnikov RB and Matusov JB 1995 *Multicriteria Optimization and Engineering*. Chapman & Hall. ISBN: 0412992310.

Stattenberger G and Braun T 2003 Performance of a Bandwidth Broker for DiffServ Networks. *Proceedings of the Fachtagung Kommunikation in Verteilten Systemen (KiVS 2003) (Symposium on Communication in Distributed Systems)*.

Stein C 1992 *Approximation Algorithms for Multicommodity Flow and Shop Scheduling Problems*. PhD thesis, Department of Electrical Engineering and Computer Science, Massachusetts Institute of Technology.

Steinmetz R and Nahrstedt K 2004 *Multimedia Systems*. Springer. ISBN: 3540408673.

Steinmetz R and Wehrle K (eds) 2005 *Peer-to-Peer-Systems and-Applications*, Lecture Notes in Computer Science 3485. Springer-Verlag, ISBN: 354029192X.

Stephens D, Bennett J and Zhang H 1999 Implementing Scheduling Algorithms in High Speed Networks. *JSAC, Special Issue on High Performance Switches/Routers* **17**(3), 1145–1159.

Stevens WR 1994 *TCP/IP Illustrated, Volume 1: The Protocols*. Addison-Wesley.

Stiliadis D and Varma A 1996 Design and Analysis of Frame-Based Fair Queueing: A New Traffic Scheduling Algorithm for Packet Switched Networks. *Proceedings of the Conference on Measurement and Modeling of Computer Systems (SIGMETRICS 1996)*, pp. 104–115.

Stoica I 2000 *Stateless Core: A Scalable Approach for Quality of Service in the Internet*. PhD thesis, Carnegie Mellon University, Department of Electrical and Computer Engineering.

Stoica I and Zhang H 1998 LIRA: A Model for Service Differentiation in the Internet. *Proceedings of the 8th International Workshop on Network and Operating Systems Support for Digital Audio and Video (NOSSDAV 1998)*, pp. 115–128.

Stoica I and Zhang H 1999 Providing Guaranteed Services Without Per Flow Management. *Proceedings of the ACM Special Interest Group on Data Communication Conference (SIGCOMM 1999)*, pp. 81–94.

Stoica I, Shenker S and Zhang H 1998 Core-Stateless Fair Queueing: Achieving Approximately Fair Bandwidth Allocations in High Speed Networks. *Proceedings of the ACM Special Interest Group on Data Communication Conference (SIGCOMM 1998)*, pp. 118–130.

Stoica I, Zhang H and Ng TSE 1997 A Hierarchical Fair Service Curve Algorithm for Link-Sharing, Real-Time and Priority Service. *Proceedings of the ACM Special Interest Group on Data Communication Conference (SIGCOMM 1997)*, pp. 249–262.

Stoica I, Zhang H and Shenker S 2002 Self Verifying CSFQ. *Proceedings of the IEEE Conference on Computer Communications and Networking (INFOCOM 2002)*, pp. 21–30.

Sun Microsystems 2000 The Sun Service Provider Initiative. http://solutions.sun.com/embedded/sp.html.

Suri S, Varghese G and Chandranmenon G 1997 Leap Forward Virtual Clock: A New Fair Queuing Scheme with Guaranteed Delays and Throughput Fairness. *Proceedings of the IEEE Conference on Computer Communications and Networking (INFOCOM 1997)*, p. 281.

Suri S, Waldvogel M, Bauer D and Warkhede PR 2003 Profile-Based Routing and Traffic Engineering. *Computer Communications* **24**(4), 351–365.

Tanenbaum A 2002 *Computer Networks*, 4th edn. Prentice Hall PTR. ISBN: 0130384887.

Tangmunarunkit H, Govindan R, Jamin S, Shenker S and Wollinger W 2001 Network Topologies, Power Laws, and Hierarchy. Technical Report USC-CS-01-746, University of Michigan, Michigan, USA. http://www.isi.edu/%7Ehongsuda/publication/USCTech01_746.ps, Shorter Version appeared as SIGCOMM 2001 Poster.

Tebaldi C and West M 1998 Bayesian Inference on Network Traffic using Link Count Data. *Journal of the American Statistical Association* **93**, 557–576.

Teitelbaum B and Shalunov S 2002 Why Premium IP Service Has Not Deployed (and Probably Never Will). *Internet2 QoS Working Group Informational Document*. http://qos.internet2.edu/wg/documents-informational/20020503-premium-problms-non-architectural.txt.

Teitelbaum B, Hares S, Dunn L, Narayan V, Neilson R and Reichmeyer F 1999 Internet2 QBone: Building a Testbed for Differentiated Services. *IEEE Network Magazine* **13**(5), 8–16.

Terzis A, Wang L, Ogawa J and Zhang L 1999b A Two-Tier Resource Management Model for the Internet. *Proceedings of the IEEE Global Communications Conference (Globecom 1999)*, pp. 1779–1791.

Terzis A, Ogawa J, Tsui S, Wang L and Zhang L 1999a A Prototype Implementation of the Two-Tier Architecture for Differentiated Services. *Proceedings of the Fifth IEEE Realtime Technology and Applications Symposium (RTAS 99)*.

Tiers 2004 Tiers Topology Generator. http://www.isi.edu/nsnam/ns/ns-topogen.html#tiers.

Tutschku K and Tran-Gia P 2005 Traffic Characteristics and Performance Evaluation in Peer-to-Peer Systems, *Peer-to-Peer Systems and Applications*, Lecture Notes in Computer Science 3485. Springer-Verlag, pp. 383–397. ISBN: 354029192X.

Vardi Y 1996 Network Tomography: Estimating Source-Destination Traffic Intensities from Link Data. *Journal of The American Statistical Association* **91**, 557–576.

Virtamo J 2003 Queueing Theory Lectures Notes. http://www.netlab.hut.fi/opetus/s38143/luennot/english.shtml.

Visual Numerics 2004 JMSL Numerical Library for Java Applications. http://www.vni.com/products/imsl/jmsl.html.

Viswanathan A, Feldman N, Wang Z and Callon R 1998 Evolution of Multi-Protocol Label Switching. *IEEE Communications* **36**(5), 165–173.

Waldvogel M 2000 *Fast Longest Prefix Matching: Algorithms, Analysis, and Applications*. Shaker Verlag, Aachen, Germany. ISBN: 3826573129.

Weiss M and Shin S 2002 Internet Interconnection Economic Model and its Analysis: Peering and Settlement. *Netnomics (Kluwer)* **6**(1), 43–57.

Westerinen A, Schnizlein J, Strassner J, Scherling M, Quinn B, Herzog S, Huynh A, Carlson M, Perry J and Waldbusser S 2001 Terminology for Policy-Based Management, RFC 3198.

Whitt W 1995 Variability Functions for Parametric Decomposition Approximations for Queueing Networks. *Management Science* **41**(10), 1704–1715.

Wolsey LA and Nemhauser GL 1999 *Integer and Combinatorial Optimization*. John Wiley & Sons. ISBN: 0471359432.

Wrege D and Liebeherr J 1997 A Near-Optimal Packet Scheduler for QoS Networks. *Proceedings of the IEEE Conference on Computer Communications and Networking (INFOCOM 1997)*, pp. 576–583.

Wroclawski J 1997 Specification of the Controlled Load Network Element Service, RFC 2211.

Yaron O and Sidi M 1993 Performance and Stability of Communication Networks via Robust Exponential Bounds. *IEEE ACM Transactions on Networking* **1**(3), 372–385.

Yavatkar R, Pendarakis D and Guérin R 2000 A Framework for Policy-based Admission Control, RFC 2753.

Yeom I and Reddy ALN 1999 Realizing Throughput Guarantees in a Differentiated Services Network. *Proceedings of the IEEE International Conference on Multimedia Computing and Systems*, Volume 2, pp. 372–376.

Zander S and Armitage G 2004 Empirically Measuring the QoS Sensitivity of Interactive Online Game Players. *Australian Telecommunications Networks & Applications Conference (ATNAC)*.

Zander S, Nguyen T and Armitage G 2005 Self-learning IP Traffic Classification based on Statistical Flow Characteristics. *Proceedings of the Passive and Active Measurement Workshop (PAM 2005)*.

Zegura EW, Calvert K and Donahoo MJ 1997 A Quantitative Comparison of Graph-Based Models for Internet Topology. *IEEE/ACM Transactions on Networking* **5**(6), 770–783.

Zhang H and Ferrari D 1993 Rate-Controlled Static Priority Queueing. *Proceedings of the IEEE Conference on Computer Communications and Networking (INFOCOM 1993)*, pp. 227–236.

Zhang L 1990 Virtual Clock: A New Traffic Control Algorithm for Packet Switching Networks. *Proceedings of the ACM Special Interest Group on Data Communication Conference (SIGCOMM 1990)*, pp. 19–29.

Zhang H and Ferrari D 1994 Rate-Controlled Service Disciplines. *Journal of High Speed Networks* **3**(4), 389–412.

Zhang ZL, Duan Z and Hou Y 2001 On Scalable Design of Bandwidth Brokers. *IEICE Transactions on Communications, Special Issue on Internet Technology* **E84-B**(8), 2011–2025.

Zhang ZL, Duan Z, Gao L and Hou YT 2000 Decoupling QoS Control from Core Routers: A Novel Bandwidth Broker Architecture for Scalable Support of Guaranteed Services. *ACM SIGCOMM Computer Communication Review* **30**(4), 71–83.

Zhang Y, Breslau L, Paxson V and Shenker S 2002 On the Characteristics and Origins of Internet Flow Rates. *Proceedings of the 2002 Conference on Applications, Technologies, Architectures, and Protocols for Computer Communications (SIGCOMM 2002)*, pp. 309–322.

Zhang Y, Roughan M, Duffield NG and Greenberg AG 2003a Fast Accurate Computation of Large-Scale IP Traffic Matrices from Link Loads. *Proceedings of the International Conference on Measurements and Modeling of Computer Systems (SIGMETRICS 2003)*, pp. 206–217.

Zhang Y, Roughan M, Lund C and Donoho D 2003b An Information-Theoretic Approach to Traffic Matrix Estimation. *Proceedings of the ACM Special Interest Group on Data Communication Conference (SIGCOMM 2003)*, pp. 301–311.

Zimmermann H 1980 OSI Reference Model – The ISO Model of Architecture for Open Systems Interconnection. *IEEE Transactions on Communications* **28**(4), 425–432.

All webpages cited in this work were checked in August 2005 or later. Owing to the dynamic nature of the WWW, webpages can change.

Index